Remote Sensing and Climate Change

Role of Earth Observation

Springer
London
Berlin
Heidelberg
New York
Barcelona
Hong Kong
Milan
Paris
Santa Clara
Singapore
Tokyo

Arthur P. Cracknell

Remote Sensing and Climate Change

Role of Earth Observation

Springer

Published in association with
Praxis Publishing
Chichester, UK

PRAXIS

Editor:
Professor Arthur P. Cracknell
Department of Electronic Engineering and Physics
University of Dundee, Dundee

This book is based on the 11th Remote-Sensing Summer School held in Dundee in 1999

SPRINGER–PRAXIS BOOKS IN GEOPHYSICAL SCIENCES
SUBJECT *ADVISORY EDITOR*: Dr Philippe Blondel, C.Geol FGS, Ph.D., M.Sc, Senior Scientist, Department of Physics, University of Bath, UK

ISBN 1-85233-321-9 Springer-Verlag Berlin Heidelberg New York

British Library Cataloguing in Publication Data
Cracknell, A. P. (Arthur Philip), 1940–
Remote sensing and climate change : the role of earth observation. –
(Springer-Praxis books in geophysical sciences)
1. Climatic changes – Detection 2. Climatic changes – Computer simulation
3. Climatology – Observations 4. Climatic changes – Remote sensing
5. Artificial satellites in remote sensing
I. Title
551.6′354
ISBN 1-85233-321-9

Library of Congress Cataloging-in-Publication Data
Remote sensing and climate change : the role of earth observation /
[edited by] Arthur P. Cracknell.
p. cm. – (Springer-Praxis books in geophysical sciences)
Includes bibliographical references and index.
ISBN 1-85233-321-9 (alk. paper)
1. Climatic changes – Remote sensing. I. Cracknell, Arthur P. II. Series.
QC981.8.C5 R438 2001
551.6 – dc21 00-067966

Printed by MPG Books Ltd, Bodmin, Cornwall, UK

Cover design: Jim Wilkie
Typesetting: Originator, Great Yarmouth, Norfolk, UK

Printed on acid-free paper supplied by Precision Publishing Papers Ltd, UK

Contents

Preface

At first sight the title of this book might seem a little surprising. Earth observation, in the form of remotely sensed data from satellites, has only become available in recent years. The very first polar-orbiting meteorological satellite (*TIROS-1*) was launched in 1960, giving us a maximum length of archived dataset of 40 years, while most archives of remotely sensed data cover a much shorter period than that. Nevertheless, remotely sensed data provide input to climatology studies in two ways. The first involves the use of remotely sensed data to provide input data for climatological models. The second involves the search of remotely sensed datasets to find evidence of climage change or of the effects of climate change, mostly likely at local or regional scale rather than at global scale. We addressed the first of these questions in a summer school in 1992 and the proceedings of that summer school were published as *Remote Sensing and Global Climate Change* (Vaughan and Cracknell 1994). At that stage we did not consider the second question, namely the possibility that remote-sensing data in the various archives might actually contain evidence of global change. Thus the element that is new this time is the attempt to see what evidence there is in the archived datasets from spaceborne remote-sensing systems for (a) the existence of global change and (b) the existence of human influence in any such changes. We shall not repeat everything that was covered previously in the 1992 summer school; nevertheless, it is necessary to recapitulate some of the basic materials from that earlier book.

The material presented in this book is based on the lectures presented at a summer school on *Remote Sensing and Long-term Global Datasets for Climate Studies* that was held at Dundee University from 2–13 August 1999. A considerable time has elapsed between that event and the publication of this book. It would be misleading to describe this as the proceedings of that summer school for two reasons. First, the material presented in the lectures has been very considerably revised and updated over the 18 months following the summer school; indeed one chapter underwent quite major revision even at proof stage. The book should therefore not be open to the accusation of being already out of date at the time of publication. Second, there were various components of the summer school which were rather ephemeral and which are not described here. These included some short seminars by participants presenting some of their own research work (which doubtless will be incorporated in theses, conference proceedings or journals in due course). There was also some time devoted to information-retrieval studies. For this purpose the participants were split into five working groups and each group chose, or was assigned, one of the following topics:

- clouds;
- vegetation/desertification;
- sea-surface temperatures;
- glaciers;
- passive microwave data (SSMR, SSM/I, ...).

Each task involved trying to find (mainly from the World Wide Web (WWW) and possibly from the Library):

1. What are the traditional sources of data on this topic?
2. What archives of satellite data exist that are relevant to this topic?
3. What is the length of time for which these archives exist?
4. What is the availability of data from these archives (a) on the WWW or, if not on the WWW, then (b) to order, (c) cost, (d) delivery time?
5. Any written material, articles, reports, etc. related to evidence of climate change in long-term remote-sensing datasets.

Each working group was asked to produce a report and to make a short oral presentation to everyone on its findings. It was decided not to include these reports in this book, principally because so much of the material referred to websites and, in our experience, much of the material on these sites, and even some of the sites themselves, seems to be somewhat ephemeral. As part of the learning process, both the practice involved and the experience gained by the participants in the working groups were extremely important.

There was also an opportunity for a hands-on session on image-processing packages and readily available satellite data. This was devoted to an exposition of the image-processing packages that we have available in Dundee, along with a consideration of the sets of remotely sensed data that are available from the WWW.

I should like to acknowledge the assistance of several colleagues in the organization of the summer school, in particular Ms Sheila Newcombe for overall organization assistance in relation to the summer school, Mr Nick Kirby for assistance with electronic media and communications, Mr Mark Slater for organizing the sessions of hands-on experience of image-processing packages and readily available satellite data, Mrs Pat Cunningham for secretarial and database assistance, Mr George Whyte for financial accounting support and Mr Ian Durajczyk, Mr Russell Gall, Mrs Marina Lawrence and Mr Jo Roden and for technical assistance.

The summer school was supported financially under contract number ERBFMMACT970276 from the European Commission and by the British National Space Centre. Finally I would like to pay tribute to the lecturers and to the participating students. The lecturers, as always, were busy people and I am very grateful to them for taking the time to prepare their material, to come to Dundee to present their material and to contribute to the revision of that material for the purpose of publication. As for the participating students themselves, we have organised ten remote-sensing summer schools over the last 20 years and this time the number of participants was the smallest we have ever had attend. But, by contrast, they were the most enthusiastic and committed group of participants that have ever attended any of our summer schools.

Abbreviations and satellites (in italics)

AATSR	Advanced Along-Track Scanning Radiometer
ACE	Atmosphere Climate Experiment
ACSYS	Arctic Climate System Studies
ADEOS	Satellite
AEM	Applications Explorer Mission
Aeolus	Atmospheric Dynamics Mission
AGCM	Atmospheric General Circulation Models
AIRS	Atmospheric Infrared Sounder
AMI	Active Microwave Instrument
AMSU	Advanced Microwave Sounding Unit
Argos	Argos Data Collection and Platform Location System
ARTEMIS	Satellite
ASAR	Advanced Synthetic Aperture Radar
ASCAT	Advanced Scatterometer
ASTER	Advanced Spaceborne Thermal Emission and Reflection Radiometer
ATSR	Along-Track Scanning Radiometer
AVHRR	Advanced Very High Resolution Radiometer
AVNIR	Advanced Visible and Near Infrared Radiometer
BAHC	Biospheric Aspects of the Hydrological Cycle
BrO	Bromine species
BUV	Backscattered Ultraviolet
CBD	Convention on Biological Diversity
CCD	Charge-coupled device
CEOS	Committee on Earth Observation Satellites
Ceosat	Satellite
CERES	Clouds and Earth's Radiant Energy System
CFCs	Chlorofluorocarbons
CH_4	Methane
CHEM	Satellite

Clark	Satellite
CNES	Centre National d'Etudes Spatiales
CO_2	Carbon dioxide
COP	Conference of the Parties
CPU	Central Processing Unit
Cryosat	Satellite
CSD	Commission on Sustainable Development
CSIRO	Commonwealth Scientific and Industrial Research Organization (Australia)
CZCS	Coastal Zone Color Scanner
DAAC	Distributed Active Archive Centres
DGVIII	Directorate General VIII (European Commission)
DGXI	Directorate General XI (European Commission)
DIS	Data and Information System (IGBP)
DISCover	DIS land-cover dataset
DMS	Dimethylsulphide
DMSP	(US) Defense Meteorological Satellite Program
DORIS	Doppler Orbitography and Radio positioning Integrated by Satellite
DTM	Digital Terrain Model
EarlyBird	Satellite
EARSeL	European Association of Remote Sensing Laboratories
Earth Explorer	Satellite
Earth Watch	Satellite
EC	European Commission
ECMWF	European Centre for Medium-Range Weather Forecasting
EDC	EROS Data Center
EEA	European Environment Agency
EIT	Economies In Transition
Envisat	Satellite
EOPB	Earth Observation Programme Board
EOS	Earth Observing System; a US programme of planned satellites
EOS/ALT	Satellite
EOS/COLOUR	Satellite
EOSP	Earth Observing Scanning Polarimeter
EPS	EUMETSAT Polar System
ER	Earth Radiation
ERB	Earth Radiation Budget
ERBE	Satellite
EROS	Satellite
ERS	Satellite (European Remote Sensing)
ESA	European Space Agency
ESAC	Earth Science Advisory Committee
ESE	Earth Science Enterprise
ESF	European Science Foundation

ESM	Earth System Models
ESSC	European Space Science Committee
EU	European Union
EUMETSAT	European Organization for Meteorological Satellites
FAO	Food and Agriculture Organization (UN)
FCCC	Framework Convention on Climate Change
FFT	Fast Fourier Transform
FIFE	First International Satellite Land Surface Climatology Field Experiment
FIRE	First International Satellite Cloud Climatology Regional Experiment
FR	Full Resolution
FTIR	Fourier Transform Infrared
FTP	File-Transfer Protocol
G3OS	Global climate, ocean and terrestrial observing systems (combination of GCOS, GOOS and GTOS)
GAC	Global Area Coverage
GAIM	Global Analysis, Interpretation and Modelling
GARP	Global Atmospheric Research Programme
GCEC	Global Change and Ecological Complexity
GCMs	General Circulation Models
GCOS	Global Climate Observing System
GCTE	Global Change and Terrestrial Ecosystems
GERB	Geostationary Earth Radiation Budget Experiment
GEWEX	Global Energy and Water Cycle Experiment
GFSSOC	Gravity Field and Steady State Ocean Circulation mission
GMS	Satellite
GMT	Greenwich Mean Time
GOCE	*Earth Explorer* gravity field and steady-state gravity circulation core mission
GOCM	Global Ocean Circulation Models
GOES	Geostationary Operational Environmental Satellites
GOEZS	Global Ocean Euphotic Zone Study
GOME	Global Ozone Monitoring Experiment
GOMOS	Global Ozone Monitoring by Occultation of Stars
GOOS	Global Ocean Observing System
GOSMOS	Global Ozone Monitoring by Occultation of Stars
GOSSP	Global Observing Systems Space Panel
GPCP	Global Precipitation Climatology Project
GRAS	Global Navigation Satellite System Receiver for Atmospheric Sounding
GTOS	Global Terrestrial Observing System
H_2S	Hydrogen sulphide
HALOE	Halogen Occultation Experiment Solar Occultation
HFCs	Hydrofluorocarbons

HIRS	High Resolution Infrared Radiation Sounder
HNO_3	Nitric acid
HRCs	High Reflective Clouds
HRDI	High Resolution Doppler Imager
HRG	High Resolution Geometry
HRV	High (Haute) Resolution Visible sensor
IASI	Infrared Atmospheric Sounding Interferometer
ICPD	International Conference on Population and Development
ICSU	International Council for Science
ICT	Internal Calibration Target
IEA	Institute of Economic Affairs
IFOV	Instantaneous Field of View
IGAC	International Global Atmospheric Chemistry project
IGBP	International Geosphere–Biosphere Programme
IGBP-DIS	IGBP Data and Information System
IGFA	International Group of Funding Agencies
IGOOS	Integrated Global Ocean Observing System
IGOS	Integrated Global Observing Strategy
IGY	International Geophysical Year
IHDP	International Human Dimensions Programme
IKONOS	Satellite
Insat	Satellite
IOC	Intergovernmental Oceanographic Commission
IPCC	Intergovernmental Panel on Climate Change
IQSY	International Quiet Sun Year
IRIS	Satellite
IRS	Satellite
ISAMS	Improved Stratospheric and Mesospheric Sounder
ISCCP	International Satellite Cloud Climatology Project
ISRO	Indian Space Research Organization
ISS	International Space Station
Jason	NASA/NOAA/CNES satellites planned as successors to *Topex-Poseidon* as part of the EOS programme
JERS	Satellite
JGOFS	Joint Global Ocean Flux Study
LAI	Leaf Area Index
Landsat	Satellite
LCWG	Land Cover Working Group
LISS	Linear Imaging Self-Scanning Sensor
LOICZ	Land–Ocean Interactions in the Coastal Zone
LRR	Laser Retro-Reflector
LSPI	Land Surface Processes and Interaction mission
LUCC	Land Use/Cover Change
LUCF	Land-Use Change and Forestry
MCP	Mission Communication Package

MERIS	Medium Resolution Imaging Spectrometer
Meteosat	Satellite
METOP	Satellite
MHS	Microwave Humidity Sounder
MIMR	Multifrequency Imaging Microwave Radiometer
MIPAS	Michelson Interferometer for Passive Atmospheric Sounding
MISR	Multi-angle Imaging Spectro-Radiometer
MLS	Microwave Limb Sounder
MLW	Mid-Latitude Winter
MODIS	Moderate Resolution Imaging Spectroradiometer
MOPIT	Measurements of Pollution in the Troposphere
MSC	Meteorological Satellite Centre
MSG	*Meteosat* Second Generation
MSS	Multi-spectral Scanner
MSU	Microwave Sounding Unit
MSU-SK	A Soviet mechanical scanning imaging radiometer
MWR	Microwave Radiometer
MWS	Microwave Sounder
N_2O	Nitrous oxide
NASA	National Aeronautics and Space Administration
NASDA	National Space Development Agency (Japan)
NCDC	National Climate Data Center
NCEP	National Centre for Environmental Prediction
NDVI	Normalized Difference Vegetation Index
NESDIS	National Environmental Satellite Data and Information Service
NGDC	National Geophysical Data Center
Nimbus	Satellite
NIMBY	Not in my backyard syndrome
NIST	National Institute of Standards and Technology
NO_2	Nitrogen dioxide
NO_3	Nitrate radical
NO_x	Oxides of nitrogen
NOAA	National Oceanic and Atmospheric Administration; also satellite
NODC	National Oceanographic Data Center
NRA	NASA Radar Altimeter
NWP	Numerical Weather Prediction
NWPM	Numerical Weather Prediction Models
O_3	Ozone
OCTS	Ocean Colour and Temperature Sensor
OECD	Organization for Economic Cooperation and Development
OGCM	Ocean General Circulation Models
OLR	Outgoing Long-wave Radiation
OMI	Ozone Monitoring Instrument

OOPC	Ocean Observing Panel for Climate
OPEC	Organization of Petroleum Exporting Countries
Orbview	Satellite
PAGES	Past Global Changes
PFCs	Perfluorocarbons
PI	Principal Investigators
PRARE	Precise Range and Range-rate Equipment
PRT	Platinum Resistance Thermometer
Quickbird	Satellite
R&D	Research and development
RA	Radar Altimeter
Radarsat	Satellite
RDED	Rio Declaration on Environment and Development
RESURS	Satellite
RFI	Radio-Frequency Interference
S&R	Search and Rescue
SAGE	Stratospheric Aerosol and Gas Experiment
SAM	Statospheric Aerosol Measurement
SAMS	Stratosphere and Mesosphere Sounder
SAR	Synthetic Aperture Radar
SBSTA	Subsidiary Body for Scientific and Technological Advice
SBUV	Solar Backscatter Ultraviolet
SCIAMACHY	Scanning Imaging Absorption Spectrometer for Atmospheric Cartography
SDS	Strategic Data Sets
Seasat	Satellite
SeaWiFS	Sea-Viewing Wide-Field-of-View Sensor
SEM	Space Environment Monitor
SETI	Search for Extraterrestrial Intelligence
SEVIRI	Spinning Enhanced Visible and Infrared Imager
SF_6	Sulphur hexafluoride
SGP97	Southern Great Plains 1997 hydrology experiment
SI	Système International d'Unités
SMMR	Scanning Multi-channel Microwave Radiometer
SMOS	Soil Moisture and Ocean Circulation mission
SO_2	Sulphur dioxide
SPARC	Stratospheric Processes and their Role in Climate Change
SPIN	Satellite
SPOT	Système Pour l'Observation de la Terre; satellite
Sputnik	Satellite
SSM/I	Special Sensor Microwave/Imager
SST	Sea Surface Temperature
SSU	Stratospheric Sounding Unit
START	Global Change System for Analysis, Research and Training
STIB	Stratosphere–Troposphere Interactions and the Biosphere

SWARM	Name of an Earth Explorer Opportunity Mission
SWIFT	Stratospheric Wind Interferometer for Transport Studies
TES	Tropospheric Emission Spectrometer
TFIS	Tropical Forest Information System
TIROS	Satellite
TM	*Landsat* Thematic Mapper
TOGA	Tropical Oceans and Global Atmosphere
TOMS	Total Ozone Mapping Spectrometer
TOPC	Terrestrial Observing Panel for Climate
Topex-Poseidon	Satellite
TOVS	*TIROS* Operational Vertical Sounder
TRMM	Tropical Rainfall Measuring Mission
UARS	Upper Atmosphere Research Satellite
UGCM	UGAMP Global Circulation Model
UKMO	United Kingdom Meteorological Office
UNCED-2	Second United Nations Conference on Environment and Development
UNEP	United Nations Environment Programme
UNESCO	United National Educational, Scientific and Cultural Organization
UNFCCC	United Nations Framework Convention on Climate Change
USGS	United States Geological Survey
UV	Ultraviolet
VISSR	Visible Infrared Spin Scan Radiometer
VWG	Validation Working Group
WCC-1	First World Climate Conference
WCED	World Commission on Environment and Development (the Brundtland Commission)
WCP	World Climate Programme
WCRP	World Climate Research Programme
WCU	World Conservation Union
WGCV	Working Group on Calibration and Validation
WGISS	Working Group on Information Systems and Services
WINDII	Wind Imaging Interferometer
WMO	World Meteorological Organization
WOCE	World Ocean Circulation Experiment
WRI	World Resources Institute
WWW	World Weather Watch; World Wide Web
WWWGOS	World Weather Watch Global Observing System

1. Ms Agnes Merza
2. Dr Maria Zoran
3. Ms Annalisa Martini
4. Ms Suwannee Narukatpichai
5. Dr C. R. Nagaraja Rao
6. Dr George Ohring
7. Mr Jawad Al-Bakri
8. Mr Minzheng Duan
9. Mr Mikkel Tamstorf
10. Ms Nicole Buschmann
11. Dr W. Paul Budgell
12. Mr Tilo Keller
13. Dr Giulia Abbate
14. Dr Peter Bissolli
15. Mr Michael David Knudsen
16. Dr Pierre-Philippe Mathieu
17. Mr Bam de Bot
18. Dr Jean-François Cretaux
19. Ms Sara Giorgi
20. Ms Anne Lucas
21. Mr Henning Hoffmann
22. Mr José Antonio Gómez-Sańchez
23. Ms Victoria Kravtsova
24. Mr Johannes Hohnhorst
25. Mr Nikolaj Kamstrup
26. Prof. Arthur P. Cracknell
27. Ms Liis Sipelgas

Participants at the 11th Remote-Sensing Summer School held in Dundee

Entitled 'Remote sensing and long-term global datasets for climate studies' the summer school took place at the University of Dundee, Dundee, Scotland, UK, between 2–13 August 1999 (numbers in parens relate to the key to the photograph of the participants).

Dr Giulia Abbate (13), ENEA Environment Department, c.r. Casaccia s.p. 77, via Anguillarese 301, Roma 00100, Italy.
Tel.: 39-06-3048-4576
Fax: 39-06-3048-4925
Email: giulia.abbate@casaccia.enea.it

Mr Jawad Al-Bakri (7), Environmental and Remote Sensing Department, Silsoe Collage, Cranfield University, Bedford MK45 4DT, UK
Tel.: 44-01525-863-633
Fax: 44-01525-863-099
Email: jtalbakri@excite.com

Dr Alan Belward, Institute for Remote Sensing Applications, JRC, 21020 Ispra, Italy.
Tel.: 39-0332-789-289
Fax: 39-0332-789-073
Email: alan.belward@jrc.it

Dr Peter Bissolli (14), Deutscher Wetterdienst, German Weather Service, FE24, Postfach 10 04 65, D-63004 Offenbach, Germany.
Tel.: 49-69-80-62-29-36
Fax: 49-69-80-62-29-93
Email: pbissolli@dwd.d400.de

Dr W. Paul Budgell (11), Institute of Marine Research, Marine Environment Centre, Nordnesgate 50, PO Box 1870, Nordnes, 5817 Bergen, Norway.
Tel.: 47-55-23-86-28
Fax: 47-55-23-85-84
Email: paul.budgell@imr.no

Ms Nicole Buschmann (10), GKSS-German Research Centre, Institut für Atmosphärenphysik, GKSS Forschungszentrum, Geesthacht G, Postfach 1160, Geesthacht 21494, Germany.
Tel.: 49-4-152/87-2001
Email: nicole.buschmann@gkss.de

Prof. Arthur Cracknell (26), Department of APEME, University of Dundee, Dundee DD1 4HN, UK.
Tel.: 44-1382-34-4549
Fax: 44-1382-34-5415
Email: a.p.cracknell@dundee.ac.uk

Mr Chris Cracknell, Openmind Software Ltd, 54 Hamilton Street, Broughty Ferry, Dundee DD5 2RE, UK.
Tel.: 49-40-77-4787
Fax: 44-1382-34-5415
Email: chris.cracknell@akamail.com

Dr Jean-François Cretaux (18), CNES/GRGS, 18 Ave Ed Belin, Cedex 4, Toulouse 31401, France.
Tel.: 33-5-61-33-29-89
Fax: 33-5-61-25-32-05
Email: jean-francois.cretaux@cnes.fr

Mr Bram de Bot (17), Joint Research Centre, JRC-SAI-EGEO-MECOM, TP 250, 21020 Ispra, Italy.
Tel.: 39-0332-789-773
Fax: 39-0332-789-469
Email: bram.de-bot@jrc.it

Dr Vera Djepa, Department of APEME, University of Dundee, Dundee DD1 4HN, UK.
Tel.: 44-1382-34-5025
Fax: 44-1382-34-5415
Email: r.a.vaughan@dundee.ac.uk

Mr Minzheng Duan (8), Institute of Atmospheric Physics, Chinese Academy of Sciences, Beijing 100029, China.
Tel.: 86-10-6204-0676
Fax: 86-10-6202-8604
Email: duan@mail.iap.ac.cn

Ms Sara Giorgi (19), University of Genoa, Institute of Geography, Via A. Bensa, 16121 Genoa, Italy.
Tel.: 39-0102-099-604
Fax: 39-0102-461-520
Email: srgiorgi@tin.it

Mr José Antonio Gómez-Sánchez (22), Camino de la Fuentes, 28, INTA, Marchamalo (Guadalajara), 19180 Spain.
Tel.: 34-91-520-1992
Fax: 34-91-520-1945
Email: gomezsj@inta.es

Mr Henning Hoffmann (21), Geographisches Institut, Georg-August-University, Abt. Kartographie und Fernerkundung, Goldschmidtstr. 5, Göttingen 37077, Germany.
Tel.: 49-5-51-39-80-03
Fax: 49-5-51-39-80-06
Email: hhoffma1@gwdg.de

Mr Johannes Hohnhorst (24), GKSS-German Research Centre, Institut für Atmosphärenphysik, GKSS-Forschungszentrum, Geesthacht GmbH, Max-Planck-Str., Geesthacht 21502, Germany.
Tel.: 49-87-2016
Fax: 49-87-2020
Email: johannes.hohnhorst@gkss.de

Mr Nikolaj Kamstrup (25), N.J. Fjordsalle 11, 4.TV, Frederiksberg, Institute of Geography, University of Copenhagen, Copenhagen DK-1957, Denmark.
Tel.: 45-353-73-597
Email: kamstrup@ostervold.dk; kn160173@geogr.ku.dk

Mr Tilo Keller (12), Meteorologisches Institut, University of Bonn, Auf dem Huegel 20, Bonn 53121, Germany.
Tel.: 49-228-735-198
Email: tilo.keller@uni-bonn.de

Mr Nick Kirby, Department of APEME, University of Dundee, Dundee DD1 4HN, UK.
Tel.: 44-1382-34-4933
Fax: 44-1382-34-5415
Email: n.e.kirby@dundee.ac.uk

Mr Michael David Knudsen (15), Bragesgade 26E, 1.TV, University of Copenhagen, Copenhagen DK-2200, Denmark.
Tel.: 45-35-81-64-01
Email: km170274@geogr.ku.dk

Ms Victoria Kravtsova (23), Donetsk State University, 29/4 Murmanskaya Str., g. Makeyevka 339021, Donetsk Region, Ukraine.
Tel.: 380-62-335
Fax: 380-62-335-3
Email: kravtsova@yahoo.com

Ms Anne Lucas (20), Department of Geography, University of Bergen, Breviksveien 40, 5045 Bergen, Norway.
Tel.: 47-55-959-659
Fax: 47-55-959-393
Email: anne-lucas@nhh.no

Ms Annalisa Martini (3), ESA/ESRIN, Via G. Galilei, 00044 Frascati, Rome, Italy.
Tel.: 39-06-941-807-06
Fax: 39-06-941-802-02
Email: amartini@esrin.esa.it

Dr Pierre-Philippe Mathieu (16), Joint Research Center, Space Applications Institute, JRC Marine Environment Unit – TP690, 21020 Ispra, Italy.
Tel.: 39-332-78-99-43
Fax: 39-332-78-96-48
Email: pp.mathieu@altavista.net

Ms Agnes Merza (1), Satellite Research Laboratory, Hungarian Meteorological Service, PO Box 39, Budapest 1675, Hungary.
Tel.: 36-1-290-01-63
Fax: 36-1-290-73-87
Email: merza@met.hu

Dr C. R. Nagaraja Rao (5), NOAA-NESDIS, E/RA1, WWB, Room 810, 5200 Auth Road, Camp Springs, MD 20746, USA.
Tel.: 301-763-8136 ext. 138
Fax: 301-763-8108
Email: nrao@nesdis.noaa.gov

Ms Suwannee Narukatpichai (4), 32/289 Chaengwattana Road, Soi 43, AIT, Pakkred, Nonthaburi 11120, Thailand.
Tel.: 66-2-955-0462
Fax: 66-2-955-0466
Email: suwannee@betagro.com

Prof. Alan O'Neill, GCAM, Department of Meteorology, University of Reading, PO Box 243, Early Gate, Reading RG6 6BB, UK.
Tel.: 44-118-931-8315
Fax: 44-118-931-8316
Email: anne@met.reading.ac.uk

Dr George Ohring (6), Chief, Climate Research and Applications Division, NOAA-NESDIS, E/RA1, WWB, Room 712, 5200 Auth Road, Camp Springs, MD 20746, USA.
Tel.: 301-763-8078
Fax: 301-763-8108
Email: gohring@nesdis.noa.gov

Dr Jean Palutikof, Climate Research Unit, University of East Anglia, Norwich NR4 7TJ, UK.
Tel.: 44-1603-593-647
Fax: 44-1603-507-784
Email: j.palutikof@uea.ac.uk

Mr Liis Sipelgas (27), Paldiski Road 1, Estonian Marine Institute, Tallinn 10137, Estonia.
Tel.: 372-2-45-68-19
Fax: 372-2-63-11-069
Email: helgi@physsea.ee

Mr Mark Slater, Department of APEME, University of Dundee, Dundee DD1 4HN, UK.
Tel.: 44-1382-34-5025
Fax: 44-1382-34-5415
Email: mtslater@rs.dundee.ac.uk

Mr Mikkel Tamstorf (9), Department for Arctic Environment, Tagensvej 135, 4th Floor, Copenhagen DK-2200, Denmark.
Tel.: 45-35-82-14-15
Fax: 45-35-82-14-20
Email: mpt@dmu.dk

Dr Robin Vaughan, Department of APEME, University of Dundee, Dundee DD1 4HN, UK.
Tel.: 44-1382-34-5025
Fax: 44-1382-34-5415
Email: r.a.vaughan@dundee.ac.uk

Dr Steven Wilson, Earth Observing Programme Manager, NERC, ITE, Monkswood, UK.
Tel.: 44-1487-773-381
Fax: 44-1487-773-518
Email: stwi@wpo.nerc.ac.uk

Dr Maria Zoran (2), Environmental Remote Sensing Department, Institute of Atomic Physics, Bucharest Magurele, Atomistilor Street 1, PO Box MG5, 76900 Bucharest, Romania.
Tel.: 40-1-42-01-001
Fax: 40-1-42-31-650
Email: maria@ifaz.sfos.ro

List of figures, tables and plates

1

Background

A. P. Cracknell

1.1 INTRODUCTION

This chapter will concentrate on general background. Background material specifically on satellite remote-sensing systems will be found in Chapter 3.

1.1.1 Why is climatology important?

Over the millions of years of the Earth's existence the climate has varied. Recently (on geological timescales) the human population has expanded and human activities have increased in such a manner that the climate may well now be influenced not only (as before) by natural events but also by human activities as well. We know, from the media, about the ozone hole in Antarctica, we know about CFCs (chlorofluorocarbons), we know about increases in CO_2 (carbon dioxide) concentrations and global warming and so on. We hear dire warnings that because of burning fossil fuels to heat our houses, to drive our motor cars, to fly our aeroplanes, to run our industries, etc. we are producing vast amounts of CO_2, which is causing the temperature of the Earth to rise significantly. It is important to try to ascertain the climatic effects of human activities so that, if possible, remedial action can be taken before we wreck the planet and exterminate ourselves. It is also important to study climate change in order to enable decisions to be made about changes to agricultural practices, about changes to building regulations and practices and about modifying manufacturing processes to take account of changes in the climate.

We could take an alarmist viewpoint and say that we must stop using up fossil fuels and live solely on windpower or hydroelectricity, or even on nuclear power. We could also be complacent and, as some people do, say that a sort of Gaia hypothesis will save us – God will somehow look after us and help us to find a way to make sure 'it is alright' in the end. The first of these would be a very long-term project requiring enormous capital investment and, in the case of nuclear power, raising political and social issues related to safety. The second of these views, the Gaia hypothesis, is something on which it might be unwise to rely.

1.1.2 Global warming

In recent years the evidence has been accumulating that on the timescale of decades there is a global warming (i.e. the global annual mean surface temperature is increasing). There is also accumulating evidence that part of this increase is a consequence of human activities. The evidence is largely statistical. Within this trend there are bound to be temporal fluctuations and spatial variations. If next summer is very hot that is not an argument in favour of global warming; equally if next summer is cold and wet that is not an argument against global warming. There is also a considerable amount of politics involved in the subject of (human-induced) global warming. Anyway, our task is to consider the scientific evidence related to two aspects (a) the existence of global warming and (b) evidence for an anthropogenic component. In particular, we shall have to see what role long-term remote-sensing datasets play in this. Inevitably we shall be led into social and political considerations as well, see Chapters 2 and 9; we cannot ignore these aspects, although they are not our primary concern.

Leakey and Lewin (1995) wrote 'Until a mere ten thousand years ago, humans across the globe lived in small bands, subsisting as hunter-gatherers over a huge range of environments, just as their ancestors had for at least 100,000 years', possibly 150,000 years. This highly successful mode of existence had its origins with the evolution of the genus *Homo* sometime prior to two million years ago. What happened about 10,000 years ago with the development of settled communities, which began to develop settled food production or agriculture, was something of tremendous significance. It was, perhaps, the beginning of a vast expansion of our species in terms of numbers. And, more especially, we reached a point at which human activities came to be able to change the landscape significantly. We look around now and virtually nowhere do we see the 'original' or 'natural' or 'virgin' landscape. We can still find a few pieces of virgin tropical rainforest or we can go to the polar regions, but virtually everywhere else on the surface of the Earth has been changed by human activities. This has various consequences, but the consequence that I wish to mention here is the following.

We know that over the history of the Earth the climate has changed all the time and this has, obviously, been as a result of natural causes. But over the last few thousand years, human beings have begun to change the face of the Earth so much that the effects of our actions have begun to become sufficiently large that they are now capable of causing climate change. When our ancestors were just small bands of hunter-gatherers, the climate was certainly changing. But this was from natural causes and, of course, these natural causes are still there. But what we have now is a situation where natural causes and human-induced causes are comparable and we have the tricky problem of trying to separate the two. If we are to be able to assess sensibly the effects of cutting, or of not cutting, our increase in CO_2 production we must be able to separate the effects of our increase in CO_2 production from the consequence of natural causes. Thus, the scientific problem that has to be addressed is the difficult question of how to disentangle human-induced effects in climate change from the consequences of natural causes. If we are to argue that we should control ourselves and our activities, in order to preserve this planet as a habitable environment for future generations, we need to have some scientific knowledge of what are the effects of our present activities. As is so often the case when scientific matters become relevant to sociological or political matters science sometimes ceases to be objective and it sometimes becomes distorted or misquoted for political ends.

In all this we cannot ignore the question of population expansion. Human population has grown from half a billion in 1600, to a billion by 1800, by 1940 it had reached almost 3 billion and now it is about 6 billion. It is not just a question of food; it is also a question of energy. The energy consumption per person in North America or Western Europe is something like the equivalent of 10 t of coal per year; in India or China it is about the equivalent of 0.2 t or 0.3 t of coal per year. As the developing countries attempt to industrialize and to achieve a standard of living comparable with that in Europe or North America – and who is to say they should be prevented from achieving that? – the increasing energy demands, even with no further increase of population, are frightening. If their increasing energy demands are to be met by burning fossil fuels, then the rate of increase in CO_2 production is going to be tremendous.

1.1.3 The importance of climate studies

Why is the study of climate important? First, as we have already said, people in general are becoming quite conscious of human effects on the climate – the related scientific question is 'How do we separate natural changes from human – induced changes?' Related to this is the question of what, if anything, we should do to attempt to control anthropogenic changes or, indeed, to control natural changes as well. Second, there is the fact that climate is a temporal average taken over a long timescale, whereas reliable detailed observations of the weather (from conventional records or from satellite data) have only been made over a very short and recent period of time. Archaeological or geological evidence that we have for longer time periods does exist but is of a much less detailed nature. Consequently it is difficult to determine the extent to which humankind's activities over the last century or two are actually already affecting the climate significantly. Third, even though there is a shortage of detailed experimental evidence about the climate itself and a lack of unequivocal direct evidence regarding the extent to which humankind's activities are already affecting the climate, it is extremely important to try to construct models in which we make use of the environmental parameters (such as the increase in CO_2 concentration) that can be quantified and use them to calculate predicted effects on climate. Constructing such models is possible, but the really serious difficulties begin to arise when we try to use these models. This is because the atmosphere/Earth system is very complicated and one is trying to extrapolate its behaviour over a very long time period, when we cannot even predict the weather more than a few days in advance with any degree of reliability! It is, perhaps, not surprising that different groups of people working with different models often produce quite different predictions of climate change; nevertheless, as we shall see, there is a large enough measure of agreement that a number of clear conclusions, even now, have been obtained.

1.2 SCALE IN TIME AND SPACE

In the Proceedings of our 1992 summer school (Cracknell 1994a) we stressed the point that life is very fragile. It happens that in the very thin layer near the Earth's surface all the necessary conditions for the development and maintenance of life have been satisfied. But quite small changes in the conditions could lead to the partial or total destruction of life. A change of only a few degrees in average temperature leads either to ice ages or to melting of the polar ice and either of these would destroy much of our present civilization. Smaller changes lead to productive agricultural land being turned into a desert. A rise or fall of a few tenths of a degree in the mean temperature may correspond to quite a large change in environmental conditions. The difference in mean temperature between now and one of the recent ice ages is only about 3.5°C.

1.3 CAUSES OF CLIMATE CHANGE

The causes of climate change can be separated into natural events on the one hand and results of human activities on the other hand. Before we go into any details of this we should, perhaps, comment on the greenhouse effect before we come to discuss the question of the enhancement of the greenhouse effect as a result of human activities.

It is important to realize that, from the point of view of life on Earth, the greenhouse effect is good, not bad. The average temperature of the surface of the Earth is about 15°C and in the absence of any greenhouse gases in the atmosphere it would be reduced by about 33°C (i.e. to −18°C). A temperature of −18°C may not sound too cold and many of us will have experienced that temperature briefly. But as an *average* temperature it would make the Earth a rather inhospitable place and any semblance of modern agriculture would only be possible in a very few locations near the equator. In past ice ages the drop in average temperature has been less than 10°C (not 33°C). However, what is worrying people at the moment is the possibility of a substantial global warming as a result of changes in the greenhouse effect induced by human activities. We shall identify these activities. However, it is not easy to determine the consequences of these activities because it is difficult to separate their effects from the natural factors that lead to changes all the time.

It is convenient to subdivide the natural events again so that we have three categories of event that may affect the climate, (i) events that occur outside the

Earth, (ii) natural events on the surface of, or within, the Earth and (iii) human activities.

1.3.1 Events that occur outside the Earth

The source of virtually all the energy that drives the atmosphere is the Sun. Thus, any changes in the intensity of solar radiation arriving at the Earth will clearly have a significant effect on the weather and on the climate. The intensity of the radiation arriving at the top of the atmosphere will depend on (i) variations in the intensity of the radiation emitted by the Sun, (ii) changes in the transmission properties of space between the Sun and the Earth and (iii) changes in the Sun–Earth distance. Further details can be found in Cracknell (1994a).

It is important to have accurate measurements of the intensity of solar radiation (the solar 'constant') over a period of years but, like all other measurements of parameters related to climatology, it has to be remembered that in climatological terms any period over which we are able to acquire measurements is infinitesimally short. Another important thing is to determine the effects of the Earth's magnetic field on the incoming solar radiation. There are two other external factors which should be mentioned. First, the gravitational effect of the differential rotation of the planets, including the Earth, gives rise to tidal effects on the surface of the Sun. Second, the variations in the gravitational effect of the Sun and the Moon on the surface of the Earth, apart from causing the familiar ocean tides, gives rise to stresses on the land-surface areas as well so that there is a greater chance of volcanic eruption when these stresses are at their maximum.

1.3.2 Natural events on the Earth

The natural events on the surface of the Earth which have an effect on climate include: (i) plate tectonics, (ii) variations in the polar ice caps, (iii) volcanic eruptions and (iv) ocean circulation. These factors have various timescales associated with them and they have been discussed earlier by Cracknell (1994a) and later in this book in Chapter 8.

1.3.3 Human activities

Human activities which now may affect the climate include: (i) the increase in concentration of carbon dioxide in the atmosphere, largely from the consumption of fossil fuels, (ii) the increase in the concentration of other greenhouse gases in the atmosphere, (iii) the depletion of the ozone layer in the stratosphere, (iv) the development of land areas and (v) other human activities.

1.3.3.1 Carbon dioxide (CO_2)

As a result of media coverage it is the increasing concentration of CO_2 and the depletion of O_3 (ozone) in the stratosphere that spring most immediately to our minds when we think about possible effects of human activities on weather and the climate.

The concentration of CO_2 in the atmosphere has risen from about 280 ppm at the time of the Industrial Revolution to over 350 ppm at the present time. At present the concentration of CO_2 in the atmosphere is rising fast because of the rapid increase in the consumption of fossil fuels and the destruction of forests leading to a reduction in the removal of CO_2 from the atmosphere by photosynthesis. The main effect of increasing the concentration of CO_2 in the atmosphere is to change the balance between incoming radiation from the Sun and outgoing radiation from the Earth and, therefore, to warm the Earth. The CO_2, along with water vapour and some of the other gases in the atmosphere, acts like the glass or transparent plastic in a greenhouse.

1.3.3.2 Other greenhouse gases

We have already mentioned the substantial increase in CO_2 concentration in the atmosphere, as a result of human activities, and the greenhouse effect associated with the atmospheric CO_2. In terms of concentration CO_2 is by far the most important of the greenhouse gases (after water vapour). However, CO_2 is not the only important product of the combustion of fossil fuels. Other gases, principally SO_2 (sulphur dioxide) and oxides of nitrogen (denoted generally by NO_x) are produced by the combustion of fossil fuels. Some fuels are cleaner than others in terms of the pollution they produce. Thus gas and oil can be treated so as to remove sulphur before they are burned and this therefore prevents SO_2 from being produced. Coal, being a solid, cannot be treated in this way before it is burned and so expensive scrubbing facilities have to be used to remove the SO_2 from the gases produced by the combustion process. The production of oxides of nitrogen is more difficult to prevent because the nitrogen comes from the air used in the combustion process and not from the fuel. When considering the combustion of fossil fuels to produce greenhouse gases, particularly NO_x, which lead to the destruction of O_3 in the stratosphere, we should perhaps give particular attention to high-flying aircraft since they are injecting these gases directly into the stratosphere, whereas other combustion processes occur at ground level and produce gases which subsequently travel up to the stratosphere.

The effect of NO_x as far as O_3 is concerned actually appears to depend on the height in the atmosphere. At low altitudes the oxides of nitrogen favour the production of ozone whereas at higher altitudes they lead to the destruction of ozone. Apart from their role as greenhouse gases SO_2 and NO_x are also undesirable products of combustion because through various reactions with water they produce mineral acids; these lead to acid rain which has catastrophic effects on the vegetation in some areas. As far as SO_2 is concerned, it should be pointed out that not all the gaseous sulphur-compound emissions into the atmosphere arise from anthropogenic causes. SO_2 is liberated by volcanoes, H_2S (hydrogen sulphide) and DMS (dimethylsulphide) are liberated by plants and DMS is liberated by plankton in the ocean. It has been suggested that, apart from producing acid rain, the DMS emissions lead to the formation of sulphate particles that constitute hygroscopic nuclei for cloud drops and thereby significantly alter the cloud patterns and, thereby, also the radiation balance.

In addition to the combustion products (CO_2 and oxides of nitrogen) and the sulphur compounds, there are the CFCs (better known for their effect on the ozone layer) and there is also methane (CH_4). Methane is produced by a whole variety of sources, some of which are natural and some of which are not (see table 4 of Cracknell 1994a).

1.3.3.3 Ozone O_3

Probably the second most well-known effect of human activity on the atmosphere is the depletion of the ozone layer. The story of the discovery of the ozone 'hole' in the Antarctic, of the identification of its cause as being attributable to oxides of nitrogen and the decomposition products of CFCs (principally $CC1_3F$ and $CC1_2F_2$) is well known. Ozone depletion does not occur only in the Antarctic; it has been occurring everywhere else as well, though less dramatically. The importance of ozone is that in the stratosphere it absorbs ultraviolet (UV) from incoming solar radiation. Without the ozone layer the UV intensity at ground level would be so high as to render life impossible for many species, including *Homo sapiens*. Even a modest reduction in the concentration of the ozone layer would lead to significant increases in skin cancers and to various other medical problems as well. The international response, once the role of CFCs was identified in the problem, was swift and international agreements (the Montréal Protocol) were made to phase out the use of CFCs and to replace them by (hopefully) innocuous substitutes.

1.3.3.4 Development of land areas

Human beings and domestic animals have been affecting the surface of the Earth for a long time; it is not just a recent situation, nor is it confined to tropical rainforest. Tickell (1977) quotes something like 20 per cent of the total area of the continents as having been drastically changed as a result of human activities. The cutting down of trees for settlement and agriculture, the slash-and-burn method of cultivation in primitive societies, the overgrazing of land by such animals as goats and cattle, and the overuse, impoverishment and erosion of topsoils have in the past affected the heat and water balance in various areas. The amount of solar radiation absorbed by the grass, crop or, in some cases, desert land is less than that absorbed by the forests which were there before. More energy is reflected back into space, less moisture is evaporated and so less rain is produced and of the rain which does come more is run off the land. The inevitable increase of dust blown up from the surface can produce effects similar to those of volcanic eruptions or sandstorms.

From the point of view of food production, or general agricultural production, the most important effects have been a decline in rainfall in some areas. Tickell cites the progressive aridity in historical times of the swathe of land from the Mediterranean to northern India. This area was once covered by dense forest and later became the site of successive civilizations; its increasing aridity seems to have been caused mostly by human destruction of the natural environment. Most of the grain-growing parts of the area which once supported large populations, including those of the Roman empire, are now largely scrub or desert. Estimates are that about 7 per cent of the Earth's surface are man-made desert and the process of progressive desertification is continuing. This inevitably contributes to some extent to changing weather and climate.

1.3.3.5 Other human activities

In more recent times other human activities have become important. For instance, there are now situations in which human intervention occurs to modify the weather rather directly (rainmaking by cloud seeding). The short-term local effects may be desired and beneficial. The effects elsewhere, or in the longer term, may be less obvious and may not necessarily be beneficial. Another human influence is interference with the hydrology of the Earth. The creation of artificial lakes for irrigation or hydroelectric purposes will affect the local heat balance and the evaporation rate. Changing the flow of rivers for irrigation purposes and the consequent change in land use and land cover will also affect the local heat balance and evaporation rate as well. The development of large cities in recent years is also important. Large cities are well known to form heat islands with temperatures significantly higher than their surroundings. This arises from two factors: (i) the replacement of vegetated surfaces by more highly reflecting road and roof surfaces and (ii) the heat generated by human activities within the cities. Locally cities' effects on the heat balance will affect the weather although, in global terms, their contribution to the heat balance is relatively small.

There is also the question of above-ground nuclear explosions. A few decades ago, in the era when such tests were common, people got very worried about the effects of these tests on the weather and the climate. The effects of those tests were probably not very different, as far as weather and climate are concerned, from those of volcanic eruptions. The effects depend on the size and altitude of the explosion and the amount of dust and other particles released into the atmosphere. A major nuclear war would be much more disastrous. It has been suggested that the use of less than 1 per cent of available nuclear weapons on 100 or 200 cities would surround the Earth with a thick cloud of dust and soot and plunge the world into such darkness that it would create a prolonged Arctic or nuclear winter. The long-term consequences of a nuclear winter cannot be assessed; but they could well be even more devastating for life than the direct effects of increased radioactivity.

1.4 PRINCIPLES OF CLIMATE MODELLING

It would be most useful if it were possible to identify the human activities that affect the climate and to make reliable predictions of the direct effects of these human activities on the future climate. However, this cannot be done directly because of the natural changes that occur all the time. It is therefore necessary to model the climate, taking into account all the natural phenomena, and predict the future behaviour of the climate in the absence of any effects of human intervention. Then, it is necessary to model the effects of human activities as well as the natural phenomena, and make predictions of the future climate on the basis of both the natural effects and the human activities. Then, the effect of the human activities can be determined by subtraction. Unfortunately, the system is very complicated: our present knowledge is somewhat uneven and our historical knowledge is very sparse indeed. However, in spite of all the difficulties, a great deal of effort has gone into climate modelling in recent years and some very useful results have been obtained.

There are two approaches which we can take to climate prediction. One is the palaeo-analogue approach, which is based on attempting to estimate future climatic change from reconstructions of past climates using palaeoclimatic data. The other is to use a numerical model, a general-circulation model (GCM). Such models are derived from weather-forecast models and are based on the same physical mechanisms and equations as are used in weather-forecast models; however, they include representations of additional elements of the climate system (using ocean models, land-surface models, etc.) with varying degrees of sophistication.

1.4.1 Weather-forecast models

What we have seen over the last decade or two has been a transition from the traditional methods of weather-forecasting to forecasts based on models run on computers. The objective of a weather-forecast model is to be able to predict the weather at a particular place and time. The idea is that we write down the mathematical equations that describe the various physical processes that occur and try to solve those equations for some future time. In an atmospheric model the behaviour of the atmosphere is represented by the values of appropriate parameters specified on a three-dimensional grid of points. The parameters involved are indicated in Figure 1.1. The dynamics included in this model involves the horizontal-momentum equations, the hydrostatic equation (any vertical acceleration is neglected) and the equation of continuity (i.e. matter is not destroyed or created). The other physics included in the model involves the equation of state, the thermodynamic equation and parametric descriptions of processes – such as evaporation, condensation, formation and dispersal of clouds, etc. – of the radiative and convective processes within the atmosphere and of the exchange of momentum, heat and water vapour with the Earth's surface. Further allowance has to be made to account for motions which occur on scales smaller than the model's grid size. We suppose that the results of meteorological observations allow the values of the atmospheric parameters to be specified at some starting time, $t = 0$. The forecasting problem then is to solve the various equations describing the physical principles and processes that we have just mentioned for a subsequent time t (i.e. to determine new values of these atmospheric parameters at a subsequent time, in practice up to a few days ahead). Because there are several of these equations, and especially because they are non-linear equations, the problem has to be tackled by numerical methods using a computer. Hence the terms 'computer model' or 'numerical model'.

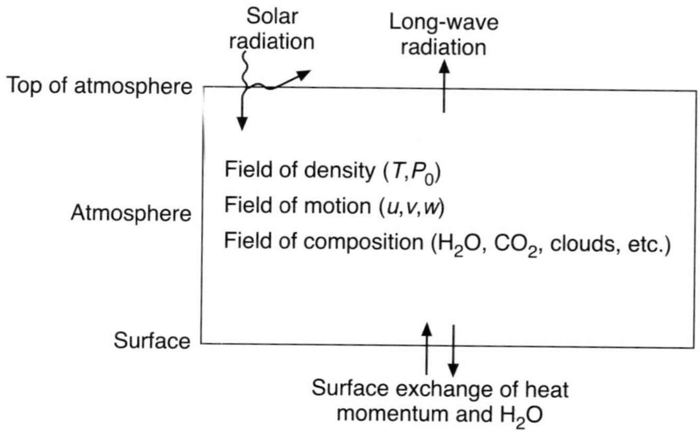

Figure 1.1. Diagram illustrating the parameters and physical processes involved in atmospheric models (from Houghton 1991).

The operational use of numerical models in weather forecasting began 30–40 years ago. Although their usefulness was originally limited, there have been rapid developments since then and within 10 years the models were able to provide better forecasts of the basic motion field than could be achieved by an unaided human forecaster. These improvements have come about for two reasons. First, the power of the computing facilities available has increased enormously. Secondly, the amount of data available to describe the present atmospheric conditions has been greatly increased in recent years. In addition to conventional surface measurements and radiosonde measurements, there are also satellite observations including satellite soundings and satellite-derived wind speeds obtained from cloud tracking from geostationary satellite images. If at present the weather forecast models are successful in providing forecasts a few days ahead then one can ask what improvements we can expect to see in the future. The European Centre for Medium-Range Weather Forecasting (ECMWF) in Reading (UK) produces forecasts for 4 weeks ahead.

1.4.2 Climate forecasting and chaos theory

The use of models along the lines we have outlined is basically a deterministic approach. We have the present conditions, we have a set of mathematical equations describing the various atmospheric processes and we rely on the mathematics to predict the future conditions. These forecasts are quite reliable for a few days ahead; the specification of what one means by a 'few' days is not very easy but, at present, a figure of 5 days might be a reasonable one. It is not a simple cut-off situation, where forecasts up to a certain time ahead are accurate and those beyond that time are not. Rather one should say that accuracy, as specified by r.m.s values in the errors of forecasts of particular parameters, is improving all the time. The question arises as to whether this improvement can go on indefinitely. It is held to be likely that the ultimate limit of forecasts from deterministic models is about 2–3 weeks ahead.

In the light of the above, we are entitled to ask the question as to whether there is any hope of ever being able to make meaningful climate-forecast predictions over the scale of a century or two if one has no hope of being able to make reliable forecasts for more than a few weeks ahead. This question is particularly pertinent since we are saying that if we take the numerical-modelling approach to climate prediction then we are using GCMs of the atmosphere which have been derived from weather-forecast models and which are based on the same physical mechanisms and equations as are used in the weather-forecast models. This brings us to the ideas of chaos theory. If it is surprising that any success can be achieved with climate models then it may help to appreciate this if we recall that in looking at climate we are looking at an average of the weather over a reasonably long time (e.g. 1 year or 10 years). The chaotic elements, or the statistical fluctuations, in the climate are the weather systems. In looking at climate change we are looking at long-term changes in this average. Viewed over a century or a millennium we see that climatic parameters (temperature, rainfall, etc.) are basically stable and vary only slowly. It is the nature of these slow long-term variations that are of concern to us in climate studies after the local short-term fluctuations have been smoothed out. It is the stability of the long-term components that makes climate prediction possible.

In a climate model we consider the effect of various forcing conditions on a given mean state. In spite of randomness or of chaos at the small-scale level we can nevertheless make useful and meaningful calculations at the macroscopic level based on a transition from one mean (or quasi-equilibrium) state to another. Thus we commonly assume in climate prediction that the climate system is in equilibrium with its forcing; that means, as long as its forcing is constant and the slowly varying components alter only slightly in the timescale considered, the mean state of the climate system will be stable and that, if there is a change in the forcing, the mean state will change until it is again in balance with the forcing. The timescale of the transition period until an equilibrium state is re-established is determined by the adjustment time of the slowest climate component (i.e. the ocean). The stable ('quasi-stationary') behaviour of the climate system gives us the opportunity to detect changes by taking time averages. Because the internal variability of the system is so high, the averaging interval has to be long compared with the chaotic fluctuations if one expects to be able to detect a statistically significant signal which can be attributed to the external forcing.

1.4.3 Climate-forecast models

Studies of the completed change from one mean state to another are called equilibrium-response studies. Studies of the time evolution of the climate change as a result of an altered forcing, which might also be time dependent, are called transient-response experiments.

The main interest in climate modelling probably lies in predicting the effects of human activities, principally the release of large amounts of CO_2 into the atmosphere, over timescales that, in climatic terms, are quite short (i.e. decades or one or two centuries). This does not mean that climatologists are not interested in other causes of climate change. It just means that CO_2 and the reasonably short-term effects are the aspects which are of great interest to society at large and, therefore, to politicians.

The climate system consists of five components:

- atmosphere;
- the oceans;
- the cryosphere;
- the biosphere; and
- the geosphere.

There are different timescales associated with these components. While the atmosphere reacts very rapidly to changes in its forcing (on a timescale of hours or days), the ocean reacts more slowly (on timescales ranging from days [at the surface layer] to millennia in the greatest depths). The ice cover reacts on a timescale of days for sea-ice regions to millennia for ice sheets. The land processes react on a timescale of days up to months, while the biosphere reacts on timescales from hours (plankton growth) to centuries (tree growth). The ideas of numerical modelling of the climate may be regarded as a development from numerical weather-forecast models (see chapter 3 of Houghton *et al.* 1990 or, for more detailed descriptions, the books by Houghton 1984, Washington and Parkinson 1986 and Trenberth 1992). Figure 1.2 is a much-reproduced diagram illustrating the atmosphere–ocean–ice–land climatic system. The solid arrows in this figure indicate external processes while the open arrows represent internal processes. The fundamental processes driving the global climate system are the heating by incoming electromagnetic radiation of rather short wavelength and the cooling by the loss of longer wavelength radiation into space. The heating is strongest in tropical latitudes and the cooling is predominantly in winter in the polar regions. The temperature gradients between equatorial and polar regions drive both atmospheric circulation and oceanic circulation; these circulation mechanisms then transport large quantities of heat from equatorial to polar regions. The five components of the climatic system listed above have been considered briefly by Cracknell (1994a) and there is no point in repeating that discussion here. However, it is useful to include some additional discussion of the handling of the geosphere in a model because it illustrates just how complicated things are and it indicates how much we have to simplify things by making approximations and taking averages (see, in particular, chapter 5 of Trenberth 1992).

Land has a dual role in the climate system. First, it acts as a lower boundary for about 30 per cent of the atmosphere; it is involved in the exchange of moisture, momentum and heat. Second, from the practical point of view for human requirements the land is crucial.

In some ways the land may be considered to be less important, in terms of general circulation, than the oceans – or the cryosphere. Its area is smaller than that of the oceans. It provides less heat storage than the oceans and it provides negligible horizontal transport. But, on the other hand, the land is much more

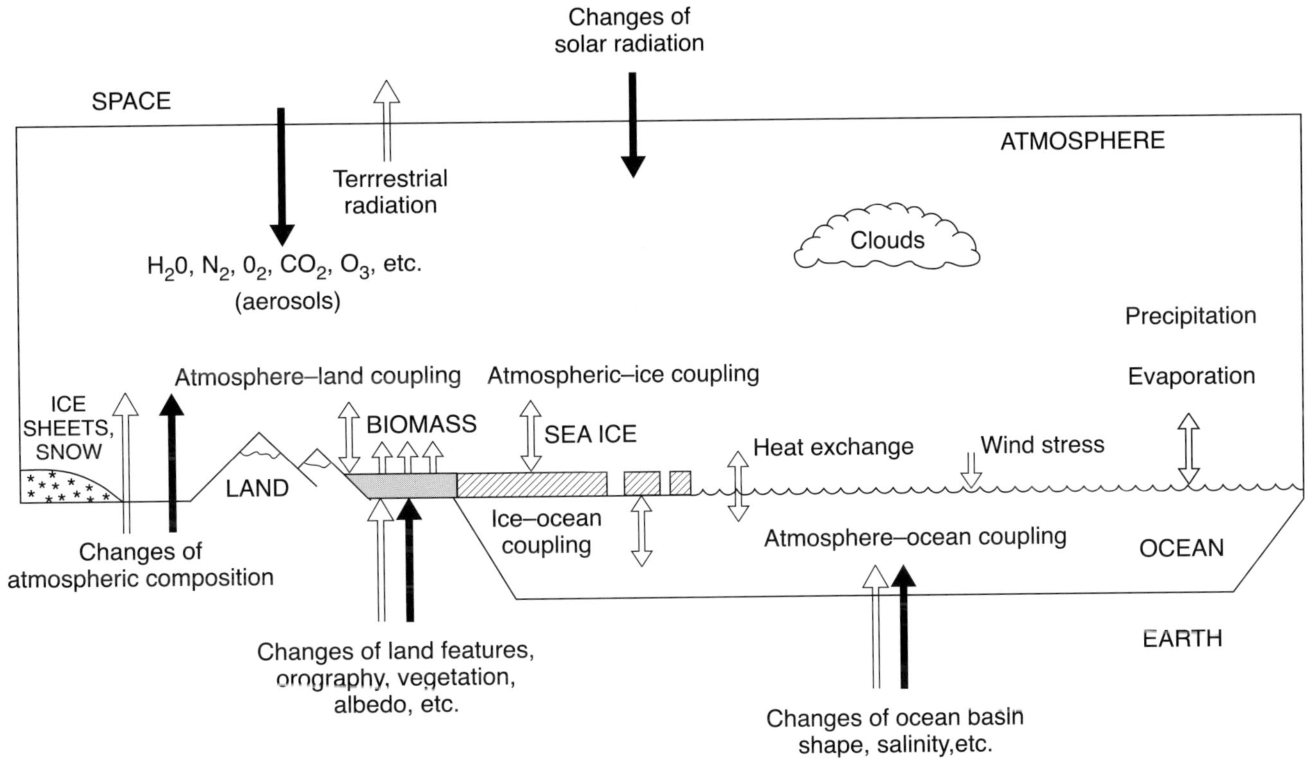

Figure 1.2. Diagrammatic representation of the climate system (from Houghton 1984).

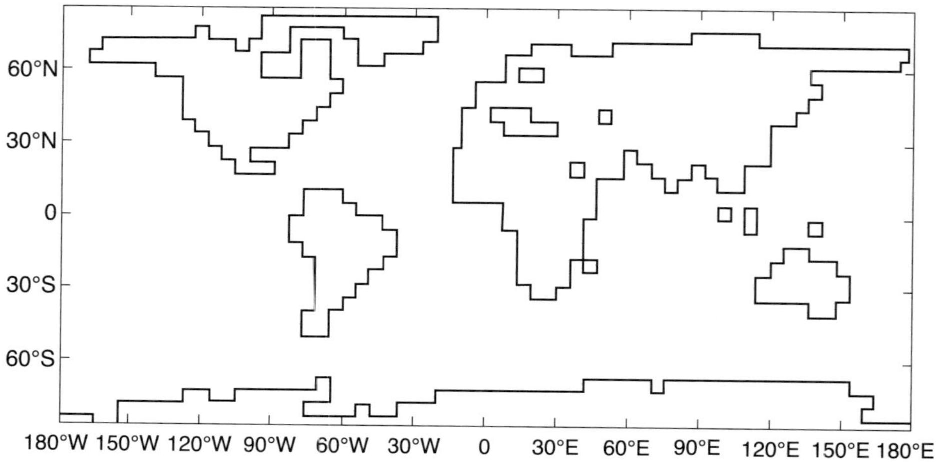

Figure 1.3. A simple land–sea mask for a climate model.

difficult to model than, say, the oceans. It has a complicated topography and relief, while the surface cover is complicated, both geometrically and in terms of its physical properties. Its nature changes very rapidly spatially. For a climate model we are concerned with grid spacings of 300 km to 1,000 km in the two horizontal directions. Thus, we have to have one value of a parameter, such as the mass of water exchanged, or the albedo, or the energy flux, or the momentum transfer, which is (a) a spatial average over the whole area and (b) a temporal average, probably over the whole year. With regard to (b) the best we are likely to do is to have a few (e.g. four) seasonal values to use. This coarse spatial resolution of a climate model also means that the land–sea boundary is only very roughly defined too (see Figure 1.3).

The fraction of solar radiation reflected (i.e. the albedo), varies with the land-surface cover. Thus any value used to describe the albedo, or any other surface parameter, is clearly a crude average. Moreover, these values are also going to vary with time as well. Temporally the land surface is far more variable than the oceans – and on a number of different timescales too. Because of the relatively low thermal capacity of the land the local thermal conditions are much more responsive to net radiation from or through the atmosphere than are the oceans. The presence or absence of clouds has a substantial effect. The temperature of the land surface varies by a much greater amount during the diurnal cycle ($\sim 10^\circ$ or 20°C or more) than is the case for the ocean.

Consider the exchange of moisture. When the land is wet it can give up water to the atmosphere more rapidly than the oceans because of its greater roughness. But when it is dry it gives up nothing. The mechanism for the transfer of water from the land surface is complicated. Some of it is by evaporation, from a wet concrete surface, from wet soil or from a river or lake. Some of the water is lost by run-off into the sea. But a great deal of it is by evapotranspiration; that is, the water is absorbed from the soil via the roots of a plant, finds it way through the plant and finally finds its way into the atmosphere via the leaves. The rate of evapotranspiration will clearly depend on the actual processes within the plant.

Water is supplied from the atmosphere by precipitation – as water, ice (hail) or snow. Soils act as reservoirs for this water but, in the long run, the precipitation is balanced by the evaporation, the evapotranspiration and the run-off. Climatologists have often viewed run-off as simply the residual after evapotranspiration requirements have been satisfied. Hydrologists have often perceived run-off as a direct response to precipitation, with evaporation and evapotranspiration as a residual. In reality, they both affect each other. Hydrologists with their studies of run-off on local and basin scales have demonstrated the importance of the rates of water infiltration into soils and the distribution of hill slopes in determining the rate of water movement from soils into streambeds and subsoil storage. Micrometeorologists and plant scientists have established the important controls of surface boundary layers and stomatal resistance. Climatologists have shown that, under well-watered conditions, water fluxes are controlled primarily by net radiation or a temperature surrogate.

Moreover, in winter in high latitudes additional processes come into play. Seasonally or permanently frozen soil impedes the flow of water into the soil. Snow accumulates during the winter season and acts as a good insulator, allowing little heat to be exchanged between the soil and the atmosphere. It enhances the surface albedo and, by this reflection of solar radiation, delays spring warming. This effect can be weakened by the presence of overlying vegetation. The parametrization of the transfer of heat and water within the soil (e.g. the balance between evaporation and precipitation, snow melt, storage of water in the ground and river run-off), is of extreme relevance for climate-change predictions, since it shows how local climates may change from humid to arid and vice versa depending on global circulation changes. It furthermore reflects, in some of the more sophisticated schemes, the changes that could occur through alterations in surface vegetation and land use.

Most soil-moisture schemes used to date are based either on the so-called bucket method or the force–restore method. In the former case, soil moisture is regarded as being available from a single reservoir, or thick soil layer. When all the moisture is used up, evaporation ceases. In the latter method, two layers of soil provide moisture for evaporation, a thin near-surface layer which responds rapidly to precipitation and evaporation and a thick deep soil layer acting as a reservoir. If the surface layer dries out, deep soil moisture is mostly unavailable for evaporation and evaporation rates fall to small values. However, in the presence of vegetation, realistic models use the deep soil layer as a source of moisture for evapotranspiration. At any given grid point over land, a balance between precipitation, evaporation, run-off and local accumulation of soil moisture is evaluated. If precipitation exceeds evaporation, then local accumulation will occur until saturation is achieved. After this, run-off is assumed and the excess water is removed. The availability of this run-off as freshwater input to the ocean has been allowed for in ocean models only recently. Most models differ in the amount of freshwater required for saturation, while few treat more than one soil type. The force–restore method has recently been extended to include a range of soil types (Noilham and Planton 1989).

1.4.4 Feedback mechanisms

An important feature of the climate system is the presence of feedback. Feedback may, of course, be either positive or negative (i.e. it may amplify or diminish the climate response to a given forcing). One example of positive feedback arises if the surface area of the Earth covered by ice expands; the expansion of the area covered by ice leads to conditions which cause the area of ice to expand still further – until some other process becomes more significant and finally causes the expansion to come to a halt. We can look at it from the opposite viewpoint – if there is a warming of the Earth, this leads to less snow and ice cover. Then the surface of the Earth becomes (on average) less reflecting and more absorbing. This causes further warming and a further reduction in the snow and ice cover (i.e. this is also an example of positive feedback). The real situation is, of course, more complicated and eventually the amplification of the warming by positive feedback (hopefully) comes to a halt. There are two other important feedback mechanisms, one involving water vapour and the other involving clouds. We shall consider them in turn.

Let us suppose that an increase in concentration of CO_2, a greenhouse gas, occurs. This will lead to some global warming. As a result of this warming, the amount of water vapour in the atmosphere will increase. The water vapour is also a greenhouse gas and so we see that an increase in the concentration of one greenhouse gas (CO_2) leads to an increase in another greenhouse gas (water vapour) and thus provides a positive feedback mechanism. However, the question of cloud feedback is very complicated. Let us consider first the effect of clouds in the present climate before we consider changes in the cloud situation. The presence of clouds heats the climate system by about 31 $W\,m^{-2}$ by reducing the infrared emission at the top of the atmosphere. Although clouds contribute to the greenhouse warming of the climate system, they also produce a cooling because they reflect incoming solar radiation and reduce the intensity of the solar radiation reaching the surface of the Earth; this produces a cooling of about 44 $W\,m^{-2}$. The cooling dominates and so the net effect of clouds on the annual global climate system is a radiative cooling of about 13 $W\,m^{-2}$.

Although clouds at present produce net cooling of the climate system, this must not be construed as a possible means of offsetting global warming because of increased concentrations of greenhouse gases. It is far from obvious what would be the effect on the cloud systems as a result of a human-induced warming arising from increased atmospheric CO_2 concentration. Let us suppose that we start from a certain dynamic equilibrium situation in which the cooling mechanisms and the heating mechanisms balance exactly. Cloud feedback arises from changes in the net cloud radiative forcing. However, this feedback arises as a result of several separate effects and it is far from obvious whether to expect the feedback to be positive or negative. Cloud feedback is not a simple matter; it depends on a variety of factors which we shall now enumerate.

- *Cloud amount*: If the amount of cloud decreases because of global warming, as occurs in typical general-circulation model simulations, then this decrease reduces the infrared greenhouse effect attributed to clouds. Thus as the Earth warms it is able to emit infrared radiation more efficiently, moderating the warming and so providing a negative climate feedback mechanism. But there is a related positive feedback; the solar radiation absorbed by the climate system increases because the diminished cloud amount causes a reduction of reflected solar radiation by the atmosphere. Thus, there is no simple way of determining the sign of this feedback component.

- *Cloud altitude*: A vertical redistribution of clouds will also induce feedbacks. For example, if global warming displaces a given cloud layer to a higher and colder region of the atmosphere, this will produce a positive feedback because the colder cloud will emit less radiation and thus have an enhanced greenhouse effect.
- *Cloud water content*: There has been considerable recent speculation that global warming could increase cloud water content, thereby resulting in brighter clouds and hence a negative component of cloud feedback. However, it has also been demonstrated that this negative solar feedback induces a positive infrared feedback. The net result of these two feedback mechanisms of opposite sign may be positive or negative. The question of cloud feedback is clearly very complex and what actually happens in any given situation will be the result of the various factors we have just outlined.

1.5 USE OF GENERAL-CIRCULATION MODELS

1.5.1 Introduction

So far we have outlined many natural phenomena and human activities which affect the weather and the climate. We have tried to indicate what is involved in modelling the climate. The general principles are fairly clear but when it comes to implementing or representing these principles in a model there are inevitably many approximations that have to be made. Generally, the more one improves the models to reduce the approximations, then the heavier the computing requirements become. A great deal of effort has gone into climate modelling in recent years and some very useful results have been obtained. Given what we said above about weather forecasting, the difficulties of producing weather forecasts for more than a few days ahead, and the whole idea of chaos versus determinism, it is perhaps both surprising and encouraging that some success in climate modelling can be achieved.

In using an atmospheric GCM for weather-forecasting purposes the surface temperature over the oceans provides a very important set of boundary conditions on the model and it is very important to have accurate values of sea-surface temperatures as input data for the model. However, because oceanic conditions change very much more slowly than atmospheric conditions, one does not make elaborate provision to take into account the circulation of heat within the oceans in a weather-forecast mode. However, in making calculations of predicted climate change we are dealing with much longer periods of time and the circulation of water and of heat in the oceans is important. For climate modelling, therefore, one needs to use GCMs for both the atmosphere and the ocean and to allow these models to be coupled together. Because this is so much more complicated than using only an atmospheric GCM for weather forecasting, the spatial resolution used in practice in climate modelling tends to be coarser than that used in weather forecasting. Grid spacings of 90 km or even 40 km in the horizontal surface may be used for numerical models for weather forecasting. For climate modelling typical models have a horizontal resolution of 300–1,000 km and between two and nineteen vertical levels.

There are a number of groups of scientists working on climate modelling in various different laboratories around the world. Each group has its own model and there are considerable differences among the results obtained by these different groups. These differences have sometimes been emphasized by those (usually politicians) who – for whatever reason – would like to claim that there is no evidence for human-induced global warming and that, therefore, there is no need to restrain our behaviour in terms of curtailing emissions of greenhouse gases into the atmosphere.

Differences among the various models arise at various stages:

(a) modelling of the physical processes (i.e. turning what we have said qualitatively into a set of mathematical equations) – there are different ways of doing this;
(b) choice of grid spacing;
(c) boundary conditions and their parametrization;
(d) the amount of computing power available – this will affect (b) and (c) (the computer power required increases very rapidly with increasing spatial resolution);
(e) assumptions made about the future.

The parametrization of radiation is possibly the most important issue for climate-change experiments, since it is through radiation that the effects of the greenhouse gases are transferred into the general circulation. A radiation parametrization scheme calculates the radiative balance of the incoming solar radiation and the outgoing terrestrial long-wave radiation and, as appropriate, the reflection, emission and absorption of these fluxes in the atmosphere. Absorption and emission are calculated in several broad spectral bands (for reasons of economy) taking into account the concentration of different absorbers and emitters like CO_2, water vapour, O_3 and aerosols.

One sensitive part in any radiation scheme is the calculation of the radiative effect of clouds. We have already mentioned that determining the effect of clouds is very complicated. In early work with numerical GCMs clouds were prescribed using observed cloud climatologies and were not allowed to alter during the experiments with, for example, changed CO_2 concentration. Later schemes contained interactive cloud parametrizations of various sophistication, but mostly based on an estimate of the cloud amount from relative humidity experiments. Only the most advanced schemes involve calculating the variation of cloud optical properties by cloud water-content experiments.

The seasonal variation of solar insolation is included in almost all experiments, but a diurnal cycle is omitted in many simulations. Climate experiments run without a seasonal cycle are limited in scope and their reliability for climate-change experiments is therefore doubtful. The inclusion of the diurnal cycle improves the realism of some feedback mechanisms and therefore the quality of the climate simulations.

There are two important consequences of climate modelling work:

1. faced with all the approximations and averaging (over space and time) it is, perhaps, surprising that any kind of realistic predictions can come out of a model;
2. many processes occur on a scale that is quite small compared with the grid-point spacing (we can therefore never expect to obtain predictions out of model calculations that relate to these processes at these scales).

1.5.2 Model outputs

It should be apparent from the above discussion that there are many uncertainties and problems associated with any given model. It would be rather risky to rely on absolute predictions made with one particular model. We should therefore consider the manner in which these models are used and this is basically done by attempting to eliminate the effects of the uncertainties or errors and trying to establish a consensus. It was in an attempt to produce some sort of order, or consensus, out of the rapidly expanding mass of conflicting or confusing results that in 1988 the World Meteorological Organization (WMO) and the United Nations Environment Programme (UNEP) set up the Intergovernmental Panel on Climate Change (IPCC). Its purpose was to prepare a report for the Second World Climate Conference (20 October–7 November 1990 in Geneva); the report characterized the present understanding of the observed regularities of climate change and possibility to forecast the climate, its impact on the environment and human activity as well as the economic actions needed to prevent undesirable climate changes. In the event, that conference itself was attended by 747 'experts' from 120 countries. Various key political world leaders were present. The IPCC was charged with assessing the scientific aspects of the climate-change issue, assessing the environmental and socio-economic impacts of climate change and formulating response strategies. The IPCC established three working groups:

- WG1, 'The scientific assessment of climate change' (Chairman: J. T. Houghton);
- WG2, 'Climate-change implications' (Chairman: Yu. A. Israel);
- WG3, 'Climate-change policy' (Chairman: F. Berntel).

Of these, by far the most interesting from our point of view was WG1.

Particularly, the IPCC attempted to 'see the wood for the trees' in terms of establishing a consensus. Thus a model is normally used to investigate the sensitivity of climate to internal and external factors and to predict the climate change by first carrying out a control integration, with the parameters set for the present climate, in order to establish a reference mean model climatology and the necessary statistics on modelled climatic variability. The climate-change (perturbation) run is then carried out by repeating the model run with appropriately changed parameters (e.g. a doubling of atmospheric CO_2 concentration) and the difference between this and the parallel control run examined. The difference between the control and the perturbed experiments is called the response to the change in the values of the parameters used. The significance of the response must be assessed against the model's natural variability (which can be determined from the control run) using appropriate statistical tests. These enable an assessment to be made of the confidence that the changes obtained represent an implied climatic change, rather than simply a result of the natural variability of the model.

The important thing is, perhaps, not to concentrate too much on the finer spatial details of the results for one time and place with any one particular model, but rather to try to concentrate on what appears to be common to the results of all, or at least most, of the models. At the time that it was written, the first report of IPCC WG1 (Houghton *et. al.* 1990) essentially did this.

1.5.3 IPCC reports

The conclusions of the first report of WG1 of the IPCC (Houghton *et al.* 1990) may not be universally accepted. However, the production of that report involved contributions from 175 scientists from 25 countries and a further 200 scientists were involved in the peer review of the draft report. The report comes with a 'Policy-maker's Summary' of about 20 pages which was prepared 'to meet the needs of those without a strong background in science who need a clear statement of the present status of scientific knowledge and the associated uncertainties'. It is quite remarkable that from the contributions of so many scientists, who surely must have many disagreements among themselves on matters of details, there have emerged some very clear conclusions. The international response to the IPCC reports will be considered in more detail in Chapters 2 and 5.

It is worth quoting from the executive summary, notwithstanding the fact that some of its evidence and conclusions have subsequently been refined (Houghton *et al.* 1992, 1994, 1996):

> "*We are certain of the following*:
>
> - There is a natural greenhouse effect which already keeps the Earth warmer than it would otherwise be.
> - emissions resulting from human activities are substantially increasing the atmospheric concentrations of the greenhouse gases: carbon dioxide, methane, chlorofluorocarbons (CFCs) and nitrous oxide. These increases will enhance the greenhouse effect, resulting on average in an additional warming of the Earth's surface. The main greenhouse gas, water vapour, will increase in response to global warming and further enhance it.

We calculate with confidence that:

- Some gases are potentially more effective than others at changing climate, and their relative effectiveness can be estimated. Carbon dioxide has been responsible for over half the enhanced greenhouse effect in the past, and is likely to remain so in the future.
- atmospheric concentrations of the long-lived gases (carbon dioxide, nitrous oxide and the CFCs) adjust only slowly to changes in emissions. Continued emissions of these gases at present rates would commit us to increased concentrations for centuries ahead. The longer emissions continue to increase at present day rates, the greater reductions would have to be for concentrations to stabilise at a given level. The long-lived gases would require immediate reductions in emissions from human activities of over 60% to stabilise their concentrations at today's levels; methane would require 15–20% reductions.

Based on current model results, we predict:

- Under the IPCC Business-as Usual (Scenario A) emissions of greenhouse gases, a rate of increase of global mean temperature during the next century of about 0.3°C per decade (with an uncertainty range of 0.2°C to 0.5°C per decade) this is greater than that seen over the past 10,000 years. This will result in a likely increase in global mean temperature of about 1°C above the present value by 2025 and 3°C before the end of the next century. The rise will not be steady because of the influence of other factors. Under the other IPCC emission scenarios, which assume progressively increasing levels of controls, rates of increase in global mean temperature of about 0.2°C per decade (Scenario B), just above 0.1°C per decade (Scenario C) and about 0.1°C per decade (Scenario D).
- that land surfaces warm more rapidly than the ocean, and high northern latitudes warm more than the global mean in winter.
- regional climate changes different from the global mean, although our confidence in the prediction of the detail of regional changes is low. For example temperature increases in Southern Europe and central North America are predicted to be higher than the global mean, accompanied on average by reduced summer precipitation and soil moisture. There are less consistent predictions for the tropics and the southern hemisphere.
- Under the IPCC Business as usual emissions scenario, an average rate of global mean sea level rise of about 6 cm per decade over the next century (with an uncertainty range of 3–10 cm per decade), mainly due to thermal expansion of the oceans and the melting of some land ice. The predicted rise is about 20 cm in global mean sea level by 2030, and 65 cm by the end of the next century. There will be significant regional variations."

It is worth while to make one or two points about this.

The executive summary only talks about the mean global temperature – and models give far more information than just that.

The four scenarios – A, B, C and D – which are mentioned need some explanation. They are taken from the original 1990 IPCC Report. In the *Business-as-Usual Scenario* (Scenario A) the energy supply is coal intensive and on the demand side only modest efficiency increases are achieved. Carbon-dioxide controls are modest, deforestation continues until the tropical forests are depleted and agricultural emissions of methane and nitrous oxide are uncontrolled. In *Scenario B* the energy-supply mix shifts towards lower carbon fuels, notably natural gas. Large efficiency increases are achieved. Carbon-dioxide controls are stringent, deforestation is reversed and the Montreal Protocol implemented with full participation. In *Scenario C* a shift towards renewable and nuclear energy takes place in the second half of next century. CFCs are now phased out and agricultural emissions limited. For *Scenario D* a shift to renewable and nuclear in the first half of the next century reduces the emissions of carbon dioxide, initially more or less stabilizing emissions in the industrialized countries. This scenario shows that stringent controls in industrialized countries combined with moderated growth of emissions in developing countries could stabilize atmospheric concentrations. Carbon-dioxide emissions are reduced to 50 per cent of 1985 levels by the middle of the next century. In the subsequent 1992 update of the Report these scenarios were replaced by six scenarios (see Table 1.1). Consequently, six different calculated global mean temperature changes resulted from these six scenarios (see Figure 1.4). But in a sense this is a detail.

However, the executive summary continued to admit the uncertainty level, to attempt to give a balanced judgement and to make recommendations.

> "*There are many uncertainties in our prediction; particularly with regard to the timing, magnitude and regional patterns of climate change, due to our incomplete understanding of*:
>
> - sources and sinks of greenhouse gases, which affect predictions of future concentrations.
> - clouds, which strongly influence the magnitude of climate change.
> - oceans, which influence the timing and patterns of climate change.
> - polar ice sheets which affect predictions of sea level rise.
>
> These processes are already partially understood, and we are confident that the uncertainties can be reduced by further research. However, the complexity of the system means that we cannot rule out surprises.
>
> *Our judgement is that*:
>
> - Global-mean surface air temperature has increased by 0.3°C to 0.6°C over the last 100 years, with the five global-average warmest years being in the 1980s. Over the same period global sea level has increased by 10–20 cm. These increases have not been smooth with time, nor uniform over the globe.
> - The size of this warming is broadly consistent with predictions of climate models, but it is also of the same magnitude as natural climate variability. Thus the observed increase could be largely due to this natural variability; alternatively this variability and other human factors could have offset a still

Table 1.1. Summary of assumptions in the six IPCC 1992 alternative scenarios.[†]

Scenario	Population	Economic growth	Energy supplies[‡]	Other	CFCs
IS92a	World Bank 1991 11.3 billion by 2100	1990–2025: 2.9% 1990–2100: 2.3%	12,00 EJ[‖] conventional oil 13,000 EJ natural gas Solar costs fall to $0.075/kW h 191 EJ of biofuels available at $70/barrel	Legally enacted and internationally agreed controls on SO_x, NO_x and NMVOC emissions	Partial compliance with Montreal Protocol. Technological transfer results in gradual phase-out of CFCs also in non-signatory countries by 2075
IS92b	World bank 1991 11.3 billion by 2100	1990–2025: 2.9% 1900–2100: 2.3%	Same as ‘a’	Same as ‘a’ plus commitments by many OECD (Organization for Economic Cooperation and Development) countries to stabilize or reduce CO_2 emissions	Global compliance with scheduled phase-out of Montreal Protocol
IS92c	UN Medium Low Case 6.4 billion by 2100	1990–2025: 2.0% 1990–2100: 1.2%	8,000 EJ conventional oil 7,300 EJ natural gas Nuclear costs decline by 0.4% annually	Same as ‘a’	Same as ‘a’
IS92d	UN Medium Low Case 6.4 billion by 2100	1990–2025: 2.7% 1990–2100: 2.0%	Oil and gas as ‘c’ Solar costs fall to $0.065/kW h 272 EJ of biofuels available at $50/barrel	Emission controls extended worldwide for CO, NO_x NMVOC and SO_x. Halt deforestation[§]. Capture and use of emissions from coal mining and gas production and use	CFC production phase-out by 1997 for industrialized countries Phase-out for HCFCs
IS92e	World Bank 1991 11.3 billion by 2100	1990–2025: 3.5% 1990–2100: 3.0%	18,400 EJ conventional oil Gas same as ‘a’ Phase-out nuclear energy by 2075	Emission controls (30% pollution surcharge on fossil energy)	Same as ‘d’
IS92f	UN Medium High Case 17.6 billion by 2100	Same as ‘a’	Oil and gas same as ‘e’ Solar costs fall to $0.083/kW h Nuclear costs increase to $0.09/kW h	Same as ‘a’	Same as ‘a’

† The assumptions for the 1990 Scenario A are described in IPCC (1990) Annex A, pp. 331–339.
‡ All scenarios assume coal resources up to 197,000 EJ. Up to 15% of this resource is assumed to be available at $1.30/gigajoule at the mine.
§ Tropical deforestation rates (for closed and open forests) begin from an average rate of 17.0 million hectares/year for 1981–1990, then increase with population until constrained by availability of land not legally protected. IS91d assumes an eventual halt of deforestation for reasons other than climate. Above-ground carbon density per hectare varies with forest type from 16 to 117 tons C/ha, with soil C ranging from 68 to 100 tons C/ha. However, only a portion of carbon is released over time with land conversion, depending on type of land conversion.
‖ EJ is exajoules (10^{18} J).

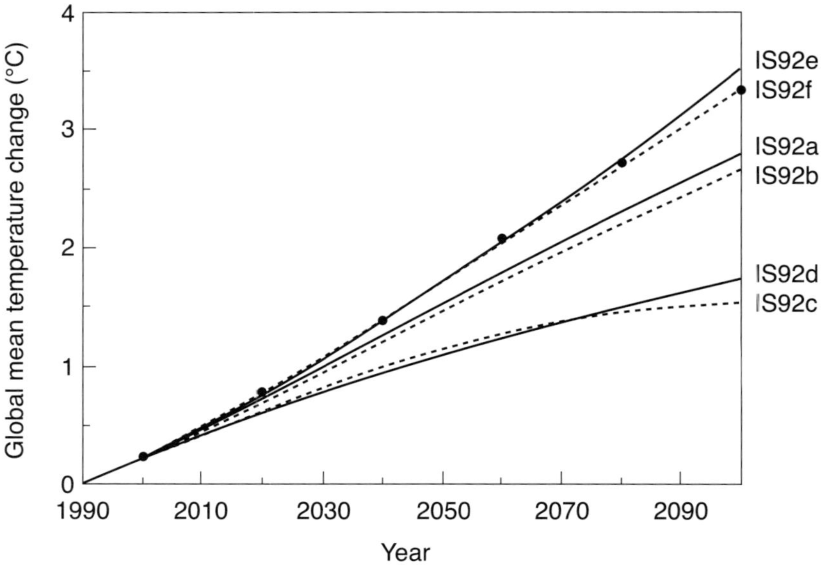

Figure 1.4. Estimates of global mean temperature change for the six IPCC 1992 scenarios (Houghton *et al.* 1992).

larger human-induced greenhouse warming. The unequivocal detection of the enhanced greenhouse effect from observations is not likely for a decade or more.

- There is no firm evidence that climate has become more variable over the last few decades. However, with an increase in the mean temperature, episodes of high temperatures will most likely become more frequent in the future, and cold episodes less frequent.
- Ecosystems affect climate, and will be affected by a changing climate and by increasing carbon dioxide concentrations. Rapid changes in climate will change the composition of ecosystems; some species will benefit while others will be unable to migrate or adapt fast enough and may become extinct. Enhanced levels of carbon dioxide may increase productivity and efficiency of water use of vegetation. The effect of warming on biological processes, although poorly understood, may increase the atmospheric concentrations of natural greenhouse gases.

To improve our predictive capability, we need:

- to *understand* better the various climate-related processes, particularly those associated with clouds, oceans and the carbon cycle.
- to *improve* the systematic observation of climate-related variables on a global basis, and further investigate changes which took place in the past.
- to *develop* improved models of the Earth's climate system.

- to *increase* support for national and international climate research activities, especially in developing countries.
- to *facilitate* international exchange of climate data."

The IPCC was attempting to establish a consensus among the results of calculations from twenty to thirty different models. The results from these various models exhibit some common features but they show very considerable differences in detail. The overall conclusions on temperature, precipitation and soil moisture in the original 1990 IPCC report were:

"1. All models show substantial changes in climate when CO_2 concentrations are doubled, even though the changes vary from model to model on a sub-continental scale.
2. The main equilibrium changes in climate due to doubling CO_2 deduced from models are given below. The number of *s indicates the degree of confidence determined subjectively from the amount of agreement between models, our understanding of the model results and our confidence in the representation of the relevant process in the model. Five *s indicate virtual certainties, one * indicates low confidence.

Temperature:

***** the lower atmosphere and Earth's surface warm;
***** the stratosphere cools;
*** near the Earth's surface, the global average warming lies between +1.5°C and +4.5°C, with a 'best guess' of 2.5°C;
*** the surface warming at high latitudes is greater than the global average in winter but smaller than in summer. (In time dependent simulations with a deep ocean, there is little warming over the high latitude southern ocean);
*** the surface warming and its seasonal variation are least in the tropics.

Precipitation:

**** the global average increases (as does that of evaporation), the larger the warming, the larger the increase;
*** increases at high latitudes throughout the year;
*** increases globally by 3 to 15% (as does evaporation);
** increases at mid-latitudes in winter;
** the zonal mean value increases in the tropics although there are areas of decrease. Shift in the main tropical rain bands differ from model to model, so there is little consistency between models in simulated regional changes;
** changes little in subtropical arid areas.

Soil moisture:

*** increases in high latitudes in winter;
** decreases over northern mid-latitude continents in summer.

Snow and sea-ice:

**** the area of sea-ice and seasonal snow-cover diminish.

The results from models become less reliable at smaller scales, so predictions for smaller than continental regions should be treated with great caution. The continents warm more than the ocean. Temperature increases in southern Europe and central North America are greater than the global mean and are accompanied by reduced precipitation and soil moisture in summer. The Asian summer monsoon intensifies.

3. Changes in the day-to-day variability of weather are uncertain. However, episodes of high temperature will become more frequent in the future simply due to an increase in the mean temperature. There is some evidence of a general increase in convective precipitation.
4. The direct effect of deforestation on global mean climate is small. The indirect effects (through changes in the CO_2 sink) may be more important. However, tropical deforestation may lead to substantial local effects, including a reduction of about 20% in precipitation."

As already mentioned, the subsequent IPCC WG1 reports refined these conclusions, but only in matters of detail. One detail was that instead of four scenarios – A, B, C and D – the 1992 report introduced predictions based on six different scenarios – labelled a, b, c, d, e and f – corresponding to different assumptions about future greenhouse-gas emissions (for the definitions of these scenarios see Houghton *et al.* 1992).

What we have been considering are equilibrium-response calculations; that is, we suppose that a change is introduced – most commonly the concentration of CO_2 in the atmosphere is doubled and the atmosphere is allowed to settle down to equilibrium again. We can ask, quite reasonably, what is the value of carrying out equilibrium calculations like this. Climate is in equilibrium when it is in balance with the radiative forcing. Thus, as long as greenhouse-gas concentrations continue to increase the climate will not reach equilibrium; moreover, even if concentrations of greenhouse-gases are eventually stabilized it would be a long time before equilibrium would be established. Thus, equilibrium calculations cannot be used directly to provide forecasts of climate changes. The first reason given for carrying out equilibrium simulations (usually with double the concentration of CO_2 in the atmosphere) is that they are done with atmospheric–oceanic mixed-layer models which ignore both the deep ocean and changes in the ocean circulation; these require less computer time than time-dependent simulations which include the influence of the deep ocean. Second, because these simulations are relatively inexpensive (in terms of computer time and therefore of money) to carry out, they can be used fairly easily for intercomparison of different models and for studying the effects of different parametrizations (e.g. of cloud). Third, it appears that, apart from areas where the oceanic thermal inertia is high, equilibrium solutions can be scaled and used as approximations to the time-dependent response.

In the models used in the equilibrium-response calculations one supposes that an abrupt change has been made in the atmospheric CO_2 concentration and that this leads to a change in the radiative forcing; this is allowed to operate in the model for a period corresponding to a few years until a new equilibrium is established. The atmospheric models used are detailed and highly developed but the treatment of the ocean, as we have already mentioned, is quite primitive. However, in reality, greenhouse-gas concentrations are changing continuously, not abruptly. The climate's reaction is not instantaneous because of the thermal capacity of the oceans. At any given time the response corresponds to only part of the equilibrium change for the corresponding instantaneous value of the forcing. For the remainder, part of the response is delayed by storage in the stable stratified layers of the upper ocean and takes several decades, or even a century, to occur. However, there is also a part of the response which remains invisible for a long time (many centuries) until the deep ocean begins to influence the surface temperature. Apart from the thermal inertia of the oceans there may also be other oceanic effects (e.g. the ocean currents can lead to a spatial redistribution of greenhouse warming). Also, even in the absence of changes in the radiative forcing the ocean–atmosphere interactions lead to fluctuations which may temporarily mask long-term climate changes.

In order to make reliable predictions of climate change under realistic scenarios of slow changes in the radiative forcing, it is necessary to take the oceanic behaviour properly into account. To do this means that we have to make use of coupled atmosphere–ocean GCMs. There is discussion of time-dependent model calculations in chapter 6 of the first IPCC report (Houghton *et al.* 1990). The use of a coupled atmosphere–ocean GCM means that the computing requirements for a calculation are very high. The experiment which is described involved a 100-year simulation. Three runs of a model were carried out with (a) no change in atmospheric CO_2 concentration, (b) a 1 per cent per annum (compounded) increase in atmospheric CO_2 concentration and (c) a 1 per cent per annum (compounded) decrease in atmospheric CO_2 concentration. The choice of 1 per cent increase in the CO_2 concentration is based on the fact that this has the equivalent effect, in terms of radiative transfer, of the combined effect of the increases in all the various greenhouse-gas concentrations at present. By comparison with the equilibrium calculations, the use of models in time-dependent response calculations is still in its early days. Some early conclusions from the results that have been obtained so far will be found in the first IPCC report (Houghton *et al.* 1990).

After the first report of WG1 of the IPCC (Houghton *et al.* 1990) was produced there were a number of subsequent attempts to update it to include the results of later work (Houghton *et al.* 1992, 1994, 1996); however, in many ways it was the first report which made the big impact and it is more in matters of detail rather than in general terms that it has been updated. The next report (Houghton *et al.* 1992) was timed to coincide with the final negotiations of the Framework Convention on Climate Change (FCCC) which was one of the five basic documents of the Rio

Convention (the Second United Nations Conference on Environment and Development [UNCED-2]) (see Section 2.4.1). This report reviewed and reaffirmed the basic conclusions of the 1990 report; it did, however, provide more detail on two sources of negative radiative forcing, namely the depletion of stratospheric ozone and the effect of aerosols derived from anthropogenic emissions. The second supplementary report (Houghton *et al.* 1994) was particularly concerned with radiative forcing and the relative importance of human and natural factors that give rise to radiative forcing as drivers of climate change. Following the success of the first report and its 1992 supplementary report, in terms of its input into the FCCC, the IPCC set itself the task of producing a second assessment and WG1 produced a report *Climate Change 1995 – the Science of Climate Change* for this purpose (Houghton *et al.* 1996). This updated the original report in many matters of detail; it also spoke rather more strongly than in the earlier report in terms of the evidence of the effects of human activities on the climate. Thus, the 1990 report said 'the observed increase (in global-mean air surface temperature) could be largely due to (the) natural variability (of the climate); alternatively this variability and other human factors could have offset a still larger human-induced greenhouse warming. The unequivocal detection of the enhanced greenhouse effect from observations is not likely for a decade or more'. The second assessment (the 1995 report) said 'the balance of evidence suggests a discernible human influence on global climate'. There is a huge wealth of detail in the second assessment but it is not appropriate to consider all that detail here.

1.6 PREDICTED CONSEQUENCES OF CLIMATE CHANGE

Some of the more direct predicted consequences of climate change are described in the later chapters of the IPCC WG1 report of 1990 (Houghton *et al.* 1990) or its various supplements (Houghton *et al.* 1992, 1994, 1996), but the best source is probably a slightly more popular book by Sir John Houghton (1997).

We see from the results of the model calculations that the likely climate change will vary a great deal from place to place (e.g. in some regions precipitation will increase, in others it will decrease). Not only is there a large amount of variability in the character of the likely change, there is also variability in the sensitivity of different systems to climate change. Houghton (1997) defines the terms sensitivity, adaptability and vulnerability:

- *sensitivity* to climate change refers to the degree to which a system responds to climate change;
- *adaptability* to climate change refers to the degree to which adjustments are possible in practices, processes or structures of systems to projected or actual changes of climate;
- *vulnerability* to climate change refers to the extent to which climate change may damage a system. It depends both on a system's sensitivity and on its ability to adapt to a given change.

Both the magnitude and the rate of climate change are important in determining the sensitivity, adaptability and vulnerability of a system. Different ecosystems will respond very differently to changes in temperature or in precipitation.

There will be a few impacts of likely climate change (i.e. warming) which will be positive as far as humans are concerned. For instance, in parts of Siberia or northern Canada increased temperature will tend to lengthen the growing season with the possibility of growing a wider variety of crops.

However, because over the centuries human communities have adapted their lives (and themselves) to the present climate, most changes in climate will have an adverse effect. If changes now occur rapidly a quick and costly adaptation to the new climate will need to be made by the affected community. The alternative of migrating to another region where less adaptation would be necessary becomes less and less of an option as the world becomes more and more crowded.

The questions addressed in Houghton's book include:

- How much will sea level rise and what effect will that have?
- How much will water resources be affected?
- What will be the impact on agriculture and food supply?
- Will natural ecosystems suffer damage?
- Will human health be affected?

The answers are far from simple. One could consider the question of sea level or of water resources fairly easily in isolation if nothing else were to change. But other factors will change. Some adaptation for both ecosystems and human communities may be relatively easy to achieve; in other cases it may be difficult, expensive or impossible.

Global warming, enhanced or reduced precipitation and changes in soil moisture are also not the only human-induced effects in the environment. There is the loss or impoverishment of soil, the over-extraction of groundwater, the damage because of acid rain, etc.

1.6.1 Sea-level rise

During the warm period before the last ice age, about 120,000 years ago, the global mean temperature was a little higher (perhaps 2°C) than at present. Average sea level was about 5 or 6 m higher than now. When the ice cover was at its maximum, about 18,000 years ago, sea level was over 100 m lower than now. The rise in sea level over the last century is shown in Figure 1.5. Estimates for the rise in sea level from the 1990 IPCC WG1 report are shown in Figure 1.6. A breakdown on the 1992 Scenario-A predictions – among different contributing factors – is shown in Figure 1.7. Note that thermal expansion is the largest single contributing factor; this will be discussed further in Section 8.2.

We are looking at rises of about 12 cm by 2030 and 50 cm by 2100. These may not sound very large, but for areas that are close to sea level they can be very significant. Half the human population lives near the coast. Particularly vulnerable areas are places such as the Netherlands, Bangladesh and many small low-lying

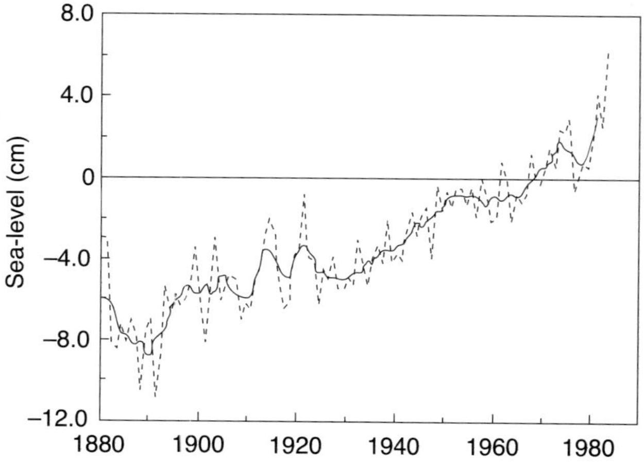

(a)

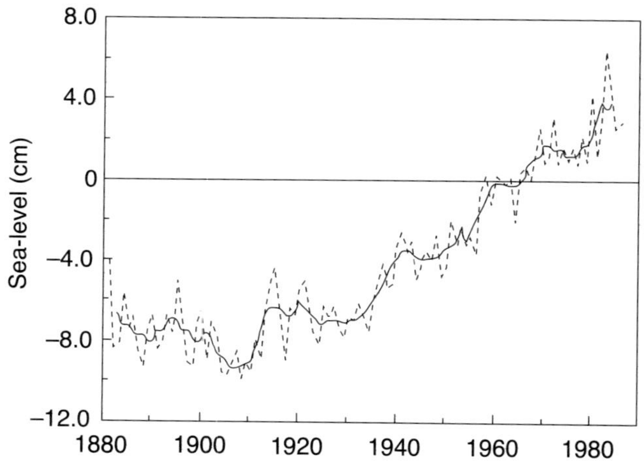

(b)

Figure 1.5. Global mean sea-level rise over the last century taken from Houghton *et al.* (1990). The baseline represents setting the average for 1951–1970 to zero; the dashed line represents the annual mean and the solid line represents the 5-year running mean.

islands. Much of the Netherlands is already below sea level and they have centuries of experience and of investment and presumably could enhance their defences to cope with a rise in sea level of 12–50 cm. But the situation is rather different for Bangladesh. It is a densely populated country of about 120 million people located in the complex delta regions of the Ganges, Brahmaputra and Meghua rivers. About 7 per cent of the habitable land and 6 million people live below the 1-m contour and 25 per cent of the land and 30 million people are below the 3-m contour (see Figure 1.8). The figures mentioned above for mean sea-level rise have to be augmented by the effects of subsidence – land movement in general plus the consequence of the removal of groundwater – giving predicted estimates of sea-level rise locally of

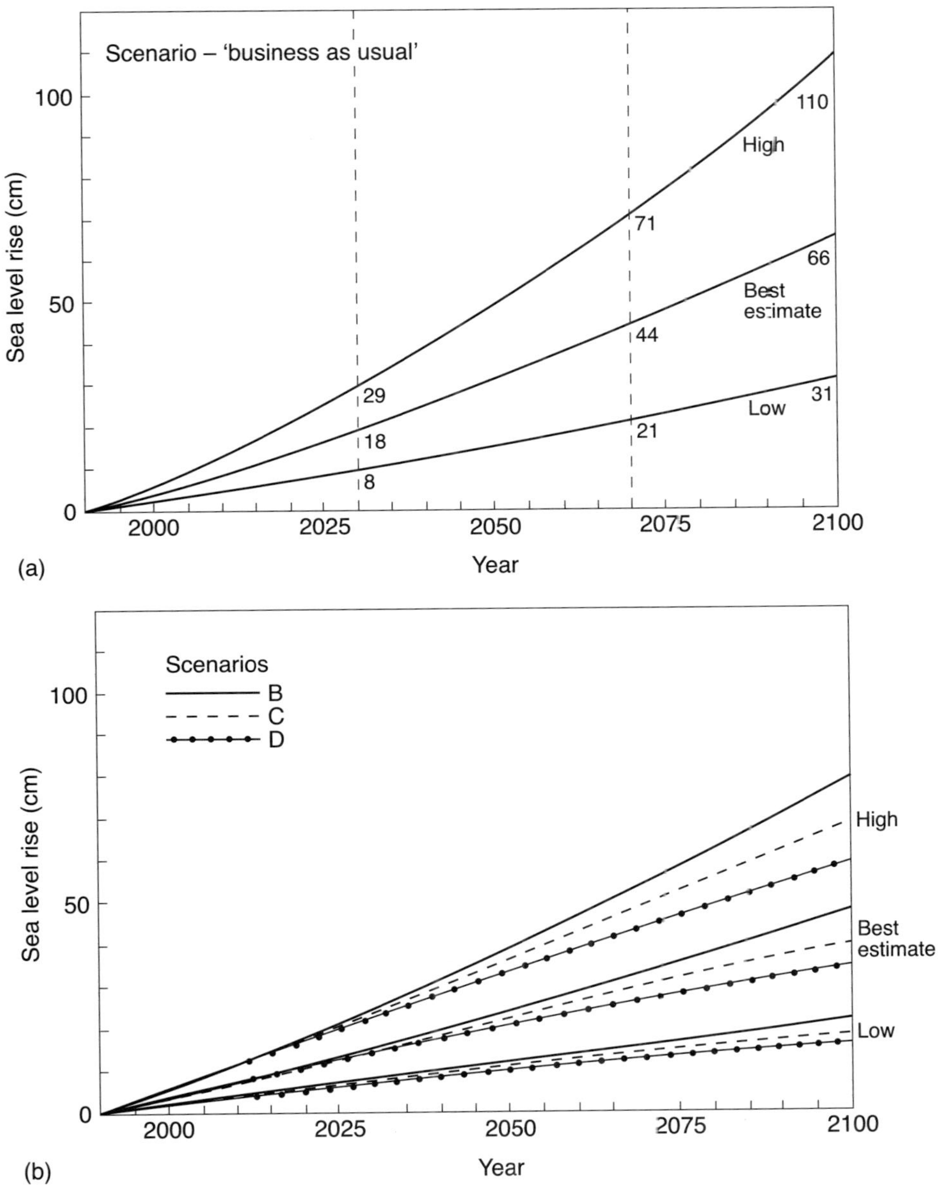

Figure 1.6. Predicted global sea-level, 1990-2100 for (a) policy scenario A, business-as-usual and (b) policy scenarios B, C and D of the IPCC 1990 Report (Houghton *et al.* 1990).

about 1 m by 2050 and nearly 2 m by 2100. The figure of 1 m is made up of 30 cm as a result of global warming and 70 cm as a result of subsidence; the 2 m is made up of 70 cm as a result of global warming and 1.2 m as a result of subsidence. It is impractical to consider full protection of the long and complicated coastline of

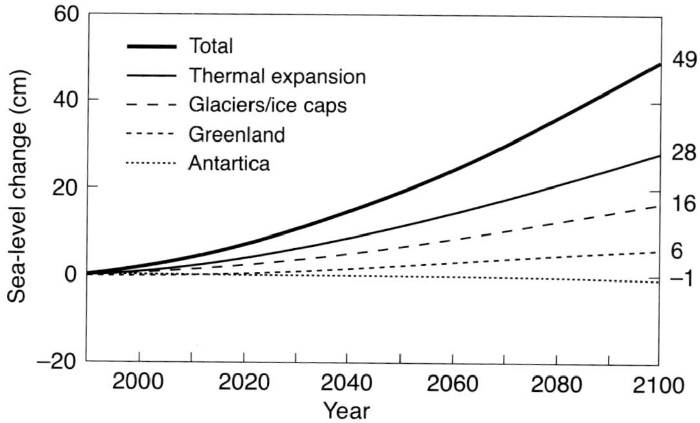

Figure 1.7. Estimates of the various contributions to the rise in global mean sea-level based on the IPCC 1992 Scenario A, business as usual (Houghton *et al.* 1992).

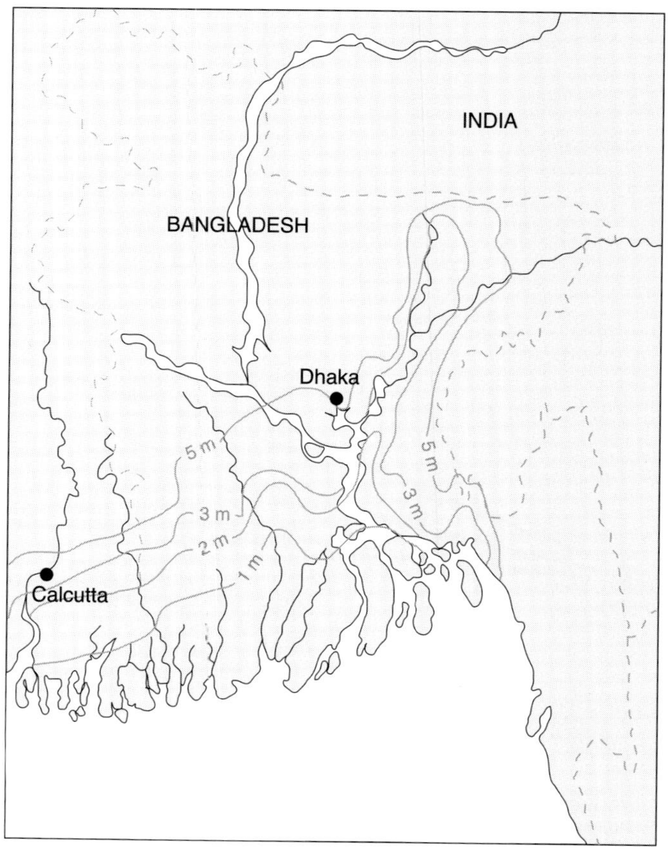

Figure 1.8. Areas in Bangladesh that would be affected by various amounts of sea-level rise (Houghton 1997).

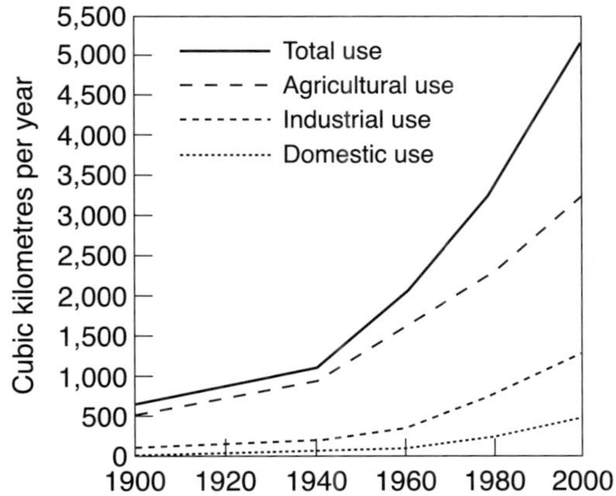

Figure 1.9. Global use of water for different purposes, 1900–2000 (Tolba and El-Kholy 1992).

Bangladesh. The most obvious effect will be the loss of good agricultural land – and this in a country where 85 per cent of the population depend on agriculture for their livelihood, with many of them living at the very edge of subsistence However, it is worse than that. Bangladesh suffers from storm surges. The storm surge of November 1970 probably caused the loss of about a quarter of a million lives, while that of April 1991 caused 100,000 deaths. Finally, the rise in sea level would lead to an increase in saline intrusion into the groundwater resources. At present it is estimated that in some parts of Bangladesh saltwater extends seasonally inland over 150 km. Similar problems to those described for Bangladesh also face the populations of other low-lying delta areas around the world.

Houghton finally makes the point that because the ocean takes centuries to adjust to an increase in surface temperature then even if the concentrations of greenhouse gases were stabilized – so that the anthropogenic cause of climate change is halted – the sea-level rise will continue for many centuries as the whole ocean adjusts to the new climate.

1.6.2 Freshwater resources

The uses to which water is put are shown in Figure 1.9. Water availability varies from 1,000 m^3 to 50,000 m^3 per head per annum in different countries. A rise in temperature means more evaporation. This may be cancelled out by increased precipitation, but in some areas there is expected to be decreased precipitation. Thus there could be less run-off. Deforestation can have an effect on rainfall; it is thought to decrease the rainfall.

1.6.3 Desertification

Drylands (defined as those areas where precipitation is low and where rainfall typically consists of small, erratic, short, high-intensity storms) cover about 40 per

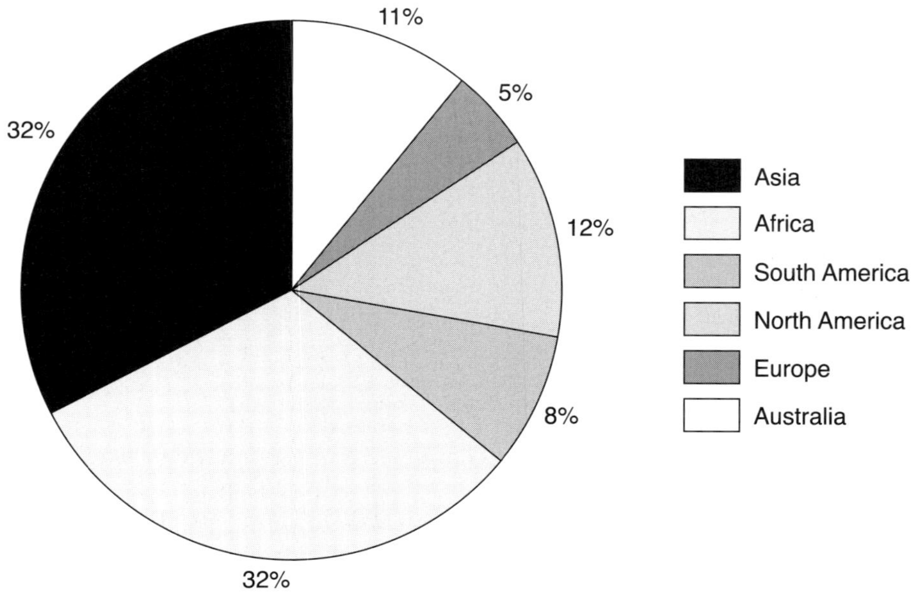

Figure 1.10. The World's drylands, by continent (Tolba and El-Kholy 1992).

cent of the total land area of the world and support over one-fifth of the world's population. Desertification in these drylands is the degradation of land because of decreased vegetation, reduction of available water, reduction of crop yields and erosion of soil. It results from excessive land use generally because of increased population and increased human needs, or political or economic pressures (e.g. the need to grow cash crops to raise foreign currency). It is often triggered or intensified by a naturally occurring drought.

The rate of desertification is currently about 60,000 km^2 per year or 0.1 per cent per year of the total area of drylands. It is a potential threat to 70 per cent of these drylands (i.e. to about 25 per cent of the world's land area). Figure 1.10 shows how these are distributed over the continents.

1.6.4 Agriculture and food supply

Temperature and rainfall are key factors in making decisions about what crops to grow. Thus, agriculture will need to adapt to changes in climate. This is not a serious problem for crops that mature over 1 or 2 years (or less). It may be a problem for trees with periods of maturing of decades or even a century or so.

The case of Peruvian farmers adjusting their choice of crops according to forecasts based on the presence or absence of El Niño is cited by Houghton (1997). There is also the carbon-dioxide fertilization effect (i.e. an increased concentration of CO_2 in the atmosphere could be expected to increase the rate of photosynthesis – with other factors assumed to be unaltered – and therefore to increase the yield of a crop). But there does not seem to be much evidence on the magnitude of this effect.

There is no simple answer to 'what is the effect of climate change on agriculture?' The answer is that agricultural success depends on human decisions, human management operating within the natural environment. Changing climate is one factor that will affect these decisions. However, if the changes are so dramatic that agricultural regions shift (because of sea-level rise, desertification, etc.) then human populations will need to move and we may see large numbers of environmental refugees.

1.6.5 Natural ecosystems

Over history, natural systems have obviously adapted to climate change. The problem is that we are now looking at quite rapid changes of climate. Changes in temperature that previously occurred over thousands of years are now occurring over decades and many ecosystems cannot change or migrate that fast. Trees, especially, are long-lived and take a long time to reproduce. They are also surprisingly sensitive to the average climate in which they develop. For the likely changes in climate in the next few decades a substantial proportion of existing trees will find themselves subject to unsuitable climatic conditions.

There has been a decline in the health of many forests in recent years and this has often been attributed to acid rain. While that is certainly a contributing factor, it now seems that climatic changes are also partly responsible. In several regions of Canada, for instance, die-back has been related to stress from a succession of warmer winters and drier summers. Sometimes, of course, it is the combined effect of pollution and of stress as a result of climate change that causes the damage.

As to marine ecosystems, while it is clear that the climate is important there is little detailed knowledge of the effects involved.

1.6.6 Impact on human health

The direct effect of global warming can be handled relatively easily. On the other hand, deaths as a consequence of heat stress in times of extremely high temperatures, or of hypothermia in times of extremely low temperature (since the amplitude of fluctuations about the mean seem set to increase) are likely to increase.

There are less direct effects as well – pollution of the atmosphere, pollution of the water supplies, poor soil (leading to poor crops and inadequate nutrition), droughts and floods, enhanced activity of carriers of diseases, etc.

1.6.7 Costs

Attempts have been made to estimate the cost of global warming and the first impression is that we might be able to buy our way out of the problem.

However, there are two factors that are important:

1. we have only tended to look at the next 50 years or so and with the hope that greenhouse-gas emissions will not go too far out of control;

2. the effects are likely to be very serious for many people currently living at subsistence level in developing countries. If as a result of desertification or sea-level rise their land becomes uninhabitable they will wish to migrate and become environmental refugees. Houghton (1997) quotes possible numbers of 3 million per year or 150 million between now and 2050. Where will they go?

1.7 CONSENSUS AND VALIDATION

What we have been talking about is largely based on the IPCC WG1 report and we have mentioned that this report was based on a consensus of the majority of the world's climatologists. But:

(a) this does not mean that it is right – there is an alternative view expressed in Chapter 9 (see also Gribbin 1990);
(b) it is still necessary to do what we can to validate the model predictions.

Let us consider briefly the question of validation. If climate models are to be used to make predictions of climatic change, whether natural change or change induced as the result of human activities, then it is obviously important to attempt to assess the accuracy of the results produced by models. To carry out the validation of climate models we need appropriate observed historical climatic data. Climate models can be tested by running them to predict historical climatic conditions from an earlier historical starting point and some considerable success has been obtained; undoubtedly there has been some fine tuning of the models in this work, but that is fair enough. The question of the validation of the outputs from climate modelling will be discussed in Chapter 8.

A long-term commitment is now needed by the world's national weather services to monitor climate variations and change. Changes and improvements in observational networks should be introduced in a way which will lead to continuous, consistent long-term datasets sufficiently accurate to document changes and variations in climate. Most sets of satellite observations are not yet of long enough duration to document climate variations. In order for these data to be most useful, it is very important that they be analysed and interpreted with existing *in situ* data. High priority should be given to the blending or integration of space-based data and *in situ* datasets in such a way as to build upon the strengths of each type of data. We are beginning to reach the stage at which calibrated long-term datasets from remote sensing (Earth observation) satellites may be considered useful in terms of seeking evidence of climate change, whether regionally or globally. There are some examples of ongoing projects that have used mixed satellite datasets, careful cross-calibration and co-ordinated data processing to produce global datasets; these include the World Climate Research Programme's (WCRP) International Satellite Cloud Climatology Project (ISCCP), the Global Precipitation Climatology Project (GPCP) and the NASA Pathfinder Program.

2

Global climate change. Socio-economic aspects of the problem

K. Ya. Kondratyev and A. P. Cracknell

2.1 INTRODUCTION

The question of greenhouse-gas emissions and human-induced global warming, including the IPCC reports, has already been considered in Chapter 1. In this chapter we try to take a wider view and refer, in particular, to two books (Kondratyev 1998; Kondratyev and Cracknell 1998); many useful references will be found in these two books. It is important to appreciate that global warming (as a result of increasing greenhouse-gas emissions) is not the only threat to the survival of the human race, or at least to our present lifestyle. There are other threats, such as population expansion, the threat to biodiversity, increasing desertification, the rapid consumption of mineral resources, the imbalance between the lifestyles of the developed and the developing world and so on. The response to the IPCC reports was largely only concerned with these other things *insofar as they were related to increasing greenhouse-gas emissions and global warming, rather than as intrinsic threats in themselves as well.* In this chapter we shall consider the World Climate Research Programme (WCRP), summarised results of the Second World Climate Conference, the Second UN Conference on Environment and Development (UNCED-2, popularly known as the Rio Summit), the related questions of population growth and sustainable development, the International Geosphere–Biosphere Programme (IGBP) and ecological monitoring.

2.2 WORLD CLIMATE RESEARCH PROGRAMME (WCRP)

In 1979 the First World Climate Conference (WCC-1) took place which produced the World Climate Programme (WCP). This programme has formed a basis for subsequent multidisciplinary developments aimed at achieving a more profound understanding of the regular features of climate formation and change, assessing the impact of climate upon human economic activity and ecosystems, and also analysing the socio-economic aspects of the problem.

The First World Climate Conference adopted a declaration on global climate problems where it has been emphasized that, although at present humankind enjoys favourable climate conditions, human economic activity is very much affected by climate variability, especially on such extremal phenomena as droughts and floods. The climatic interdependence of different countries indicates the necessity for a co-ordinated global strategy in the interests of a better understanding and rational use of climate.

At the present time mankind inadvertently is often the cause of climate change at various scales. It is reasonable to expect that further expansion of human economic activity may result in considerable regional, and even global, climate changes; hence there is an urgent need for international co-operation. Climate change induced by natural processes will continue. The scale of present-day human economic activity now makes it possible for humankind to cause (unintentional) anthropogenic climate changes on the planet that can be compared with basic, naturally occurring climate variations taking place during the lifetime of several generations.

Since we can suppose that in the future humankind will be able to introduce deliberate climatic changes on a large scale, it would be irresponsible to consider such actions until we have acquired an understanding of climate-governing mechanisms that would be sufficient to predict the potential consequences of such actions. Before projects of this kind can be implemented, it is important to reach relevant international agreements.

The basic objectives of the World Climate Research Programme (WCRP), oriented towards climate studies on a timescale of several weeks to several decades, are:

1. determining possibilities of climate forecast;
2. finding out the extent of human impact on climate.

To achieve these objectives, the following are required:

1. further development of the understanding of the regular features of regional (on the scale of the order of 100 km and more) and global climate;
2. a more reliable assessment of the regional and global climate trends;
3. development and advancement of physico-mathematical climate models capable of reconstructing the real climate and estimating its predictability;
4. study of climate sensitivity to natural and anthropogenic impacts, with special emphasis on climatic changes as a result of these impacts.

The development of the WCRP is concentrated on six main directions of research:

1. the Global Energy and Water Cycle Experiment (GEWEX), whose main objectives are studies of the interaction between cloudiness and radiation, and the water cycle in the atmosphere and on the Earth's surface;
2. the Tropical Oceans and Global Atmosphere programme (TOGA) – aimed at the study of the dynamics of the interactive atmosphere – ocean system in the

tropics which affects in the most essential way the interannual variability of the climatic system (El Niño/Southern Oscillation);
3. the World Ocean Circulation Experiment (WOCE) – focused on studying the slow changes of the climatic system which are due to the interaction between the oceans and the atmosphere;
4. the Arctic Climate System Studies (ACSYS) – taking account of the specific features of high-latitude climate-forming processes;
5. Stratospheric Processes and their Role in Climate Change (SPARC) – focused on the analysis of interactive chemical, dynamical and radiative processes in the context of global climate change;
6. a programme of numerical climate modelling (including both the rapid and slow variability of the climatic system).

These developments have a single main purpose – achieving such a level of understanding of the climatic system variability which will make it possible to make a sufficiently reliable basis for climate forecasting on timescales of seasons to a century.

Although for the WCRP objectives timescales of several weeks to several decades are of primary interest, palaeoclimatic reconstructions are also very desirable, particularly for the verification of the reliability of climate models. The range of spatial scales varies from regional (of the order of 1,000 km) to global. The WCRP involves three approaches to the solution of problems formulated in the programme: (1) study of climatically important processes; (2) development of physico-mathematical models of the climatic system; (3) statistical analysis of the observed climate parameters.

2.3 SUMMARIZED RESULTS OF THE SECOND WORLD CLIMATE CONFERENCE

The Second World Climate Conference (WCC-2) was held in Geneva during the period 29 October–7 November 1990. This conference collected 747 experts from 120 countries. The IPCC reports, which have already been discussed in Chapter 1, were prepared as working documents for this conference. The Conference's main purpose was summarizing the results of the work performed during the previous decade (i.e. since the First World Climate Conference in 1979). It was concerned with discussing the perspectives of future scientific efforts, with special emphasis on the human dimension of the problem (i.e. the necessary urgent political decisions capable of preventing the dangerous [not to say catastrophic] impact of climate change on human life and economic activity).

WCC-2 comprised two stages: (1) a scientific discussion based on the IPCC reports, involving about 1,000 experts from 70 states, and summarizing the results of the implementation of the World Climate Programme (WCP); (2) a ministerial conference aimed at the working out of political recommendations (this second stage is the characteristic feature of WCC-2 which makes it different from the previous conference, WCC-1, which was basically a scientific conference). There were speeches by numerous important world leaders. But it should be noted at once that, although sensible and substantial, all speeches lacked an important element, namely a clear concept of long-term (e.g. on the scale of a decade) perspectives of investigations within the framework of the World Climate Programme (WCP) and its scientific component, the World Climate Research Programme (WCRP), and also other closely related efforts (first of all, the IGBP).

The conference can be criticised for becoming too obsessed with numerical climate modelling (possessing sophisticated global models whose realization is possible only with the help of the most powerful computers) and a concentration on the question of greenhouse-gas emissions. But, while this is important, it is not the whole problem. One can note the speech of Margaret (now Baroness) Thatcher, ex-Prime Minister of the UK, at the second stage of the conference. According to her, it could not be said with definiteness that the increase in global mean temperature had been caused by human economic activity. She also emphasized that the global carbon cycle had not been adequately studied. She also rightly stressed the fact that the problem of climate cannot be considered on its own and that other urgent ecological aspects of our time should be taken into account, such as the explosive population growth, the loss of soil fertility, the growing pollution of the ocean, the intensive use of fossil fuel, deforestation (especially in the Tropics).

The scientific results of WCC-2 were summarized by twelve working groups which prepared relevant recommendations finalized in the form of a statement. The estimates contained in this statement are based on the conclusions of the IPCC report which has already been mentioned. In particular, it has been emphasized that the problem of climate is broader than the framework of atmospheric and oceanic sciences, and that it has already touched upon every aspect of life on our planet. The importance of the problem of climate increases incessantly because it determines, to a great extent, the future environment and well-being of humankind.

To avoid the risk of undesirable climatic changes, it was recommended to begin the implementation of measures aimed at the reduction of greenhouse-gas emissions. In order to stabilize the level of the CO_2 (carbon dioxide) concentration by the middle of the twenty-first century at a figure that exceeds the amount of the pre-industrial era by 50 per cent, it would be necessary to achieve an annual reduction in CO_2 concentration of 1–2 per cent. In the opinion of the majority of conference participants, all states possess technically available and economically acceptable possibilities for reducing CO_2 emissions; however, for some countries this view is not acceptable. The conclusion about the necessity of a 1–2 per cent reduction in CO_2 emissions annually is based on the IPCC report and this, in turn, is based on certain models of the global carbon cycle.

The statement by the ministerial conference of WCC-2 defined the following top-priority directions in climate research:

1. clouds and the hydrological cycle;
2. greenhouse gases, the global carbon cycle and other biogeochemical cycles;
3. oceans: physical, chemical and biological processes; interaction between the atmosphere and the oceans;

4. palaeoclimatic studies;
5. polar ice sheets and sea ice;
6. terrestrial ecosystems.

An important item is the statement that the ‘present observational systems for monitoring the climate system are inadequate for operational and research purposes’. Hence, the following conclusion is drawn: There is an urgent need to create a Global Climate Observing System (GCOS) modelled on the World Weather Watch Global Observing System [WWWGOS] and the Integrated Global Ocean Observing System [IGOOS] and including both space-based and surface-based observing components’. GCOS should be designed to meet the needs for:

1. climate-system monitoring, climate-change detection and response monitoring, especially in terrestrial ecosystems;
2. data for application to national economic development;
3. research towards improved understanding, modelling and prediction of the climate system.

Further development and implementation of the GCOS concept should be pursued urgently by scientists, governments and international organizations. All countries must ensure a full-scale and open exchange of datasets needed for climate-system research, process and impact studies and modelling.

With regard to socio-economic aspects of the problem, the statement indicated that ‘immediate steps should be taken including:

1. national and regional analyses of the impacts of climate variability and change on society, and study of the range of response and adaptation options available,
2. closer co-operation and communication among natural and social scientists, to ensure that climate considerations are accounted for in development planning,
3. significant increases in resources to carry out impact/adaptation studies.

Improvements in energy efficiency and non-fossil-fuel energy technologies are of paramount importance, not only to reduce greenhouse-gas emissions, but to move to more sustainable development pathways’.

Regrettably, these reasonable statements were not accompanied by the necessary analysis of the impact of political, social, economic and other factors on the environment and climate. Hence, a paradoxical situation was created. On the one hand, the efforts of the Intergovernmental Panel on Climate Change (IPCC) WG2 and WG3 were devoted to this very range of problems; on the other hand, they were practically focused only on a detailed analysis of the potential climate-change impact on human economic activity (water resources, agriculture and food, oceans, fishing industry and coastal zones; energy production; land use and city planning; forestry) and human health.

During the final stage of the first part of the WCC-2 great attention was devoted to the problem of greenhouse-gas emissions and to whether recommendations for their reduction had been justified or not. At the same time, the key aspect of the whole problem of the interactivity of the human society – environment system (i.e. the impact of socio-economic factors on the biosphere, environment and climate) was actually ignored. That is why the discussions themselves and the statement and declaration adopted on their basis were distorted and obviously biased towards the task of reducing the emissions of CO_2 and other greenhouse gases.

If the programme of the conference had contained presentations concerning the biosphere dynamics and the socio-economic roots of threats to the environment (the continuing growth of the world population, the increasing poverty in many of the developing countries as a result of the gigantic pumping of their resources to the developed industrial countries, etc.), each problem would presumably have found its proper place, including that of greenhouse-gas emissions.

The ministerial declaration adopted during the second part of the WCC-2, with the participation of representatives from 137 countries, including those of the European Community, expressed approval of the IPCC conclusions and a high appreciation of the activity relevant to the implementation of the WCP during the previous decade. The global strategy has been formulated in the document as follows: '... we consider that a global response, while ensuring sustainable development of all countries, must be decided and implemented without further delay based on the best available knowledge, such as that resulting from the IPCC assessment', and further: '... we stress that special efforts must be directed to the areas of uncertainty as identified by the IPCC. We maintain that there is a need to intensify research on the social and economic implications of climate change and response strategies'.

Then a statement follows which appears to be very important, viz.: 'We note that per capita consumption patterns in certain parts of the world along with a projected increase in world population are contributing factors in the projected increase in greenhouse gases'. This is open to the criticism that it sees the population problem only in relation to greenhouse gases. In what follows it is stressed that the lack of scientific certainty must not delay taking economically justified measures to stop environmental degradation. Furthermore, it is stated: 'We acknowledge, however, that those developed countries with as yet relatively low energy consumption (measured on a per capita or other appropriate basis) which can be reasonably expected to grow, and some countries with economies in transition, may establish targets, programmes and/or strategies that accommodate socio-economic growth, while improving the energy efficiency of their economic activities. We urge developed countries, before the 1992 UN Conference on Environment and Development, to analyse the feasibility of and options for, and, as appropriate in light of these analyses, to develop programmes, strategies and/or targets for a staged approach for achieving reductions of all greenhouse gas emissions not controlled by the Montreal Protocol, including CO_2, CH_4 [methane] and N_2O [nitrous oxide], over the next two decades and beyond'. Among measures which are of importance to achieve sustainable socio-economic development, the following have been noted: increase in the efficiency of power production, the maximum possible use of ecologically harmless technologies, the preservation of the world's forests and (taking account of the role of forests in the maintenance of biological diversity), protection of soil stability and of the hydrological system).

It should be recognized that the WCC-2 was doubtless an outstanding event both from the point of view of its acutely important thematic coverage, and the authoritative composition of its technical and intergovernmental stages. The IPCC report which served as the basis for scientific discussions and political decisions should, on the whole, be evaluated positively.

The most important points may be summarized as follows:

1. the principal significance of the WCC-2 is connected with the fact that after many years of overemphasizing the role of anthropogenic impacts it has been clearly stated that what we need is to study interrelationships between internal variability of the climatic system (the basic uncertainty is here) and impact of growing concentrations of greenhouse gases – only such an approach will give us an opportunity to explain causes of global warming during the last century;
2. it is important to overcome the stereotype of greenhouse warming and overemphasis on the necessity to reduce the emissions of greenhouse gases because otherwise it looks like a panacea against all environmental dangers;
3. the global climate-change problem should be considered in the context of global change (hence the significance of the biosphere as an interactive component of the climate system, and to be more concrete the necessity of a coupled consideration of at least the global carbon cycle which is still far from being properly understood).

One cannot avoid preoccupation about the fact that the conference did not properly analyse the socio-economic roots of global ecological problems, including that of climate change: doubtless, today the problem of climate should be given consideration in the broader context of the entire range of ecological problems.

2.4 UNCED-2

On 14 June 1992, in Rio de Janeiro, the Second United Nations Conference on Environment and Development (UNCED-2) was concluded. Preparations for the Conference were started a few years before and took a large amount of time, effort and money. A large secretariat was formed and positioned in Geneva, Switzerland. Workshops devoted to the development of the framework conventions on climate and biodiversity were organized, and four long sessions of the Preparatory Committee were held. The result was a huge amount of various materials and documents. It is important to analyse the results of UNCED-2 which has become a unique event from the viewpoint of the range of the problems discussed and the high level of representation (which included presidents and heads of governments). Doubtless, UNCED-2 has played an important role as a worldwide forum devoted to problems of global ecology which put forward many closely interrelated subjects proceeding from socio-economic development and environmental dynamics. Closing the conference, Dr M. Strong, the UNCED-2 Secretary General, stated: ‘First, we have carried out successfully the largest intergovernmental Conference ever held on this planet. The right people have been here and the people of this planet have watched and wondered. The world will not be the same after this Conference. Diplomacy will not be the same. The United Nations will not be the same. And prospects for our Earth cannot ... must not ... be the same ... Finally, this decade must be a time of transition that firmly moves us onto a new pathway and a new economy, one that is truly Eco-nomic system'.

The scale and political level of the UNCED-2 were certainly unprecedented. Rio de Janeiro welcomed 114 heads of governments, diplomats from 178 countries, representatives of 1,600 non-governmental organizations, and a huge number of journalists who represented diverse mass media. The conference produced five basic documents:

- Framework Convention on Climate Change (FCCC);
- Convention on Biological Diversity (CBD);
- Rio Declaration on Environment and Development (RDED);
- Agenda 21;
- the Forest Principles.

The first two of these agreements, the Framework Convention on Climate Change and the Convention on Biological Diversity, were legally binding conventions which had been negotiated independently of the main UNCED-2 process but on schedules which had been set so that they could be signed at UNCED-2 in Rio. The states that have signed these conventions are identified on page 15 of Grubb *et al.* (1993). On the other hand, the other three agreements, the Rio Declaration on Environment and Development, Agenda 21 and the Forest Principles, which were signed by nearly all the participating states, were defined as documents that are not legally binding on the signing parties. Indeed, the full title of the Forest Principles is the 'Non-legally binding authoritative statement of principles for a global consensus on the management, conservation and sustainable development of all types of forests'.

Rio's greatest achievement has become the process itself, the recognition of the very fact that the problems of the environment and development cannot be treated separately. UNCED-2 has demonstrated most convincingly the organic interrelationship between the problems of poverty and underdevelopment, on the one hand, and ecological problems, on the other hand, including the difference between the standards of living in the North (or developed/industrialized countries) and in the South (or developing countries). Energy consumption and climate, the tropical-wood trade, the problem of poverty, the increasing pressure of population growth, desertification – all these aspects of global and regional ecology have been given consideration on a far broader scale than ever before. An event of special importance was the Global Forum which was held in Rio simultaneously with UNCED-2 and which was supported by numerous non-governmental organizations (about 9,000 organizations, with a total number of 29,000 participants, 450,000 visitors and about 1,000 meetings of various kinds).

2.4.1 Framework Convention on Climate Change (FCCC)

The FCCC established principles that climate change is a serious problem, that action cannot wait until the scientific uncertainties have been resolved, that developed countries should take the lead and that developed countries should compensate developing countries for any additional costs incurred in taking measures under the convention.

We can regard the IPCC as having spent the two years from 1998 to 1990 in establishing and publicising the scientific issues and having brought into the discussions on climate change a range of countries that had hitherto been uninterested and ill-informed. This work of the IPCC, together with continuing public concern, established the political conditions for the start of the formal negotiations which eventually led to the FCCC. In December 1990 the UN General Assembly took control of the negotiating process directly (unlike the ozone negotiations and the IPCC, which had been under the auspices of the United Nations Environment Programme [UNEP] and the World Meteorological Organization [WMO]) and negotiations began in February 1991. According to the analysis of Grubb *et al.* (1993) the gulf between the negotiators was enormous. Some were already convinced about the seriousness and urgency of the problem and wanted a convention which contained binding commitments to limit greenhouse gases. Others recognized the problem, but wanted to tread more cautiously, providing first a general legal framework, leaving any binding measures to subsequent negotiations. Some developing-country participants even feared the whole issue as a conspiracy by Northern (or developed or industrialized) countries to impede their development, and their main instructions seemed to be to avoid any commitments. The final draft which was eventually produced as the basis for the final negotiations represents a snapshot of an evolving political process, in some respects incoherent and in many details undefined or deliberately ambiguous.

In the analysis of Grubb *et al.* (1993) the FCCC represents a considerable achievement. It succeeded in one of its primary aims – that of providing an international legal framework and set of principles which was acceptable to almost all the countries involved. It accepts that climate change is a serious problem and that addressing it is primarily a responsibility of the industrialized countries, given their relative wealth and the fact that they have been responsible for the bulk of the accumulated greenhouse-gas emissions to date. The FCCC commits the developing countries to nothing unless the developed countries pay for it. The weakness of the FCCC lies in the fact that, although it accepts the existence of the problem and despite accepting the value of a precautionary principle, there is no commitment to any quantitative reductions in greenhouse-gas emissions. The FCCC merely indicates that industrialized countries should aim as a first step to return greenhouse-gas emissions to 1990 levels by 2000.

The FCCC which formulates certain responsibilities taken by the 154 States which signed it (except Malaysia, which called it meaningless because of its lack of binding targets for greenhouse-gas reductions) was a product of long evolution which was finally completed (on a working level) at the fifth session of the

Intergovernmental Negotiating Committee on the FCCC. Regrettably, the final version of the FCCC approved by UNCED-2 (the FCCC took effect on 21 March 1994) appears to be the worst possible variant, being vague and ambiguous even within the 'greenhouse' concept of the document. Berreen and Meyer (1992) have stated, quite correctly, that the basic proposals formulated in the FCCC are nothing else but vague declarations concerning the undesirable nature of excessive greenhouse-gas emissions. Serious criticism of the FCCC can be made and two issues of particular importance can be considered.

1. The two conventions (FCCC and CBD) and the statement on Forest Principles were prepared by different working groups as independent, non-linked documents. The preparation of the FCCC is a spectacular example of the inconsistency of such an approach. The point is that the solution of environmental problems is impossible without taking due account of the interactivity of its components. The ecological well-being of the Earth is determined by the presence of a slightly disturbed biosphere that is capable of recovering itself. The biosphere, which functions on the basis of the Le Chatelier principle, ensures the compensation of disturbing forcings and the regulation of the environmental properties to a great extent, owing to the biodiversity existing in nature. This reveals a close interrelationship between the problems of climate, biodiversity and forest dynamics. The principle of the unity of nature is a fundamental one, and by ignoring it one can only find oneself in trouble. Hence the conclusion: further efforts in the development of the documents under discussion should be based on a close and efficient co-ordination, which can only be constructive if the key aspects of the interactivity of the problems considered are agreed upon at the preliminary stage.
2. The 1992 supplement to the IPCC report states, with regard to the global climate warming observed during the last century (within 0.3–0.6°C) (Houghton *et al.* 1992):
   - '... the size of this warming is broadly consistent with predictions of climate models, but it is also of the same magnitude as natural climate variability. Thus the observed increase could be largely due to this natural variability; alternatively, this variability and other human factors could have offset a still larger human-induced greenhouse warming;
   - the unequivocal detection of the enhanced greenhouse effect from observations is not likely for a decade or more'.

By the time of the IPCC 1995 report (Houghton *et al.* 1996) there were two apparent developments. One was a shift towards the view that the observed warming trend during the twentieth century 'is unlikely to be entirely natural in origin'. The other was the result of more extensive studies of the effects of aerosols; it was concluded that tropospheric aerosols of anthropogenic origin contribute a negative component to climate warming (i.e. they have a cooling effect that, to some extent, can be expected to offset the warming due to the increase in greenhouse-gas concentrations).

Thus, it clearly follows from the various IPCC documents that there are no grounds for a one-sided support of the stereotype of greenhouse warming and for putting forward the problem of greenhouse-gas-emission reduction as the central problem if we are preoccupied with the task of preventing anthropogenically induced global climate change. It is most surprising to see that this kind of concept has been made the basis of the FCCC. Since this subject has been treated at length by Kondratyev (1998, 1999), it will suffice here just to mention that a vital factor of anthropogenic impact on global climate is the destruction of the biosphere which leads to increased emissions of CO_2 in the atmosphere by the continental biosphere. This is a consideration of primary importance; hence, we must first of all be concerned about the preservation of the biosphere, which is the main factor of global ecological safety. The problem of forest dynamics should be considered in this context.

2.4.2 Convention on Biological Diversity (CBD)

The CBD aims to preserve the biological diversity of the planet, through the protection of species and ecosystems, and to establish terms for the associated uses of biological resources and technology. It affirms that states have 'sovereign rights' over biological resources on their territory the fruits of which should, however, be shared in a 'fair and equitable' way on 'mutually agreed terms'. The CBD requires that countries develop plans to protect biodiversity and submit some information on them.

Biological diversity, or biodiversity, is the global composite of genes, species and ecosystems. The CBD originated with UNEP. The rapid acceleration of loss of species and ecosystem degradation has caused rising concern among scientists, policy makers and the public. UNEP estimates that there are about 30 million species on the Earth, of which only about 1.5 million have ever been described and, moreover, that about a quarter of the Earth's species risk extinction over the next 30 years. Consequently in 1987 governments were called on by UNEP to examine the possibility of establishing an international agreement on the conservation and sustainable use of biodiversity. UNEP set up a series of meetings which started in 1998 and eventually evolved into the Intergovernmental Negotiating Committee for a Convention on Biological Diversity. Negotiations soon focused on an argument between countries over the exchange of biological resources in return for access to technologies, particularly biotechnology. Countries with the greatest diversity of species, most of which are developing countries, insisted on sovereign rights over genetic resources within their territories, as well as access to benefits resulting from the use of such resources in the shape of biotechnology. These demands were unacceptable to certain countries, which regarded biodiversity as a common heritage of humankind and biotechnology to be the concern of industry and best advanced without too much interference by governments. The final draft which was produced for UNCED-2 in Rio was inevitably a compromise document.

As is the case with other documents, the CBD is the result of continuous efforts in the adjustment of contradictory interests of various states. The U.S.A. refused to sign the convention in contrast with 153 other states who did sign it; President Bush protested that it posed a threat to the US biotechnology industry and put American jobs at risk. However, the USA did eventually sign the convention in 1993.

The CBD commits the participating states to take substantive action in many key areas, including:

- the development of national plans, strategies or programmes for the conservation and sustainable use of biodiversity;
- to construct inventories and to monitor components of biodiversity and of processes adversely affecting it;
- the development and strengthening of *in situ* mechanisms for biodiversity conservation both within and outside protected areas;
- the development of *ex situ* mechanisms for biodiversity conservation as a complement to *in situ* approaches;
- the restoration of degraded ecosystems and recovery of endangered species;
- the regulation of the release of genetically modified organisms;
- the preservation and maintenance of indigenous and local systems of biological resource management and equitable sharing of benefits with local communities;
- the promotion of research, training and public awareness;
- the assessment of impacts on biodiversity of proposed projects, programmes and policies.

Miller and Barber (1992) expressed the view that the CBD contains numerous unconvincing and/or contradictory statements which can be reduced to the following:

1. The provisions on intellectual property rights and the transfer of technology are exceptionally confusing, and several industrial countries have expressed their discomfort with these sections, to the extent that we have already noted that one – the U.S.A. – refused to sign the convention on the grounds that the property rights of biotechnology companies (and their patented technological processes) would be compromised and the growth of that industry stunted. This may require a separately negotiated protocol to elaborate more precise statements in the CBD.
2. An important financing mechanism is the Global Environmental Facility established as an interim structure. Its status requires revision, with special emphasis on the governance of the Global Environmental Facility.
3. Policy and institutional issues that the Global Biodiversity Strategy (launched by the World Conservation Union [WCU], the World Resources Institute [WRI] and UNEP in February 1992) identified as key ones for conserving biodiversity and reducing the Third World debt, were treated weakly or not at all in the CBD.
4. The CBD does not take into account completely the possibilities of non-governmental organizations.

As a result of the prolonged arguments over access to genetic diversity and technology transfer, the CBD concentrated too heavily on these issues and on financing, at the expense of focusing on methods to conserve biodiversity. For instance, there is no article in the CBD requiring the construction of lists of globally important areas and species on which to focus international attention; this was largely because developing countries saw them as potential threats to their sovereignty. The CBD is also open to the criticism that it places too much emphasis on biotechnology and genetic modification as solutions to reverse the loss of biodiversity in nature. While the CBD acknowledges the importance of patent protection for modern biotechnology, it fails to give concrete protection to cultural diversity, traditional knowledge and inventions of indigenous peoples and rural communities.

In addition to these problems one important point should be stressed: the CBD does not reflect the fact that biodiversity is the basis of global ecological safety because it is a factor securing the sustainability of ecosystems and the biosphere as a whole. In this connection, an issue of great concern is the low level of biodiversity studies, in which further progress is most urgently needed and which must include documenting of the structure, composition, global distribution and analysis of the entire hierarchy of biosystems from molecules to the biosphere, and also the development on this basis of rational measures of biodiversity protection on local, regional and global levels, with necessary accounting of the unique nature of every form of life on the Earth. The solution of these problems requires multidisciplinary studies on the basis of an adequate information base and modern systems of monitoring, taking proper account of the hierarchy, dynamics and spatial distribution of biosystems.

Perhaps Grubb *et al.* (1993) should have the last word on the CBD: 'with all its limitations the Convention on Biological Diversity is still a considerable achievement. But it is best viewed as the first step in the climb of a very steep mountain'.

The Intergovernmental Committee on the Convention on Biological Diversity has adopted a recommendation for the necessity of regular conferences (beginning with 1993) for the co-ordination of efforts aimed, in particular, at the promotion of the necessary studies and projects. We would like to hope for provision in the future of a much broader interdisciplinary base for working out relevant recommendations than was the case previously.

2.4.3 The Rio Declaration on Environment and Development (RDED)

Initially, the basic document of the Rio Summit was intended to be the so-called 'Earth Charter'. The vision of the UNCED Secretary-General, Maurice Strong, was of a short, uplifting, inspirational and timeless expression of a bold new global ethic. But, to quote from Grubb *et al.* (1993), 'under the guidance of delegates who felt anything but uplifted and inspired, it became a distillation of many of the conflicts and political differences which infused the whole of the (UNCED-2) conference agenda'.

Since the Preparatory Committee had not managed to reach a consensus with regard to the contents of the 'Earth Charter' (a well-balanced statement concerning priority issues in ecological studies), it was replaced by a far more vague and rhetorical document entitled 'The Rio Declaration on Environment and Development'. This declaration comprises twenty-seven principles for guiding action on environment and development. Many address development concerns, stressing the right to and need for development and poverty alleviation. Principles concerning trade and environment are ambiguous and in some tension; others concern the rights and roles of special groups.

The following two theses of the RDED appear to be very significant:

> All states and all people shall co-operate in the essential task of eradicating poverty as an indispensable requirement for sustainable development, in order to decrease the disparities in standards of living and better meet the needs of the majority of the people of the world.
>
> To achieve sustainable development and a higher quality of life for all people, states should reduce and eliminate the unsustainable pattern of production and consumption and promote appropriate demographic policies.

According to Grubb *et al.* (1993) the RDED is far from the original idea of an inspirational Earth Charter. 'It is predominantly socio-political in content, and not as ecological as some may have wished. It is a typical piece of UN negotiation, which some claim reduces every principle to the lowest common denominator. In other words it is not a pointer into the future, but a snapshot of history ...'. It is hardly worthy of further consideration here.

2.4.4 Agenda 21

Agenda 21 is intended to set out a master plan or an international programme of action for achieving sustainable development in the twenty-first century throughout the world. One can hardly doubt that the most important but surely most difficult thing in securing sustainable socio-economic development is the necessity of adaptation to the resources of the global biosphere. The statement adopted by the reconvened World Commission on Environment and Development (WCED 1987) (the Brundtland Commission) focuses general attention on the following:

1. Broadening of discussions on population to include the economic and political causes of poverty. Poverty, environment and population can no longer be dealt with as separate issues.
2. The necessity of writing off the commercial bank debt of the Third World which reached the figure of $1.3 thousand billion. The economic policy, standards of production and consumption of industrialized countries put a limit on the capacity of the developing countries to achieve a sustainable development and eliminate poverty.
3. Substantiation of the polluter-pays principle and working out of measures for the realization of this principle including prices which reflect full environmental costs.
4. Implementation of the official development assistance target of 0.7 per cent of GNP of developed countries by 2000.

The Brundtland Commission also made the very timely recommendation for the creation of a new, independent international body outside the UN system for the monitoring, analysis and assessment of the state of the environment.

Agenda 21 seeks to be comprehensive in its scope, and to make recommendations on the measures which should be taken to integrate environment and development concerns. To this end, it provides a broad review of issues pertaining to sustainable development, including statements on the basis for action, objectives, recommended activities and means of implementation. These are based on experience and analysis of the issues, combined with the interests of different parties, including interests brought forward by both developing and industrialized countries. Just about any issue related to environment and development receives mention.

Agenda 21 is an immense document, of 40 chapters running to over 500 pages, outlining an action plan for sustainable development, covering a wide range of specific natural resources and the role of different groups, as well as issues of social and economic development and implementation. It effectively integrates environment and development concerns; it is strongly oriented towards 'bottom-up', participatory and community-based approaches in many areas, including population policy. It is not simply the product of a committee of experts or of a commission; it is a negotiated document agreed among more than 150 governments. However, it is not a legally binding agreement on those states which have signed it. Governments are not required to follow every detail of each recommendation. This, and the fact that it was prepared through a rather broad participatory process, allowed a different flavour to emerge from that which is typically found in international conventions. It shows more acceptance of market principles, within appropriate regulatory frameworks, than is usual in UN agreements. Compared with most international conventions Agenda 21 contains far more about the nature of the problems, aims, possible approaches and desirable policies, often expressed in rather general and non-committal ways.

An extensive chapter-by-chapter analysis of the Agenda 21 document will be found in Chapter 9 of Grubb *et al.* (1993). A few of the more general points will now be summarized. One underlying theme which emerges clearly is a 'bottom-up' approach which places the emphasis on people, communities and non-governmental organizations. For example, the chapter on poverty emphasizes the need to address the basic needs of the poor, to improve their education and strengthen their opportunities and participation in local governance. This contrasts with previous emphasis on international sponsorship for grandiose public projects intended to boost local economies and generate wealth and employment through the expenditure and resulting supply of cheap resources, such as big electricity projects. Disillusion with such grant project-oriented planning is evident throughout Agenda 21. Agenda 21 hardly treats such important issues as the population and consumption patterns, Third World debt, the role of the military, the export of hazardous wastes, deforestation, etc.

As stated in the preamble, governments carry the primary responsibility for implementing Agenda 21. They agreed the document and have the means to develop the regulatory measures required, and are the relevant authorities to decide upon national and international financial disbursement. In practice, however, it is inevitable that although nearly all governments did participate in the negotiations and agree the final text, they will vary considerably in the seriousness with which they take the exercise, and in their readiness to accept and act upon the ideas embodied in Agenda 21. Yet many other actors will be involved – local authorities, industry, non-governmental organizations, etc. And a range of UN and other international bodies will be expected to take full account of relevant recommendations for action. Those who pushed so hard to change attitudes and language in Agenda 21 will have to continue their efforts if the language of the document is to be translated into reality. Annual expenses required for the implementation of Agenda 21 recommendations have been estimated at $21 billion.

The Commission on Sustainable Development (CSD) was created within the United Nations to oversee the implementation of Agenda 21. CSD is responsible for co-ordination within the UN system and reports to the General Assembly through the Economic and Social Council of the UN on the implementation of the programme of Agenda 21. However, the CSD cannot 'enforce' the agreement in any legal sense. The real impact of Agenda 21 will depend on the extent to which national governments and all the various groups discussed in the document – from local councils and trade unions to scientific groups, business and industry – absorb and pursue the recommendations therein, influenced also by the continued efforts of environment and development groups.

2.4.5 Non-legally binding authoritative statement of principles for a global consensus on the management, conservation and sustainable development of all types of forests (in short, the Forest Principles)

The 'Forest Principles' form the rump of blocked attempts to negotiate a convention on forests. It emphasizes the sovereign rights of states to exploit forest resources along with various general principles of forest protection and management. It was not until the early 1980s that serious concerns were voiced about the state of the world's forests. Assessments by the Food and Agriculture Organization (FAO) of the United Nations showed that the tropical forests were disappearing at a rate of over 11 Mha per annum. The effects of deforestation could be seen in economic, social and environmental terms, but research showed that a range of factors were responsible, including poverty, debt, trade, high population growth rates, landlessness, inappropriate policy planning and poor forestry practices. To solve the problem would therefore need a broad-based and concerted effort.

During the early 1980s a number of unilateral and multilateral responses were proposed. However, in spite of these efforts it became increasingly clear in the late 1980s that the attempts to reduce the rate of tropical deforestation were not being successful. In early 1990 the FAO repeated its tropical forest survey. The results showed that, despite the various efforts which had been made, the rate of tropical deforestation had increased by 50 per cent since the 1980s.

It was then argued that a more heavyweight organization would be needed to make countries accountable for protecting their forests. Also towards the end of the

1980s there was a movement away from talking about tropical forests in isolation as it became more clearly apparent that the world's natural temperate and boreal forests were also under threat (and retreating). This led to the idea that what was needed was a 'global forest agreement' which would require all the world's governments to commit themselves to looking after their own forests and ensuring that their national policies did not promote the misuse of forests overseas.

In 1990 several factors contributed to the idea that since efforts were being made to develop binding agreements on climate change and on biodiversity for signature at UNCED-2 in Rio, then another proposed binding convention should be put forward as well on tropical forest protection. However, governments of the tropical nations began to become worried that the efforts to draw up a global forest convention (so strongly promoted by the Northern nations) were little more than a mechanism to prevent developing nations from using their forests in the way that they wished to do. Developing nations also viewed the proposals for a forest convention as a mechanism for the North to divert attention away from the main issue of reducing greenhouse-gas emissions. There was widespread and open opposition throughout the tropical countries to any form of legally binding agreement on forests. Throughout the UNCED-2 process it was clear that there was no consensus on how to manage the world's forests wisely. The initial attempts to launch negotiations on a convention were rapidly transformed into discussion on a set of 'principles for forest management', and repeated attempts to elevate this into negotiations on a convention failed in the face of very strong opposition from several key developing countries. The deadlock was never really broken in Rio, and the resulting document on Forest Principles, with its most unwieldy title, bears the scars of what became a very difficult negotiation. The discussion was not really about forests, forestry or even sustainable management, but was much more to do with international politics and sovereignty. Thus, the negotiations on the Forest Principles highlight the formidable obstacles to achieving an effective convention on forests, but nevertheless do represent a modest step forward towards an international regime for forest management.

The text of the Forestry Principles is discussed in some detail by Grubb *et al.* (1993) and their general conclusions are worth noting. UNCED-2 failed to reach a 'minimum' agreement on a forward-looking statement of principles that would have recognized the interconnections between forests and issues such as biological diversity and climate change, as well as recognizing the role of forests within a broader land-use context. Nevertheless, many believe it was a miracle that the Forest Principles were agreed in Rio, considering the range of views between states and the short time period available for negotiation. Agreement was eventually achieved by a process of avoiding specific commitments or contentious principles and incorporating a wide range of generalized observations, recommendations and goals, sometimes in themselves ambiguous or perhaps contradictory. The result is a document with something for everyone, without any clear message or direction.

Although to many the Forest Principles are disappointing in that they fail to provide a legal instrument for stimulating the protection and improved management of the world's forests, they do provide a framework which can be used to hold national governments to a number of principles. While paving the way for future efforts, the Forestry Principles quite appropriately focuses our attention on the fact that '... the programmes for agriculture and rural development, and for water resource development seek to combine a basic development concern with sound environmental management'. Furthermore, the statement declares that 'It is necessary to recognise the key role of all types of forests in the maintenance of ecological processes and balance on local, national, regional and global scales through protection of unstable ecosystems, water catchment areas and fresh water resources, and also as accumulators of biodiversity and biological resources and sources for genetic material for biotechnological production, as well as for carbon.' Mere acceptance of the need for some form of international co-operation on forests was for some developing countries a substantial pyschological step.

2.4.6 UNCED-2 Conclusion

UNCED-2 has doubtless played an important role in attracting public attention towards ecological problems of vital value. However, to achieve the desired results, additional, more strenuous efforts are required. The scale and complexity of the whole problem are unprecedented.

A significant book on the outcome of UNCED-2 was written by Al Gore who was the US Vice President. In this book Gore (1993) suggested that we start the implementation of a 'Global Marshall Plan' using as a model the Marshall Plan implemented in Western Europe after World War II was over. This would include the development of a Strategic Environment Initiative. Gore offered a detailed characterization of measures necessary to implement the Strategic Environment Initiative in various fields: agriculture and forestry, energy, transport, communications, waste reduction and recycling, etc. Gore proposed a comprehensive and ubiquitous change in the economic 'rules of the road' by which we measure the impact of our decisions on the environment. He suggested concrete measures that require the attention and discussion at an ecological 'summit' of world leaders.

In this context, Gore's opinion is that the task for the USA as a leader in the free-market economy is the development of efficient measures to use the market power for the saving of the global environment. Gore extends criticism of President Bush's administration concerning its inability to achieve any success along this path, in spite of the existence of some realistic opportunities (e.g. the realization of the principle 'let the polluter pay' [a special tax for polluting the environment]). In this connection Gore formulated a number of specific proposals relevant to the USA. Gore proposed the negotiation and approval of a new generation of international agreements concerning environmental problems (the Montreal Protocol on Ozone can be the prototype of this kind of agreement). He also proposed the establishment of a co-operative plan for educating the world's citizens about our global environment.

An integrating objective should be the creation (both in developing countries and in developed countries) of social and political conditions that might ensure the transition to the community of sustainable development throughout the world (the notion of 'sustainable development' should, however, be much more precisely defined). Central here must be an effort to convince people everywhere on our planet that the global environment is part of their 'backyard', as it really is. Regrettably, the famous 'not in my backyard' syndrome (NIMBY) has often been used as a powerful political force. Gore's final words in his book are as follows: 'For civilization as a whole, the faith that is so essential to restore the balance now missing in our relationship to the Earth is the faith that we do have a future. We can believe in that future and work to achieve it and preserve it, or we can whirl blindly on, behaving as if one day there will be no children to inherit our legacy. The choice is ours; the Earth is in the balance.'

The former US Vice President, Al Gore, has published a very important book written on a high professional level, where two most important (and so often underestimated) problems have been analysed in a most convincing manner:

- the Earth's balance (in its various manifestations) as the basis for global ecological safety;
- the necessity for the prevention of further non-controlled population growth (few politicians have demonstrated enough courage to face this vital, though painful and acute, problem).

A more extensive discussion of the contents of Gore's book is given by Kondratyev (1998).

2.5 KYOTO PROTOCOL

In Section 2.4.1 we noted the vagueness of the FCCC that was signed in Rio in 1992, which caused a sharp response on the part of Malaysia, and its inadequate conceptual basis. It was realized that further serious efforts were needed to improve the FCCC and so it was decided to continue the functioning of the IPCC. However, no purely mechanical continuation of the previous activity will be a success. What is necessary is the broadening of the group of experts involved in relevant developments which must be enriched with regard to ideological aspects. A considerable increase in the efficiency of international co-operation is also absolutely necessary (see further in Kondratyev and Donshenko 1999). In recent years annual conferences devoted to discussions of the FCCC problems took place. The most important of these was the conference in Kyoto in December 1997 at which participants discussed and approved the so-called Kyoto Protocol which specified levels of recommended reductions in greenhouse gas emissions to the atmosphere (UNFCCC 1997). This rather controversial document has been discussed in detail by Kondratyev and Donshenko (1999) and by Grubb *et al.* (1999). While the FCCC provided the legal and political foundations for international action, it did not, however, actually specify levels of reductions of greenhouse-gas emissions. In 1995 it was decided to open new negotiations.

US proposals for specific binding quantified commitments, implemented with flexible mechanisms including international market instruments, defined the core structure of the Kyoto Protocol, while pressure from the EU provided an impetus for stronger commitments than would otherwise have been possible. Renewed if fragile consensus among the developing countries to back strong industrialized-country commitments, alongside measures to ameliorate adverse impacts, removed the last hurdle in principle, but enormously divergent perceptions and interests remained. The Protocol was adopted in December 1997 under intense pressure from the deadline and desire for a meaningful agreement.

The heart of the Kyoto Protocol consists of specific limitations on the emissions of greenhouse gases from each industrialized country (listed in Annex I of the Convention) for the period 2008–2012, see Table 2.1 (Grubb *et al.* 1999). The aggregate aim is to reduce these emissions to about 5 per cent below 1990 levels, which is roughly the level already achieved in 1995 because of reductions in the economies in transition (i.e. the economies of the various countries of the former Soviet bloc). The reductions for Russia and Ukraine were set at zero in the Kyoto Protocol because of the large decline in heavy industry, and the consequent large decline in greenhouse-gas emissions in those countries over the period 1990–1995 following the collapse of communism. The headline commitments for all the OECD (Organization of Economic Cooperation and Development) countries, except possibly Australia and Iceland, would require strong action, particularly if applied to domestic CO_2 emissions in isolation. However, the collective Annex I goal is far weaker and the basic OECD commitments are modified by flexible mechanisms that serve both to weaken and redistribute the effort much more widely. The Protocol contains no proposed limitations on the emissions from developing countries. Eighty-four countries signed the Protocol, including all the industrialized countries in Annex I, except Hungary and Iceland.

Source and sink flexibility is likely to mean that total industrial CO_2 emissions exceed the collective commitment by several percentage points (circa 5 per cent with a wide margin of uncertainty):

- the inclusion of five other greenhouse gases, notably CH_4, eases significantly the commitment, as compared with CO_2 only, for most industrialized countries other than Japan and Iceland; across Annex I these other gases account for about 20 per cent of base year emissions;
- the inclusion of carbon sinks – 'direct human-induced changes' in land use initially limited to 'afforestation, reforestation, and deforestation', which lead to the absorption of CO_2 – will be very important for some countries and may amount to a few per cent of Annex I emissions overall, depending on the detailed rules adopted.

International flexibility allows national emissions to diverge widely from the initial national commitments as a product of resource transfers and foreign investment. The three mechanisms for international emission transfers (joint implementation and emissions trading within the Annex I group of countries, and the clean development mechanism involving the developing countries) represent international

Table 2.1. Emissions and commitments in the Kyoto Protocol (from base year) (after Grubb *et al.* 1999).

Country	Commitment (% from base year)	1995 emissions (% from base year)	Other greenhouse gases† (%GWP in base year)
OECD			
Australia	+8	+4.8	33.8
Canada	−6	+9.3	22.9
European Union‡	−8	−3.4	21.6
Iceland	+10	+4.3	25.7
Japan	−6	+10.7	8.0
Liechtenstein	−8	n.a.	n.a.
Monaco	−8	+25.2	2.7
New Zealand	+1	+0.5	65.1
Norway	0	+1.7	35.6
Switzerland	−8	−1.8	16.1
United States	−7	+5.8	17.6
EITs			
Bulgaria (1988)§	−8	−35.7	28.8
Croatia‖	−5	n.a.	n.a.
Czech Republic	−8	−21.5	13.9
Estonia	−8	−44.4	7.2
Hungary (1985–7)§	−6	−24.0	17.7
Latvia	−8	−46.2	30.6
Lithuania	−8	n.a.	23.3
Poland (1988)§	−6	−22.2	15.5
Romania (1989)§	−8	−38.1	28.8
Russian Federation	0	−29.3	22.0
Slovak Republic	−8	−25.2	17.2
Slovenia‖	−8	n.a.	27.5
Ukraine‖	0	n.a.	22.7

† Share of the greenhouse gases other than industrial CO_2, but excluding LUCF (Land-Use Change and Forestry), in the base-year emissions, which account for a differing trend by 1995 compared with CO_2 only.
‡ The fifteen countries of the European Union are listed as each having a target of −8%. These targets were subsequently redistributed under the 'bubbling' provision.
§ EIT (Economies In Transition) that use a base year other than 1990.
‖ Base-year data not agreed, data refer to 1990.

market instruments that are unprecedented, controversial and poorly defined. Much will hinge on specific rules to be developed through further negotiation. Trading is envisaged in general among Annex I countries. Moreover, as far as the EU is concerned there was a 'bubbling' provision that enables that group of countries to redistribute their emission commitments in ways that preserve their collective total. The different units will be subject to separate regimes of monitoring, certification, accountability, etc.

The Kyoto Protocol provides a menu of possible and mandated specific actions that will serve as hooks for future negotiations:

- a list of specific policies and measures will serve as a checklist for national reviews and peer pressure but is unlikely to gain a stronger role for some years;
- bunker fuels for international shipping and aviation, which are not included in national inventories, will come under rapidly increasing pressure for separate action;
- developing countries including oil-exporting countries will use key references in the Protocol to further debate on ways of minimizing any adverse impacts upon them.

Provisions concerning developing countries will support a significant expansion of their involvement in the process and capacity building, including further development of national inventories, programmes and reporting, and a new phase of negotiations on ways of promoting appropriate technology transfer. Originally there was a proposed article on the adoption of voluntary commitments to emission reductions by developing countries – the quid pro quo for including emissions trading and the clean-development mechanism – but this was lost. This leaves the process by which new countries might be brought into the structure of specific commitments, at least in the first period, legally and procedurally unclear. This represents the only significant failure of core US objectives in the negotiations. In all other respects the structure of the Protocol is very much as sought by the US administration and represents a testament to the dominance of US power, and the continuing weakness of foreign policy in the EU, Japan and elsewhere, at the close of the twentieth century.

The political and economic circumstances in which the Kyoto Protocol was negotiated led to some strange emission allocations, and an overall pattern that may induce large-scale emissions trading, principally from the transition economies to the United States, Canada and Japan. The allocations of Russia and Ukraine, in particular, could represent a surplus amounting to 5–10 per cent of total Annex I assigned amounts. Transferring this surplus to other parties would undermine the legitimacy of emissions trading and set dangerous precedents for the subsequent evolution of commitments and the trading system.

Grubb *et al.* (1999) estimate that, if emissions trading is wholly unrestricted and conducted purely to minimize costs in the first commitment period, the energy-sector CO_2 emissions during the first commitment period could be:

- 25–30 per cent above 1990 levels in Australia and some of the other smaller OECD countries;
- 10–15 per cent above 1990 levels in the United States, Canada and Japan;
- around 1990 levels in the EU, with increases in most of the EU offset by reductions principally in Germany and the UK;
- 25–35 per cent below 1990 levels in Russia and Ukraine, with wide variation across the other transition economies.

In this scenario the outcome could differ from the headline commitments to a far greater extent than the negotiators ever conceived and would clearly be inadequate to lead efficiently towards atmospheric stabilization below a doubling of pre-industrial CO_2 levels, or to induce necessary technological changes. Indeed, under some scenarios we find real resource-abatement costs to be negligible, although Russia and Ukraine could still extract a price as effectively duopoly suppliers. The implied scale of West–East transfer of both emissions and finance would also lead to considerable political debate. Extensive use of sinks or the clean development mechanism would further reduce the level of effort required in Annex I countries, and would allow Annex I emissions to grow from the 5 per cent reduction already achieved by 1995, violating the stated collective aim of the commitments.

The clean development mechanism for trading between industrialized countries and developing countries is the most unexpected and original of all Kyoto's innovations. It is a leap into *terra incognita* that is unlikely to work simply as a way to distribute abatement efforts globally at least cost. The difficulty of ensuring that crediting reflects real and additional emission reductions is compounded by the paradoxes that the most 'cost-effective' projects may be the least 'additional' and that strict project additionality would give perverse policy incentives. Other eligibility criteria will introduce several factors other than just greenhouse-gas limitation.

Specific design of the clean-development mechanism faces dilemmas of integrity and size. Rigorous rules and strong governance imply high administration costs, thereby deterring investment. Weak rules and governance may attract more investment at the risk of corrupting and discrediting the system, as well as undermining the criteria of sustainable development and the objectives of the Protocol, since 'maximising emission credits' under the clean-development mechanism is equivalent to 'minimising action within Annex I'. At best, the clean-development mechanism could encourage large flows of foreign investment towards sustainable and low-greenhouse-gas development. At worst, it could become not just a source of spurious emission credits, but a sink for the intellectual as well as some physical resources of the developing world, and a distraction from the fundamental goals of sustainable development. However, for both good and bad, the clean-development mechanism, as generally conceived, is likely to be constrained by the inherent limitations of credit-based, project-based systems, and for some countries it may be primarily a bridge towards adopting commitments and joining the international trading system.

However, the clean-development mechanism may also evolve to perform a more fundamental and long-standing role. The defining purpose of the clean-development mechanism should be to help direct foreign corporate investment towards goals of sustainable development in its many forms and interpretations, according in part to national preference. In this it should be complementary to the Global Environment Facility. Indeed, if appropriately designed, these institutions should correspond to the natural complementarity between the private and public sectors at a global level.

Developing countries, as formally represented by the G77 and China, have firmly resisted efforts to open discussions on quantified emission commitments and have blocked procedural routes to voluntary accession to the Kyoto Protocol, fearing that individual countries would be pressured into joining by the might of the OECD. The status of countries seeking to join the trading system without undertaking full Annex I obligations remains unclear. Assuming procedural issues can be addressed, accession of new countries individually is one route towards developing a global regime. This would tend further to inflate the trading system. Thus, one path may be for the Kyoto regime to evolve by accession and inflation, and loose development of allocation and trading rules to attract new entrants. This is as much as seems plausible for first-period commitments. For subsequent periods an alternative might involve negotiations around principles of 'emissions convergence' towards more equitable long-term distribution of assigned amounts. A third approach could involve criteria for the evolution of commitments based around threshold indicators but this may be less appealing for the G77. The scope for compromise between the different models is not obvious, but possibilities include the concept of a 'convergence corridor' towards per capita and perhaps other proposed indices of equity and performance.

The Kyoto Protocol, which was concluded in December 1997, is an extraordinary and unprecedented achievement in international affairs, which may have a profound impact on industrial countries' production and consumption habits. It is not a perfect international agreement, but it is the best compromise the international community was able to reach at that time. It was negotiated among a universal representation of the international community, including governments concerned with climate change and governments with totally different concerns. Some negotiators, particularly but not only OPEC (Organization of Petroleum Exporting Countries), were reluctant to agree measures or targets that might induce reductions in fossil-fuel consumption. Others, particularly but not only the USA, were trying to protect their lifestyles by advocating a reduction in other countries' emissions. Developing countries were defending their right to make the same mistakes that industrial countries had committed before, although real economic development may not follow the old growth patterns; on the contrary, they are 'leapfrogging' between technologies, skipping many processes which are now obsolete. Industrial countries in general behaved jealously, overprotecting their competitiveness in a context that revealed wide differences in their energy-production profiles. Nevertheless, the Protocol's inclusion of binding commitments with varied flexibility is more sophisticated than many political analysts considered possible. The many other elements – spanning global 'soft' commitments, related processes on technology transfer and financial mechanisms, policies and measures, minimization of adverse impacts, sinks and compliance mechanisms – represent a complex package designed to gain and sustain participation. The central structure of successive 5-year commitment periods provides a natural basis for the evolution of the regime over time.

The transfer mechanisms potentially establish new markets based upon resources (emission credits and assigned amounts) that are purely human constructs – harnessing the greed of nations, and their industries, in the cause of climate protection. The incentives to expand ownership and access to these resources will be immense. Lax rules and wide scope for joint initiatives and the clean-development mechanism would maximize the potential credit generation in transition economies and developing countries respectively, but lessen the value of credits and action within the OECD. The major OECD importers may seek permissive rules on both the project mechanisms and trading, so as to maximize their access to cheap-emission credits and to ease the struggle for ratification that may be long and difficult in some countries. However, the net result of all this would be massive inflation – resulting neither in significant resource transfers to anyone, nor in significant progress in tackling climate change. This may be restrained by competition between the mechanisms and regions, and by injecting a firmer sense of the overall constraints on emissions. Implementing the Kyoto Protocol without reference to long-term goals and trajectories would be like inviting treasuries to print money without having any concept of inflation.

Almost everything about the governance of these mechanisms remains to be determined. We have yet to develop a clear understanding, let alone consensus, about the international social, ethical and legal foundations upon which such instruments need to rest. Other aspects of the Kyoto Protocol, including its provisions on policies and measures, minimization of adverse impacts, technology transfer, and compliance procedures, all require much further development to become effective.

Yet the Protocol has already had two fundamental achievements. First, its adoption has persuaded the private sector that the world will indeed start getting to grips with the problem of climate change, however slowly and ponderously, and indeed that there may be varied benefits accruing to those who move first. Second, it has advanced the debate on international economic instruments from whether to adopt them to how to implement them. The Kyoto Protocol could yet be destroyed: from within, by excessive national greed leading to uncontrollable inflation; or from without, if those who do not want any agreement, acting together with those for whom Kyoto is not good enough, can block ratification in key countries. Yet this is unlikely. Governments have already made major political investments to establish this nascent regime. It offers a solid basis, and there are no credible alternatives on offer. The world has started down an ambitious and efficient road to tackle its most daunting environmental problem, and there is no turning back. At the technical level there still remain problems associated with the reliability of monitoring emissions, with a lack of standardization of the form of reporting. There also remains the question of the policing and enforcement of the agreed emissions, not to mention the question of the penalties to be imposed on defaulting countries.

2.6 POPULATION GROWTH

Once humans ceased to be nomadic hunters and settled down and developed agricultural practices, such things as vegetation cover, evapotranspiration, hydrological-flow patterns and the radiation budget at the Earth's surface began to be affected by their activities. In the early days the anthropogenic effects on the climate, at a global scale, were quite small. However, with the Industrial Revolution and the vastly expanding human population since then, the situation has changed. The effect of human activities is not now simply a minor perturbation on the climate. Human activities now constitute a significant factor in causing climate change, both at a regional scale and at a global scale. The important question which is being addressed by the scientific climatological community is that of trying to disentangle the anthropogenic contribution to climate change from the natural factors that also lead to climate change. The UN International Conference on Population and Development (ICPD) in Cairo in September 1994 has been a cornerstone event.

The consequences of the Industrial Revolution for climate change are twofold. There are the environmental consequences of the various large-scale industrial and agricultural processes that have been developed. There is also the fact that there has been a massive growth in global human population since the Industrial Revolution and this, inevitably, has had environmental consequences of a large enough scale to have effects on the climate. In recent years the discussion of anthropogenically induced climate change has often concentrated on the question of greenhouse-gas emissions. However, to some extent, the question of greenhouse-gas emissions is symptomatic and it is not really the fundamental cause of the problems. It can be argued that the biggest problem of all is that of world-population growth. A source which we shall follow is that of Buckley (1994). There is nothing particularly unique about this source except that it does focus concisely on the fundamental issues and it also has a useful collection of up-to-date figures and diagrams in it (see also Brown *et al.* 1999).

In 10,000 BC the total population of the world was about 6 million, which was probably the sustainable maximum at that time, when human beings had to rely on hunting and gathering in order to survive. Extinction was a real possibility as humans fought with the elements and with other animals to keep their species alive. Life expectancy was only about 20 years and birth rates were only marginally above death rates. At that time, humankind's grip on life was precarious and its impact on the natural environment was negligible. As settled agriculture and weaponry developed over the centuries, human beings were able to produce food more easily and to defend themselves more readily. The result was to increase their life expectancy and prolong the years of potential parenthood, leading to a growth in population.

In the early Middle Ages, when world population was less than half a billion, human activity had already transformed the landscape over large areas, in particular by cutting down trees to clear land for farming and to provide wood for fuel and building materials. Annual population growth between 1000 and 1750 was only about 0.1 per cent. By 1750 average life expectancy was still only 27 years. Half of newborn babies died before reaching the age of 5. Although fertility was high, famines, disease and war prevented the population from expanding rapidly. At that time in Europe the total population of about 100 million was probably all that could be sustained by contemporary agriculture.

The Industrial Revolution changed the technology of food production and enabled a much larger population to be supported. At the same time improvements in health care reduced infant mortality and enabled people to live much longer lives. By the early 1960s, world population was growing at 2 per cent per annum, adding 70 million people annually. World population is now 6.0 billion (by the UN calculation, the world's six-billionth human being was horn in Sarajevo, Bosnia, on

12 October 1999 (among the 370,000 babies born on the same day), which is twice as big as it was in 1960 and increased by 1 billion people in the past 12 years. It will probably have doubled again by about 2050 (the minimum assessment is 8.9 billion in 2050). No one actually knows how many people the world can support. The world's population has grown from 1 billion in 1830 to 5.6 billion in 1994. Further growth to 8.9 billion seems inevitable, with a peak of 10.7 billion appearing likely. This explosive growth in world population is a recent phenomenon; the last century alone accounts for 75 per cent of all the population growth that has taken place in the last 12,000 years. The predicted population growth rate is 78 million people per annum. Feeding, housing and employing these extra 78 million people each year presents human society with a massive problem.

In many countries the scientific and technological revolution has led, however, to tragic social consequences. At present, 750 million people are starving and over 1 billion people live in dire poverty. In developing countries, 95 per cent of children are born, live and die poor, unaware of any scientific and technological progress. Almost 2 billion people drink unsanitary water. There are reports about outbreaks of diseases which are supposed to have been substantially eradicated (e.g. malaria and cholera). New diseases have sprung up, the most dangerous being AIDS. Drug abuse, violence and terrorism are rampant in developed countries, where the number of poor and homeless keeps growing as well. There has been a sharp increase in technological accidents, genetic and blood diseases and malignant tumours.

The discrepancies between the northern and southern hemispheres are increasing. In 1950 North America and Europe accounted for 22 per cent of the world's population, while Africa accounted for only 9 per cent. Now this proportion has been reversed. The debts of the developing countries of the southern hemisphere exceed US$1.2 trillion (i.e. one-half of their gross national product). The poverty, misery, disease and hopelessness in people's lives are fraught with unpredictable conflicts and the threat of violence.

Over 90 per cent of the population growth in the next 50 years will take place in developing countries, especially in Africa, many of which have serious social, economic and environmental problems. Some of these countries are seeing rapid economic growth and becoming massive, new, consumer societies. There are enormous implications of the continuing growth in world population and we ignore this at our peril. Humankind, and civilization as we know it, is sitting on a time bomb. All species strive to proliferate, but predators, disease and starvation limit populations which get out of hand and disturb the balance of nature. The question arises as to whether humankind, with its superior powers of intellect and reasoning, can keep its own population in check before nature applies its own brutal methods.

The ICPD, which was held in Cairo in September 1994, was arguably one of the most important gatherings in history. The stakes could hardly be higher, for the next 20 years will be a crucial test of humanity's commitment to protecting its own future. Population growth is a very serious problem, especially for the Third World; if population growth is not stemmed soon, it could overwhelm efforts to move the world to more sustainable forms of economic development. In the worst scenario, it could lead to famine, war and destruction of the global ecosystem. It is sometimes argued that views such as these are unnecessarily alarmist and that human ingenuity will find ways to cope with burgeoning population. Some say that poverty is the real problem and that the main effort should be devoted to helping the economies in the developing 'South' to catch up with those in the developed 'North'.

Both these proposed solutions – a rapid technological fix and the voluntary redistribution of wealth – require more faith in humanity than seems to be justified by the facts. Hunger already afflicts 800 million of the world's poor. Expansion of global food supply is being made harder by water scarcity and soil degradation. On a per-capita basis, the area of cropland is set to drop by one-fifth by 2010. Intensive-farming practices have indeed achieved remarkable yields, but they have also caused widespread damage to soil. The need to force the soil to produce even higher yields in the future is likely to accelerate the process of environmental damage. Genetic engineering can be expected to help with food supply and, indeed, it has already helped to some extent by producing new strains of crops with much higher yields; however, it is questionable whether it can be expected to be successful in the future on the necessary scale and within the time frame needed to avert disaster. In forty-six out of ninety-five developing countries, food output per head is declining. At present at least thirty-seven countries, mostly in Africa, are unable to feed themselves from their own resources.

As far as the possible redistribution of wealth is concerned, far from getting narrower, the gap between the haves in the developed countries and the have-nots in the developing countries is actually widening and the burden of Third World debt is becoming crippling. At present the world's richest 20 per cent of the population receive 82.7 per cent of global income and the poorest 20 per cent only receive 1.4 per cent, according to the UN's 1992 Human Development report. The disparity between the richest and poorest has doubled since 1960. The developed countries show little inclination to distribute their wealth among the poor or to curb their own extravagant lifestyles. Even small steps in this direction, such as increasing foreign aid or putting taxes on energy use, are steps which are too uncomfortable for most politicians to propose and for most voters to accept. Meanwhile, the vast army of urban poor in the developing world continues to grow, increasing the dangers of social unrest and aggressive international migration.

Although experience has shown that economic improvement eventually brings declining birth rates, the world cannot afford to wait for universal affluence to solve its population problem. Indeed, it is economic wealth which has already caused so much environmental damage. The 20 per cent of people who live in the developed world are reckoned to account for 80 per cent of global pollution and resource use. The action plan agreed at the Cairo conference recognized that solving the population problem involves a complex mixture of solutions involving education, gender equality, health care, sustainable economic development and family planning.

It was inevitable that progress would seriously affect the environment, since it was based on runaway exploitation of its resources. Economic estimates ignored the amortization of, and damage to, resources, the resultant hazards to people's health, the shortage of resources left for future generations and the more indirect consequences of the biospheric destruction. Technology that destroyed resources proved to be most lucrative both under the free-market economy and in planned economic systems.

The last 40 years of economic progress have been catastrophic for the biosphere. As much as 20 per cent of the upper fertile layer of agricultural land (1.5 Gha) has been lost, 20 per cent of tropical forest has been cut down, the annual rate of desertification has now reached 6 Mha and every year forest area dwindles by 11 Mha. Over the past 40 years the concentration of CO_2 in the atmosphere has risen by 13 per cent, the ozone layer has shrunk by 2 per cent annually, and an ozone 'hole' forms every year over Antarctica. Dead lakes and damaged forests have become a ubiquitous landscape feature in industrialized countries. Most rivers in Europe and North America remain polluted all year round. The pollution of reservoirs is mainly caused by dry and wet (rain- and snow-induced) deposition of gases and aerosols emitted by industrial enterprises and by the run-off of fertilizers and pesticides from agricultural land. Soil, sediments, plants and animals become repositories of dangerous toxins such as heavy metals, products of pesticide disintegration, plastics, radionuclides, etc. More than 100 species of organisms disappear every day, mostly in the tropical forests. On the whole, human beings appropriate as much as 40 per cent of the biological products created by photosynthesis, leaving a huge number of other organisms without food.

Developing and industrialized countries differ from one another in the nature of their impact on the environment, although in each case the impact is highly destructive. In developing countries, poverty, technological backwardness and rapid population growth lead to degradation of land, forests and non-renewable resources. Industrialized countries, which have amassed a tremendous industrial and energy potential, are the main consumers of resources (energy in particular) and, similarly, the main contributors of pollutants. In the United States, for example, the per-capita energy consumption is thirty-five times greater than that of India. In 1988, the amount of carbon emitted into the atmosphere in North America was six, two and half, and nine times greater than that in Africa, Latin America and South-East Asia, respectively. Thus, the results of socio-economic development over the past 40 years testify to a high degree of social degradation and sharp aggravation of the global ecological problems.

Ecological problems have been repeatedly dealt with by many state leaders at numerous scientific conferences and in various programmes and agreements. As before, however, this is usually mere rhetoric, which is not backed up by significant actions. What is most obvious is the uncertainty of views on the approaches necessary for the solution of ecological problems, the concept of ecological security and the ways and means of sustainable development.

2.7 SUSTAINABLE DEVELOPMENT

2.7.1 Need for sustainable development

Estimates of future world population vary enormously. It is impossible to be precise about the size of the population in 50 years time, but the present number of people of reproductive age makes further substantial growth inevitable. Even at the low United Nations' level projection, world population in 2050 would have tripled since 1950. Projections as far ahead as 2150 vary wildly, from a figure as high as 28 billion to an extremely low figure of 4.3 billion, which is lower than today, because of falling fertility rates. Assuming that the medium forecast is most likely to come about, world population will eventually peak at about 11.5 billion around 2150. Even the most optimistic scenarios see growth rising to 8.9 billion by the middle of the next century, before beginning to decline. Therefore, 10 billion is a reasonable ballpark number of people to contemplate having to be accommodated on the Earth.

Protecting the environment is not just a question of reducing the growth of the population. It also depends on how people behave. At local level, growing populations may cause deforestation, soil degradation and loss of biodiversity, but the link between population and the environment is complex. Third World countries with large poor populations are releasing far fewer chemicals and gases into the environment than consumers in the developed world. However, if lower population growth in developing countries leads to greater economic prosperity, and therefore to higher consumption of manufactured goods and use of services, pollution may be increased. If North American or European consumption patterns were extended to 10 billion consumers, it would wreck the environment. One demonstration of the industrialized world's profligacy is that homeowners in the USA spend $7.5 billion a year on tending their lawns, pouring huge amounts of water and chemicals on to their personal patches of grass. The world faces a terrible dilemma, namely how to bring economic improvement to billions of people in the developing world, without destroying the quality of life for everyone on the planet.

We have already mentioned the disparity of wealth, with the richest 20 per cent of the population of the world receiving 82.7 per cent of global income and the poorest 20 per cent receiving only 1.4 per cent of global income. One can look at another aspect of the imbalance between the lifestyles of the rich and poor nations and this is to try to assess their relative impacts on the environment or on the ecosystems of the planet. It is estimated that although people in the developed world account for only about 20 per cent of the world's population, they cause about 80 per cent of the global pollution and resource depletion. A useful indicator is to consider the annual per-capita consumption of energy; this varies from

India	0.3 tCE
Africa	0.4 tCE
China	0.7 tCE

at the bottom of the scale to

USA	9.5 tCE
Canada	9.7 tCE

at the top of the scale, where the units are metric tonnes of coal equivalent (i.e. the amount of coal that would have to be burned if the energy were all to be produced by the burning of coal). We can see from these figures that the impact on the planet, either in terms of the consumption of resources or in terms of the contribution to global warming by the production of the greenhouse gas CO_2, of one person added to the population of North America is more than thirty times greater than that of one additional person in India or in Africa and may be as much as fifty times greater. However, all poor countries are striving to develop technologically and this means that their per-capita consumption of energy and of natural resources will increase and become closer to those of the inhabitants of North America.

The impact of population growth on the environment is not clear-cut. In new frontier regions, where people are clearing the jungle for farmland or cutting down forest on hillsides, there is obvious damage to the environment. In most countries, the best cropland and most suitable settlement areas are already occupied, so that further expansion has to take place in marginal areas where ecosystems are fragile and environmental damage is likely to be high. Each year, some 16–20 Mha of tropical forest and woodland are lost in the search for new farmland and settlements and the resulting loss of biodiversity is a serious problem. In some parts of sub-Saharan Africa the demand for fuel wood outruns the renewal of tree stocks by 30–200 per cent. Fertile soil is a vital resource which is also being depleted. The Global Assessment of Soil Degradation, carried out for the United Nations, found that 10.5 per cent of the Earth's vegetated surface has been 'moderately', 'strongly' or 'severely' damaged by erosion, chemical contamination, salinization or compaction since 1945. A further 6 per cent of world soil has been lightly degraded. Because most human settlements begin in areas where land is most fertile, urbanization often results in concreting over some of the best farmland.

However, even more damage can be caused by consumer demand in the rich world. The impact on rainforest destruction of, for example, the fast-food industry (which has cleared forest for pasture) and the timber industry (which offers its customers hardwood products) has been well documented. Increasing affluence may bring consumer pressure for conservation, especially at local level. Thus Japanese forests have been conserved by placing two-thirds of them under public ownership. However, consumer pressure is less intense when it comes to protecting the environment in someone else's country and Japanese customers remain heavy buyers of tropical timber.

Consumer societies all over the world are increasing in size and making ever more heavy demands on natural resources. Although known reserves of oil and most minerals have increased in recent years, this cannot go on for ever. The question is how to achieve sustainability.

The first task of the industrialized countries is not how to get agreement with the developing countries about resource shares. It is to reach consensus among themselves on a radical plan to reduce their collective impacts on the global commons – they are the debtors in terms of the equitable use of the biosphere. When this has been done they will be in a moral and technical position to help the developing countries to adopt the most efficient and pollution-free industrial strategies.

An important aspect of the contemporary situation is that about half of the global population lives in cities with worsening environmental conditions and in coastal areas subject to calamities such as inundation and tsunami. The increasing concentrations of human population in certain areas and the development of relevant infrastructures lead to increased loss of human life and increased economic losses as a result of various environmental catastrophes such as earthquakes and volcanic eruptions.

Formulating ecological priorities from the viewpoint of the most dangerous trends that determine unstable global ecodynamics, one can identify the following six: population growth, deforestation, desertification, loss of ecological diversity, environmental pollution and climatic changes. All of which emphasizes the true fact that the means for overcoming these trends and a transition to stable ecodynamics are in the hands of politicians and not the scientists.

Awareness of environmental issues is growing in the rich world. Consumer pressure is forcing companies to invent less wasteful manufacturing processes and to use fewer toxic substances. But the green revolution is taking place only slowly. It will be many years before even the most progressive countries reduce their polluting of the environment. Their efforts could be swamped by increases in pollution from newly industrialized countries, whose populations are anxious to catch up with the developed world and are likely to be less concerned with issues such as the environment. Human societies can be remarkably adaptive, but whether they can switch to sustainable ways of living quickly enough to prevent serious damage to the planet remains to be seen.

From a commercial point of view, a larger population means a bigger market and a source of cheaper labour. In recent years, nothing has excited the multinational companies as much as the opening up of China and other Asian markets. But the attraction lies in the prospective buying power of these new customers rather than in their numbers. Population is growing fastest in Africa, but far more businessmen are flying to Beijing, Jakarta and Bangkok than to Lagos and Abidjan. Many Asian countries with high economic growth have managed to reduce fertility rates to close to the replacement level (2.1 children per mother). Examples include Japan (1.5), South Korea (1.6), China (2.0) and Thailand (2.2), all of which have well-educated populations and effective family planning programmes. Their populations will keep growing until the age structure becomes evenly balanced – China alone is expected to have 1.5 billion people by 2025 – but at least demographic transition is under way there. However, in other parts of Asia – such as Pakistan, the Philippines and northern India – increases in population still outpace the growth in GDP.

2.7.2 What is sustainable development?

A great deal has been written about sustainable development, but the basic principle is simple and not very new. A dictionary definition of *development* might be 'gradual unfolding; fuller working out'. It is a process by which potential is realized. In environmental terms, it is the process by which people alter their environment to make it supply more products and services useful to humanity. Sustainable development is simply a kind of development that goes on delivering its benefits indefinitely. To quote the familiar statement by the World Commission on Environment and Development (1987) – the Brundtland Commission – it is: 'Development that meets the needs of the present without jeopardising the ability of future generations to meet their own needs.'

Superficially, development has been a success and has supported a vast expansion of the human population. However, the level of development varies greatly in different parts of the world. It is easy to draw up a list of achievements. Agriculture has greatly enhanced the capacity of the environment to support people. In recent decades, food production per head has risen everywhere except in Africa. Life expectancy worldwide has been increasing steadily (though it is twice as high in Japan, the USA and Western European countries as it is in a number of African countries). Literacy in the developed world is almost universal and it is increasing in even the poorest countries. As medicine, health care, economic growth and (female) literacy expand, so the rate of increase in human numbers is slowing. In most developed countries the industrial concerns are far less polluting and far more efficient in their use of materials than they used to be. Our lifestyle in terms of clothing, leisure activities, travel, consumer goods, etc. becomes ever more luxurious. We must, however, not let achievement blind us to failure. The world's environment is a mosaic of both achievements and failures. Even the richest countries have patches of poverty and deprivation. Of the world's 6.0 billion people, about 600 million people are chronically hungry and at least one billion suffer greater or lesser poverty. Avoidable infectious diseases still kill several million children every year. Nevertheless, population growth is outstripping resources in many areas and undermining efforts to improve the quality of life. The new industries in many developing countries use dirty technology that would not be tolerated in many developed countries. Human degradation is evident in the sprawling shanty towns that spread around many great cities – often side by side with wealthy suburbs. Those shanty towns are being swollen continually by migration from the countryside and some of them are run by drug traffickers rather than by governments. Some countries have drifted into civil war as governments lose the consent of the people and can no longer control the warring factions.

The unevenness of the development pattern, and the expansion of some of the areas of worst environmental and social degradation should be a warning. Development may have allowed many of our habitats to support more people than ever, but it does not support all of them well. Moreover, it has eroded – and continues to erode – the living diversity of the Earth, which is our life-support system and the foundation for continuing evolution to meet an uncertain future. It is doubtful whether present policies and present systems of environmental management can support, never mind give quality of life to, the 10 to 12 billion people that some demographers project will be living on Earth by 2075. Conflict between the haves and the have-nots is likely to erupt. The debate about sustainable development arises from concern over these deficiencies. The assumption is that continued development along present lines will aggravate the imbalance between people and the environment.

2.7.3 How can development be sustainable?

Holdgate (1995) gives a list of examples of inefficiencies or unsustainability;

- some 70 per cent of the world's drylands are degraded to some extent, and desertification costs over US$42 million per year, mostly in Asia and Africa;
- an area as large as India has been lost to agriculture as a result of mismanagement;
- an area three times as large as India is producing less than it might;
- some 39 per cent of the net primary production by plants on land worldwide and 2.2 per cent of that in the oceans is pre-empted or destroyed by people;
- the amount of farmland per head is falling – from a worldwide average of 0.38 ha per person in 1970 to 0.28 ha per person in 1990 and, if present trends continue, perhaps to 0.15 ha per person by 2050;
- fish catches in all of the world's seventeen fishery areas are either at or above sustainable limits – in nine areas stocks are in serious decline;
- between 2 and 13 per cent of all species on Earth may become extinct;
- pollution has enlarged from being a largely local nuisance of urban smoke and rivers dirtied by town drains to a global menace, with erosion of the stratospheric ozone layer and a probability of significant climate change in the next 50 years.

Holdgate (1995) quotes three reasons for these problems:

- some resources are being used faster than they can be replenished (forest products, fossil fuels);
- some resources are stressed so that their capacity to produce what we need is diminished (acid-rain pollution of rivers and lakes);
- some potentially productive resources are not used because of tradition or ignorance (e.g. the clear-cutting of tropical forest instead of cropping for (mainly) non-timber products, poor rangeland management).

There are no instantaneous, global or magical solutions to the problems of these inefficiencies. We could pin our hopes on top-down imposed solutions and on greenprints for the whole Earth, like Agenda 21 (see Section 2.4.4) – the vast compendium adopted by the Earth Summit held in Rio de Janeiro in June 1992. However, Holdgate argues quite strongly that sustainable development cannot be achieved by a top-down intergovernmental process. He argues that these problems can be alleviated to some extent by local decisions and actions. The collective effect of large numbers of such local actions can be to alleviate the problems substantially. It is unrealistic to argue that the science-led processes that have raised farm yields so greatly, advanced energy efficiency and pollution control in industry, and brought gains in medicine and social care, must for some reason now suddenly come to a halt. But this does not mean that the potential is infinite. There has been a tendency to believe that if we have economic growth we can provide everything that everybody needs, but this is not necessarily the case. Where the population has increased faster than economic growth there has been a decline in the quality of life and enormous pressure on the environment. Holdgate argues that 'the way ahead has to be guided by critical analysis of resource potential and environmental sensitivity and' that 'the key to action will also lie in the economic and social processes which in turn constrain or help individual choice'. He also argues that there is now a great deal of convergent thinking that is necessary for our construction of a market system which will work to optimize our use of natural resources and deliver the best attainable future. The three central issues appear to be about the economic valuation of natural resources, the concept of growth and the basis for inter-temporal, or inter-generational, comparisons. The aim should be that:

- natural resources (called by some economists 'natural capital') are correctly valued in economic assessments;
- prices for the use of these resources, and for the products they supply, correctly reflect these values;
- the system is not distorted by omitting major costs (like those of damage to natural systems outside the formal economy) or constraints (like the limits imposed by the nature and sensitivity of the physical world);
- inter-temporal comparisons are soundly based (which brings in the question of discounting).

People cannot avoid choosing the path of development. In the extreme situation development is a self-regulating process. If people use the resources at their disposal well, their living standards are likely to improve (so long as their neighbours do not appropriate their assets). If they misuse their resources, their living standards will fall, eventually to the point where their numbers are regulated by starvation and emigration, as has occurred in some parts of the world. The social challenge therefore is for communities and individuals to manage the resources at their disposal to sustain the quality of life and to do so in a manner which does not undermine the quality of life of their neighbours or of future generations.

While there is a genuine international, regional and governmental dimension to the action we need, Holdgate argues that the larger systems are an aggregation of their local and individual components. Let us illustrate this by an example from our discussion of population growth. The one-child family policy in China has achieved some success; this clearly results from the actions of individual couples, but those actions take place in a social and political system which carries a strong and clearly defined message from the government of that country. We might argue that the real task that faces governments and intergovernmental organizations is not primarily to produce legislation, particularly if that legislation is unenforceable. Rather their task should be to change the opinions and attitudes of individual people to act in an ecologically responsible fashion. On the other hand, the success which has been experienced by Japan in addressing environmental problems is partially due to social pressure expressed through citizens' movements. The improvement of the environmental situation in Japan has thus not been simply a top-down process. It is naive to suppose that people are generally informed, rational, friendly, law-abiding and altruistic; many of us need to be encouraged to behave in these ways. Moreover, we should not assume that we are immortal; we are not, neither individually nor as a species. While a top-down approach cannot be successful on its own, the matter cannot simply be left to individuals either. There is a need for a partnership between central government, on the one hand, and local and individual opinion and pressures on the other hand; the nature of the partnership is bound to be different in different countries. If we are not careful we shall see more disaster areas alongside areas of entrenched wealth. There may well sooner or later be some very destructive conflicts between the haves and the have-nots.

The question arises as to whether there is any example of truly sustainable development and the answer is probably that there is no such example. There is no country where every component of development is sustainable. However, there are examples of sustainable resource use and sustainable lifestyles within many countries and what is required is social, economic, scientific and political processes that favour sustainable developments so that overall the national lifestyle becomes more sustainable. Holdgate suggests some parameters which can be studied to attempt to monitor the success of development from the point of view of sustainability. These are:

- *the environmental quality gap* (i.e. how far the environment at the time of monitoring fell short of optimum conditions);
- *the life quality gap* (i.e. how far social conditions fell short of a reasonable level);
- *the sustainability gap* (i.e. how far we were undermining the future or failing to build it).

The study of these parameters would indicate how far present social and environmental policies are likely to deliver an acceptable and sustainable future for the community, while the measurement of the gaps in itself would provide a feedback to the continuing process of decision making in the community and in government.

2.7.4 Moral issue

On the moral front, one issue which arises in discussions of population control is the uncompromising official position of the Vatican on the question of artificial methods of birth control. However, it is unfair to see the Roman Catholic Church's teaching solely in terms of its position on artificial birth control and on abortion. There is more to it than that and there have been several messages of Pope John Paul II on ecological matters and these have largely been ignored. The encyclical 'Sollicitudo Rei Socialis' of 1987 argued that human development must include respect for 'The beings which constitute the natural world' and three points especially were made related to (a) respect for other species, (b) the limits of natural resources and (c) the harmful consequences of industrial pollution. Another encyclical 'Centesimus Annus' of 1991 links ecological problems to consumerism. A third encyclical 'Message of Pope John Paul II on the World Day of Peace' on 1 January 1990 starts by saying that now '... there is a growing awareness that world peace is threatened not only by the arms race, regional conflicts and continued injustices among peoples and nations, but also by a lack of due respect for nature, by the plundering of natural resources and by a progressive decline in the quality of life.

The sense of precariousness and insecurity that such a situation engenders is a seedbed for collective selfishness, disregard for others and dishonesty. This message goes on to stress that the present ecological crisis is not just a social or political problem but is also a moral problem and elaborates on this. In this message, various aspects of the required solutions are noted:

- the ecological crisis reveals the urgent moral need for a new solidarity, especially in relations between the developing nations and those that are highly industrialized;
- the proper ecological balance will not be found without directly addressing the structural forms of poverty that exist throughout the world;
- any form of war on a global scale would lead to incalculable ecological damage;
- modern society will find no solution to the ecological problem unless it takes a serious look at its lifestyle. In many parts of the world society is given to instant gratification and consumerism while remaining indifferent to the damage which these cause;
- the aesthetic value of creation cannot be overlooked.

It is clear, therefore, that recent papal teaching calls very clearly and forcefully for people to care for our world and to restrain the individual and collective greed and negligence that leads us to do it harm. Traditionally, the Orthodox Church has emphasized the close relationship between human beings and nature more strongly than have the Catholic and Protestant Churches.

2.8 INTERNATIONAL GEOSPHERE–BIOSPHERE PROGRAMME (IGBP)

The scientific community has been aware of the necessity for complex studies of the geosphere and biosphere for a long time. During recent decades, this awareness has revealed itself in a most spectacular way through the implementation of a large number of international scientific programmes dedicated to particular topics; for example, the International Geophysical Year (IGY), the International Quiet Sun Year (IQSY), and later the World Weather Watch (WWW) and the Global Atmospheric Research Programme (GARP) are just a few of them – there are many more. These various programmes are characteristic of a complex approach to the solution of relevant problems. However, analysis of the set of problems relevant to global ecology shows that a closer co-ordination of efforts is required and that those efforts should be oriented towards achieving the particular goals of various programmes, and the establishment of a kind of superprogramme encompassing the key aspects of geospheric studies.

It is these circumstances that have stimulated the development of the long-term IGBP. Its main goal is the study of global changes in the geosphere, especially from the viewpoint of anthropogenic impact on the biogeochemical cycles of carbon, nitrogen, sulphur, phosphorus and water, along with the dynamics of such life-supporting factors as radiation, the quality of air and natural waters, soil fertility and the interactions between biospheric and geospheric phenomena.

IGBP, which has been in its implementation phase since 10 September 1990, originally proposed seven key directions of research; these subsequently led to the formulation of six key research questions. Each of these six questions is now addressed by an established core project:

- How is the chemistry of the global atmosphere regulated, and what is the role of biological processes in producing and consuming trace gases? (International Global Atmospheric Chemistry project, IGAC).
- How will global changes affect terrestrial ecosystems? (Global Change and Terrestrial Ecosystems, GCTE).
- How does vegetation interact with the physical processes of the hydrological cycle? (Biospheric Aspects of the Hydrological Cycle, BAHC).
- How will changes in land use, sea-level rise and climate alter coastal ecosystems, and what are the wider consequences? (Land–Ocean Interactions in the Coastal Zone, LOICZ).
- How do ocean biogeochemical processes influence and respond to climate change? (Joint Global Ocean Flux Study, JGOFS).
- What significant climatic and environmental changes occurred in the past, and what were their causes? (Past Global Changes, PAGES).

In addition to these six core projects (IGAC, GCTE, BAHC, LOICZ, JGOFS and PAGES) there are a few other core projects which have been proposed or adopted, including the Stratosphere–Troposphere Interactions and the Biosphere (STIB) (which has now become part of the WCRP as SPARC), Global Change and Ecological Complexity (GCEC), Global Analysis, Interpretation and Modelling (GAIM), Land Use/Cover Change (LUCC) and the Global Ocean Euphotic Zone Study (GOEZS). There is a special place for the problem of global modelling. A very important further development is the International Human Dimensions Programme (IHDP).

2.9 ECOLOGICAL MONITORING

One might ask where Earth observation, or remote sensing from spacecraft, fits into what has been said. The exceptional complexity of modelling, and more so of forecasting global ecodynamics, requires creating an adequate system of ecological monitoring to be able to obtain sufficiently comprehensive observational data. The most promising research directly relevant to the above is presently being implemented in the USA within the programme Mission to Planet Earth, whose observational component is the Earth Observing System (EOS, 1999). In what follows, we shall give a brief characterization of efforts being undertaken in this regard. Naturally, the implementation of EOS must proceed from the analysis of the priorities of global and regional ecology. We shall examine some of its key aspects concerning the establishment of priorities and requirements for observational data.

From the viewpoint of a researcher of the geosphere–biosphere, the study of biogeophysical cycles of carbon, nitrogen, phosphorus and sulphur, which are governed by both natural and anthropogenic factors, is of special interest. The latter ones are of particular significance for the carbon cycle which has not been sufficiently investigated so far. The main sources of uncertainties are processes relevant to the contribution of the continental biomass dynamics (deforestation, etc.), and also to changes in the total productivity of ecosystems which are due to variations in the cycles of other components. This is what determines the need for satellite monitoring of the biomass dynamics on a global scale with suitable scaling to obtain assessments of detailed spatial–temporal variability in the biomass in individual regions, which can be reliable only if there are adequate control data of direct ground-based measurements.

The key aspect of the carbon cycle relates to estimates of the contribution of the sea biota and the carbon exchange between the atmosphere and the ocean. Of principal significance is the accounting of the interaction between the cycles of carbon and nitrogen. Nitrogen-cycle dynamics are determined by three processes with different time constants: long-term processes of the transport of nitrogen to and from the atmosphere; a faster nitrogen exchange inside the biosphere; the growing income of anthropogenically induced nitrogen (as a consequence of high-temperature combustion processes alone about 20 Mt of nitrogen per year is emitted to the atmosphere, while the utilization of fertilizers adds about 40 Mt of bound nitrogen per year to this amount). However, the global anthropogenic impact on the nitrogen cycle still remains negligible in comparison with the nitrogen reservoirs in the soil and in oceans. It may become substantial only after several decades or in about a hundred years.

Unlike the cycles of carbon, nitrogen and sulphur, the phosphorus cycle is independent of processes in the atmosphere. In this case, the main processes are connected with dissolved and particle-shaped phosphorus that can be found in river waters, with soil weathering and biogenesis processes in sedimentary rocks. The important problems of the contribution of river-carried phosphorus, involved in the biological cycle, and the characteristic timescales of the transport of phosphorus to the ocean have not been resolved so far. Among the basic aspects of studying the global phosphorus cycle there are:

1. mechanisms controlling the presence of phosphorus in soils, including the effect of anthropogenic factors (e.g. acid rain);
2. the transport of phosphorus to the ocean by river waters;
3. the relation between ocean bioproductivity and phosphorus concentration (in particular, the influence of phosphorus income on the fixation of nitrogen in coastal zones).

The problem of anthropogenic impact on the sulphur cycle is of great interest. There are indirect assessments that indicate that by now the emissions of gas-phase sulphur compounds to the atmosphere (as a result of fuel-burning processes) have reached the order of magnitude of the contribution of natural factors. Of special interest are:

1. estimates of biologically induced gaseous compounds of reduced sulphur, studies of processes and anthropogenic impacts determining them;
2. study of the transformation of anthropogenic emissions of SO_2 (sulphur dioxide) and transformations of gaseous sulphur compounds in the atmosphere in general, including gas-to-particle reactions involving the formation of sulphate aerosols which may affect climate to a great extent.

Since processes of anthropogenic impact on biogeochemical cycles are slow and sluggish, it is urgently necessary to be able to trace such dangerous tendencies well beforehand.

Satellite remote-sensing techniques form a very reliable basis for studying the peculiar features of the geographical distribution of ecosystems and boundaries between them, the scales and rates of anthropogenic impact on ecosystems and the effect of climate and other factors on ecosystems. These techniques are especially important from the viewpoint of monitoring the variability of biogeochemical cycles, controlling the primary biomass dynamics and chlorophyll concentration. Table 2.2 presents a brief characterization of requirements for observational data. Lidar sounding (mainly employing the principle of differential absorption) appears to be the most promising method. Very detailed considerations of the data requirements for studying the main types of ecosystems and the principal requirements for data from remote-sensing systems are to be found in section 1.12.2 of Kondratyev (1998).

Satellite remote-sensing systems are very costly and it takes a very long time to plan, design, build and launch them. There are severe restrictions concerning the weight of instruments, the power-supply system capacity and the amount of information accumulated on board a spacecraft. Besides, one faces difficulties with regard to purely scientific aspects of the problem of interpreting the data of satellite measurements: the mathematical difficulties of inverse problems (parameters of a relevant medium must be retrieved from the characteristics of emitted, reflected or absorbed radiation), uncontrollable fluctuations of the atmospheric optical characteristics, etc. Consequently great care needs to be taken in the planning of such systems in order to maximize the chances of success in terms of the ultimate aim of obtaining useful geophysical or environmental results. The idea of taking a formal and systematic approach to the relation between the requirements of individual information users, and the choice of spectral characteristics for satellite instrumentation has been developed by Kondratyev and Pokrovsky (1977, 1989); it is also presented in section 1.12.2 of Kondratyev (1998).

In the solution of the problem of planning remote-sensing experiments, two extreme cases should be avoided: excess or insufficiency of information obtained because of the poor choice of conditions for an experiment. In both cases the effectiveness of the solution of the inverse problem may be considerably reduced.

Two approaches to the solution of the problem of multi-purpose planning of remote-sensing systems have been developed, namely the information approach and the economic approach. It is assumed in both cases that information users who

Table 2.2. Requirements for data of remote observations of biogeochemical sources, sinks and fluxes.

Observed values	Observation targets	Spectral region
Ecosystem boundary	Indicators of varations caused by environmental changes	Visible or near IR
Turbidity plumes in rivers	Indicators of erosion and eutrophication	Visible or near IR
Chlorophyll content in the ocean	Indicators of primary productivity and its variation	Visible (including observations of forced fluorescence)
Atmospheric parameters: CH_4 – total content in the troposphere CH_4 – source intensity (and gradient) in individual regions	Intensity of biological and anthropogenic sources, lifetime	IR absorption or lidar sounding
N_2O – total content in the atmosphere and (desirably) gradients over individual regions	Biological and anthropogenic (agricultural) sources	IR or lidar sounding
CH_3Br – total content and (desirably) gradients	Discrimination of assumed biological sources	IR-absorption or lidar sounding
$(CH_3)_2S$, $(CH_3)SH$, H_2S – total content and (desirably) gradients	Discrimination of assumed biological sources and their duration of their functioning	IR-absorption or lidar sounding
NH_3 – gradients in the troposphere over individual regions (forests, emission regions, etc.)	Intensity of biological sources	?
Non-methane hydrocarbons – gradients in the troposphere over individual regions	Discrimination of assumed biological sources and the duration of their functioning	?

represent the interests of various areas of the economy, are able to formulate precisely their requirements for the basic characteristics of information they need (spectral channels, the geometry of the experiment, the spatial–temporal resolution, etc.). In their formulation Kondratyev and Pokrovsky (1977) identified many requirements relevant to four groups of applications: I – oceanography (102 requirements), II – hydrology (33 requirements), III – geology (36 requirements), IV – forestry and agriculture (32 requirements).

The idea of the information approach consists in singling out the most important factors and finding relevant vectors of factor loadings in a formalized space of requirements. The method of compiling a space of requirements and the modification of the scheme of factor analysis used were presented by Pokrovsky (1981). Tables giving optimal sets of spectral intervals were produced for groups of problems I – IV, see tables 1.22 and 1.23 of Kondratyev (1998). A discussion was also given of the relation between these sets of spectral intervals and the spectral intervals of the bands of a number of well-established existing instruments flown in space. The spectral bands of these instruments had been chosen on a much more empirical basis several years ago (e.g. it was shown that information obtained with the Landsat Thematic Mapper [TM] is in fair agreement with the requirements of the group of problems in geology, but not with the hydrology group). The Landsat MultiSpectral Scanner can satisfy the needs of users still less. Not surprisingly, it was shown that the satellite instruments available are not at all adequate to provide for the information requirements of potential users of satellite information in various areas of applications.

Since the future development and production of satellite remote-sensing systems must increasingly be financed by those fields of science and branches of applications that may be interested in satellite information, the question of economic profitability of planned instrument complexes arises, taking account of their long-term operation onboard space platforms. This leads to the second approach to the planning of satellite missions, namely the economic one. The multi-zonal nature of satellite information and its multi-purpose application make it possible to analyse the problem of optimal planning of relevant observational systems from the viewpoint of maximization of the integral economic efficiency index V of a multi-purpose observational system subject to the condition that the individual efficiencies for individual problems V_i $(i = 1, \dots, N)$ are regulated by special restrictions of the type of inequality from above and from below. A detailed account of the analysis in the approach will be found in section 1.12.2 of Kondratyev (1998).

2.10 CONCLUSION

It is generally accepted that global warming, as a result of greenhouse-gas emissions, and damage to the ozone layer (the ozone hole) are (largely) the result of human activities. But it is also important to realize that there are many other aspects of global change that are induced as a result of human activities. The threat to the future of humanity does not come from human-induced global warming alone; it comes from a whole range of activities that damage the environment.

The deterioration of the ecological situation in many regions of the globe and on the planet as a whole requires more intensive efforts for studying the regular features of regional and global ecodynamics. The central aspect of the problem is the proper establishment of priorities, which is so important because the problem under discussion is very complex and multidisciplinary. This becomes obvious after studying the documents produced by UNCED-2 (which was held in Rio de Janeiro in June 1992), the declarative nature of which is evident. It is the lack of clearly formulated priorities that predetermined the overemphasis on the part of the UNCED-2 participants on the problem of the necessity for reduction in greenhouse-gas emissions to the atmosphere to prevent catastrophic consequences of potential global climate warming. However, the discussion of the greenhouse gas problem has proved to be very useful; it uncovered the socio-economic roots of environmental problems, revealing the interdependence of ecology and politics. It is our view that discussion should be focused first on socio-economic development and the environment, and second on greenhouse-gas emissions and global warming – and not the other way round, as was the case with UNCED-2. It is considered that (i) the main limiting factor for further development of humankind are the resources of the biosphere and (ii) the main criterion of ecological well-being, or otherwise, is the level of consumption per capita of biospheric resources. Hence follows the acute social problem of the contemporary world, which is the fact that the level of consumption of biospheric resources in industrially developed countries is much higher than that in the Third World nations and that this has already resulted in an unacceptable extent of damage to the biosphere. This problem can partly be eliminated through the advancement of the currently existing technologies (in particular, through 'decarbonization' of energy) and the development of new ones, but a decisive factor can only be the reduction in population size and the establishment of a new and more fair global economic order. Thus, promising global ecological perspectives are dependent on the optimization in choice of acceptable ways for socio-economic development.

3

Satellite climatology – current and future European systems

R. A. Vaughan

3.1 INTRODUCTION

Space offers a unique vantage point from which to view the Earth, and satellites now play an invaluable role in the understanding and ultimate protection of our global environment. Although aircraft still have an important role to play in remote-sensing, providing complementary data, the combination of remote-sensing instrumentation and space technology now provides a sophisticated and versatile set of tools which are finding more and more applications. Traditionally, satellite remote-sensing systems have been used for global, regional and national programmes to support government activities. Now, however, with the relaxation on restrictions for ownership and operation of Earth-observing satellites, unprecedented competition has been stimulated among large aerospace companies to create new geospatial products and services aimed at markets never before envisaged – such as news gathering, farming, shipping, and conveyancing – as well as the traditional mapping and science community. At present, a number of commercial high-resolution systems, with multispectral resolutions of the order of 3–10 m and panchromatic resolutions in the range 1–5 m, are in prospect; indeed, the first was launched at the end of 1999 with about eight more to follow.

Few scientific techniques have seen such rapid development in a relatively short time as has remote sensing. This has been driven both by the recent pace of technological developments and by the current politicisation of the environmental agenda (see Chapter 9 for further information). Little more than 40 years has elapsed since the launch of the first artificial satellite *Sputnik* in 1957. The first meteorological satellite *TIROS-1* was launched in 1960, which was followed by the US (and Soviet) military space programmes and manned space programmes of the 1960s (Mercury, Gemini, Apollo). The first Earth observing satellite for land-based applications was probably *Landsat-1* launched in 1972, since when literally hundreds more satellites have been put into orbit, with a surprisingly high success rate. Not all these have been Earth observing, though. Many were for making scientific measurement of space and the atmosphere, as well as for meteorology, and others were for telecommunications and positioning.

The major sources of Earth data for the last 30 years have been the *Landsat* and *SPOT* (Système Pour l'Observation de la Terre) programmes, which will be described later, but nowadays a number of systems are available for providing specific data for many different scientific communities. This chapter will look at those systems which are particularly relevant to the study of the climate, and in particular to the monitoring of change. It will briefly review the current systems, but the main purpose is to look forward at the provisions being made for the next decade or so, particularly by the European Space Agency (ESA) which is now becoming a major player in this field. First, however, a short summary of the particular requirements for global climate monitoring will be given.

3.2 DATA REQUIREMENTS

The global climate is hugely complex. It extends over large ranges of spatial scales and consists of a great number of subsystems varying on many different timescales. On the face of it, remote sensing might be considered an ideal technique for obtaining information, as it provides a continuous and synoptic means of acquiring data. But, unfortunately, it is not the panacea that it appears. Some geophysical parameters can be determined indirectly from space, but are limited by the spatial, spectral and temporal properties of the systems. Global coverage at any reasonable spatial resolution, on a suitable timescale for climate monitoring will produce prodigious quantities of data. These problems are discussed in depth by Vaughan (1994, 1998).

The Earth system consists of the atmosphere, the hydrosphere and the lithosphere, each varying on its own timescale and on different spatial scales. Between these three systems there exist complex coupling mechanisms with different strengths and lag times. Understanding and explaining past changes and predicting future changes in the climate system can most profitably be achieved by some form of modelling, although models of any sophistication are beyond present computing capabilities. The development of the model algorithms requires an understanding of the complexities of the geophysical interactions, many of which at present are poorly understood. The input to these models consists of the measured geophysical parameters, representing the natural world, which are measured by both direct and indirect methods. Field, or *in situ*, measurements provide the most direct observations (e.g. temperatures, rainfall, insolation, wind strengths, etc), although, by necessity, the spatial and temporal coverage will be very limited. These are usually single, periodic or continuous measurements performed at fixed locations or on a grid of points. Conversely, remotely sensed parameters can be synoptic and repetitive, but are indirect, the geophysical information is derived from the measured data, usually reflected or emitted electromagnetic radiation.

Table 3.1. Classification of spatial resolutions of satellite images.

Very low	$\geq$ 300 m
Low	30 m–300 m
Medium	3 m–30 m
High	0.5 m–3 m
Very high	< 0.5m

Another difference between satellite measurements and *in situ* measurements is that satellite data consist of averages of the response over the instantaneous field of view (IFOV) of the sensor rather than the actual value at a point. Each pixel in the image depicts the average value, with no information about the spatial distribution within the pixel. The spatial resolution of images can be classified as in Table 3.1. Atmospheric sounders and passive radiometers usually have resolutions of tens or hundreds of km. The Advanced Very High Resolution Radiometer (AVHRR) has a resolution of 1.1 km, TM (Landsat Thematic Mapper) has 30 m and the new family of high-resolution satellites now provide 1-m resolution images.

Coupled with the spatial resolution is the temporal resolution. The coarser the spatial resolution the greater the swathe width and the more frequent the repeat. AVHRR repeats every 12 hours, TM every 16 days. Ideally, we would like high-resolution data at very frequent intervals. Not only is this not at present possible, but it would generate unacceptably high rates of data flow and impossibly large quantities of data to store and analyse. It is estimated that even with the present limitations on resolution, terabytes of data will be collected every day (Vaughan 1994). At present there are about 7.5 million SPOT scenes in the central catalogue, comprising nearly 300 Tbytes of data. NASA (National Aeronautics and Space Administration) currently has an archive of approximately 1,000 Tbytes, which is estimated to grow to 9,000 Tbytes by 2010 and to 14,000 Tbytes by 2014. Much of the storage media which has been used has a lifetime of only about 10 years, and the reading equipment is rapidly becoming obsolete. Archiving is now done using robotic mass-cartridge devices, but National Oceanic and Atmospheric Administration (NOAA) estimates that it would take 80 years to copy its existing environmental data archives. *SPOT* is being selective and copying only about 40–50 per cent, based on a zonal cloud-cover map of the world. Improvements in computing speeds, data-compression techniques and data-storage systems are taking place rapidly, but even these will not be able to cope with the anticipated flood of data. Selectivity and compromise will be required, at least for the immediate future.

There will also be problems with the huge amount of *in situ* data which will be required (about 3Tbytes of storage will be required just to hold the digitized 1 : 50,000 topographic map coverage of Europe – one CD-ROM holds about 600 Mbytes). Field data, image data, map data and socio-economic data will need to be co-registered and analysed. These will probably be in different formats, with differing levels of accuracy, subjectivity, resolution, classification and stratification.

The uniformity of data from different parts of the world will be affected by differences in culture, political regimes, language, population and state of development. Thus the 'global database' will not necessarily be homogeneous, and hence this will affect the accuracy and validity of models, both for explaining the present situation and for predicting the future.

The accuracy of both the data and the model algorithms are paramount. Just a few per cent change in the reflectivity of clouds generates a radiative forcing comparable with the effects of doubling the amount of CO_2 (carbon dioxide) in the atmosphere, yet neither the flux of CO_2 from the atmosphere to the ocean, nor the flux of DMS (dimethylsulphide) from the ocean to the atmosphere (DMS forms particles which control the radiative properties of clouds) can be calculated to much better than a factor of 2. The present prediction of the apocalyptic effects of a doubling of the CO_2 in the atmosphere depends on a linear extrapolation without considering the possibility of saturation (see Chapter 9 for further information). Until the relative importance of the parameters in the model are known, and the accuracy to which the important ones can be measured is improved, the predictions of models are worth very little – hence the investment in improving satellite retrieval of the parameters.

3.3 CURRENT SYSTEMS

3.3.1 International programmes

The whole field of remote sensing is a rapidly changing one. There are currently many dozens of environmental and meteorological satellites in various orbits around the world, operated by a number of different national and international agencies, and now even by commercial concerns. Perhaps the most extensive programme is that of the USA; particularly, NASA's 'Mission to Planet Earth' project, now renamed the Earth Science Enterprise (ESE). This is an Earth science programme 'aimed at providing the observations, understanding and modelling capabilities needed to assess the impacts of natural events and human-induced activities on the Earth's environment' (Lillesand and Kiefer 2000). The programme incorporates both space-based and ground-based measurement systems to provide the basis for documenting and understanding global change with an initial emphasis on climate change. The Earth Observing System (EOS) is one of the primary components which includes *Landsat-7*, carrying the Enhanced Thematic Mapper (like the TM, but only with a 15-m panchromatic band) and the recently launched (end of December 1999) *Terra* satellite carrying the first spaceborne hyperspectral instrument. This is a very large and complex system, carrying five different sensors:

- ASTER Advanced Spaceborne Thermal Emission and Reflection Radiometer;
- CERES Clouds and Earth's Radiant Energy System;
- MISR Multi-angle Imaging Spectro-Radiometer;
- MODIS Moderate Resolution Imaging Spectroradiometer;
- MOPITT Measurements of Pollution in the Troposphere.

These instruments have been designed to be used in a complementary manner, but the two of most interest here are MODIS and ASTER. MODIS is based on the AVHRR and CZCS concept, but has thirty-six wavebands with spatial resolutions of 250 m, 500 m and 1,000 m. It provides 2-day repeat global coverage with improved geometric properties and radiometric calibration, necessary for the long-term continuous observations required for monitoring small changes in global climate. ASTER (a joint US/Japanese venture) acts as a 'zoom lens' (Lillesand and Kiefer 2000). It consists of three separate instruments with fifteen wavebands in different spectral regions, with resolutions varying between 15 m and 90 m. One of the 15-m detectors is mirrored by a backward-looking one which enables stereo image acquisition as well as study of bidirectional reflectance.

AVHRR has been carried on a number of NOAA satellites. Although designed as a meteorological instrument, the data have been used extensively to monitor and measure a number of environmental parameters. Its main advantages are that it is an operational system and it has rapid global coverage and a 12-hour repeat over a swathe of more than 4,000 km. With two satellites operational at any one time, this provides 6-hourly overhead passes and at mid-latitudes up to eight passes can be obtained per day. This makes it ideal for global monitoring of events and changes. The disadvantages are its rather large IFOV (1.1 km at nadir, increasing to more than 1 km $\times$ 4 km at the extremities because of the very oblique view) and its limited spectral range (five fairly broad channels, one visible, one near-infrared, one mid-infrared and two thermal infrared, or six channels [with an extra near-infrared channel] on the latest versions of the instrument). Environmental applications have included sea surface-temperature mapping, ocean-colour monitoring and regional-vegetation monitoring. The reader is referred to Cracknell (1997) for a full description of the instrument and for a discussion of the non-meteorological applications of the data.

Brazil, China, Japan, India and several other countries have also entered the 'space race' and have already launched Earth-Observing systems or will do in the near future. The most relevant are the Indian *IRS* satellites carrying LISS-I (a *Landsat*-MSS-like sensor) and LISS-II (a *Landsat*-TM-like sensor), Russia's *RESURS-01* series carrying the MSU-SK multispectral scanner, Japan's *ADEOS* carrying the AVNIR and OCTS (Ocean Colour and Temperature Sensor) which only survived 7 months, and Japan's *JERS-1* carrying an optical sensor and a synthetic-aperture radar system. A number of others are documented in the proceedings of a previous summer school (Vaughan 1994), but it must be borne in mind that some of them will have failed by now, and most of the then-planned missions will have slipped to a greater or lesser extent or will have failed completely. Readers are referred to Kramer (1996) (now unfortunately rather dated), Rees (1999) and Lillesand and Kiefer (2000) for more up-to-date detailed information about these international satellites and payloads.

The two present trends in instrumentation are for hyperspectral and high (spatial) resolution instruments. A number of hyperspectral imagers (with anything up to 350 narrow wavebands, usually selectable) have been flown on aircraft, but, because of technical limitations, have only recently been included as part of any satellite's payload. The first, launched in December 1999, was MODIS. The next will probably be ",4 > MERIS (Medium Resolution Imaging Spectrometer) on *Envisat*. *ADEOS-2* and *Orbview-4* are also intended to carry hyperspectral instruments.

Until very recently, the highest spatial resolution available from a civilian satellite was that from *SPOT* (10 m in the panchromatic band). Recently, however, a number of commercial missions have been planned and launched giving resolutions down to 1 m. *Clark*, *EarlyBird* and *IKONOS-1* all failed on launch, in 1997, 1998 and 1999, respectively. *IKONOS-2* (now renamed *IKONOS*) was successfully launched in September 1999 and has provided excellent imagery at 1-m resolution (panchromatic) and 4-m resolution (multispectral) and the Russian *SPIN-2* has been producing 1-m resolution digitized photographs since 1998. *QuickBird*, *Orbview-3* and *EROS* are scheduled for launch in the coming year. Although such satellites cannot really be used for general environmental monitoring because of the small size of their images and lack of global coverage, their imagery may nevertheless be useful for studying particular events such as earthquake damage or localized flooding. Some national governments, particularly America and Canada, are very worried about the security risks posed by the availability of high-resolution data, which is approaching that obtained from specialized intelligence satellites, and are contemplating introducing legislation to control its availability. It is for this reason also that the US government will not allow *Radarsat II* to be launched in the United States.

3.3.2 European programme

Europe was quite late in entering the satellite remote-sensing arena, although a number of national agencies and institutions developed their own airborne scanners and SAR systems. France has led the way with its SPOT series as well as with its Ariane launcher, while most other European countries have been contributing to ESA's programme, funded through national contributions. Up till now, the main European contribution to Earth observation have been through the Meteosat programme and the two European Remote Sensing (ERS) missions although ESA's Living Planet Programme is intended to lay the foundations for Europe's Earth-Observing activities over at least the next decade.

ERS

The major current European component is the ERS programme, which originated in requirements framed in the early 1970s. Since then the requirements have become more refined as has the context within which these needs have been expressed. Early on in the mission definition, the emphasis was on commercial exploitation. By the time the mission configuration was finalized in the early 1980s the emphasis had changed with a realization of the importance of global climate and ocean-monitoring programmes. More recently the need to establish a commercial return on the data has reappeared (ESA 1998d).

ERS-1 was launched in 1991 with an expected lifetime of 3 years. It was kept operational for about a year after *ERS-2* was launched in 1995, during which time they operated in tandem, collecting pairs of SAR (Synthetic Aperture Radar) images for interferometric use. Then *ERS-1* was kept on standby until March 2000, when its on-board attitude control system failed. During its 9-year lifetime, *ERS-1* completed about 45,000 orbits and acquired more than 1.5 million individual SAR scenes, contributing to scientific understanding in many varied fields, including oceans, ice, land and atmosphere. The main feature of *ERS-1* was that it was the first satellite to use advanced active microwave techniques since the short-lived *Seasat* in 1978, other than the brief experiments in the Space shuttle programme. The main instruments on both the *ERS-1* and *ERS-2* satellites are:

- *Active Microwave Instrument (AMI).* This is a C-band instrument capable of operating both as a synthetic-aperture radar, in both image or wave modes, and as a wind scatterometer through its set of four antennae (three for the scatterometer and one for the SAR). This is a joint instrument, making common use of much of the hardware in order to reduce the payload. A consequence of this is that it is not possible to collect both types of data at the same time. Another problem is that, because of the large power requirements, the SAR instrument can only operate for a maximum of 12 min per orbit. In addition, because the data rate is too high for on-board storage, data are only acquired within the reception zones of suitably equipped ground stations. Although this instrument was designed to measure wind and wave fields over the world's oceans, it has nevertheless been used extensively over land for mapping and vegetation monitoring. Tandem images have been used to generate interferometric SAR images which are used for determining elevations and elevation changes (e.g. for volcanic studies) and structural movements (e.g. earthquake monitoring) as well as for creating digital terrain models (DTMs).
- *Radar Altimeter (RA).* This is a K_u band nadir-pointing instrument which measures the time of the return echoes from ocean and ice surfaces. This can provide information about surface elevation, significant wave heights and surface wind speeds. Over the oceans, it can measure heights to better than 10 cm. It has been used over land, but, because of the large size of the footprint (greater than 20 km), it can only give meaningful height data over reasonably flat desert areas (Vaughan and Karamjavan 1999; Mantripp, 1994).
- *Along Track Scanning Radiometer (ATSR).* This is a four-channel radiometer designed to provide sea-surface and cloud-top temperatures. It has a spatial resolution of 1 km, a swathe width of 500 km and a relative temperature accuracy of about 0.1°C. It is therefore in many ways similar to the AVHRR, but it uses a conical scanning system to obtain two looks at the surface, at nadir and at about 55° ahead to permit atmospheric correction. It also incorporates a microwave sounder (MWS), a two-channel passive radiometer, whose data are merged with the ATSR data before being transmitted to the ground. In addition

to the marine and meteorological applications for which it has been designed, many land uses have been found (e.g. vegetation and snow monitoring) as well as surface/atmosphere flux measuring (see Chapter 15 for further discussion).

Two other instruments are carried. These are the *Precise RAnge and Range-rate Equipment (PRARE)* and the *Laser Retro-Reflector (LRR)*. These provide precise orbit determination for accurate navigation of imagery and for referencing the altimeter height measurements.

SPOT

SPOT is a European satellite (a joint French, Swedish and Belgian undertaking) operated by Centre National d'Etudes Spatiales (CNES), the French space agency. *SPOT* is in a low-Earth orbit and carries the High (or Haute) Resolution Visible (HRV) sensor. This is a pushbroom scanner operating in two modes, multispectral (three wavebands in the red and near infrared) with 20-m resolution and panchromatic mode with 10-m resolution. There are two sensors, each with a swathe width of 60 km, which are steerable up to 27° on either side permitting stereo viewing and more frequent observation of a particular location. *SPOT-4*, launched in 1999, carries an improved version of the HRV – the HRVIR, incorporating a short-wave infrared band, and also carries the Vegetation instrument, with spatial resolution of 1.2 km and a swathe of 2,200 km for global vegetation monitoring.

Meteosat

Meteosat is Europe's contribution to the system of geostationary meteorological satellites which are in equatorial orbit around the world. It is situated above the Greenwich meridian at 0° latitude, 0° longitude. It is a spin-stabilized satellite carrying the Visible Infrared Spin Scan Radiometer (VISSR). This operates in three wavebands, visible, thermal infrared and mid-infrared (water vapour channel) with spatial resolutions of 2.5 m and 5 m at nadir. Images are transmitted at 30-min intervals. It is operated by EUMETSAT (a spin-off of ESA's control centre) in Germany.

Topex-Poseidon

The *Topex-Poseidon* mission is an altimetry mission conducted jointly by CNES and NASA. The mission was launched in 1992 to study the global ocean circulation from space. Two altimeters are carried. The Poseidon is an experimental lightweight single-frequency instrument operating at K_u band, and the main operational instrument is Topex, a dual frequency K_u/C band NASA Radar Altimeter. A microwave radiometer provides atmospheric water-content data and a laser retro-reflector is used for accurate positioning. At the time of writing, this mission is still flying, having completed nearly 300 ten-day repeat total Earth coverage cycles.

3.4 FUTURE SYSTEMS

In the immediate future, European missions will build on the success of these recent, past, systems. *Envisat* and *Meteosat* Second Generation will carry improved versions of some of the instruments on *ERS* and *Meteosat* respectively in addition to a number of new ones. But ESA's 'Living Planet' programme (ESA 1998c) marks a new era for European Earth observation based on smaller, more focused missions which will be user driven.

Envisat

Envisat is possibly the last of the large, multi-project satellites, and the largest European satellite so far. It will cost $2\frac{1}{2}$ billion euros and will weigh $8\frac{1}{2}$ tonnes. It will be a major technological challenge because of its complexity and the diversity of the payload data. It includes many developments from the *ERS* missions and will continue the process of data gathering initiated by these missions in 1991, so providing at least 15 years of data. Its advanced instruments will provide greater accuracy and greater quantities of data which will be more flexible for addressing a variety of complementary issues such as global warming, climate change, ozone depletion and changes in oceans, ice sheets, vegetation and atmospheric composition. It is also designed to support further commercial business based on radar imagery. It will provide continued global sea surface-temperature measurements for climate-change monitoring, improved measurements of the variability of sea surface heights, more advanced monitoring of atmospheric chemistry including ozone and new measurements of the ocean and land-surface biosphere.

Envisat is scheduled for launch in mid-2001 on an Ariane 5 launcher. It will occupy a Sun-synchronous polar orbit at a height of about 800 km with an inclination of 98°. The cycle involves a 35-day repeat of the higher resolution instruments. This mission is discussed at greater length in Chapter 13 but, for completeness, the main instruments are given below:

- AATSR — Advanced Along-Track Scanning Radiometer;
- ASAR — Advanced Synthetic Radar;
- DORIS — Doppler Orbitography and Radio positioning Integrated by Satellite;
- GOSMOS — Global Ozone Monitoring by Occultation of Stars;
- LRR — Laser Retro-Reflector;
- MERIS — Medium Resolution Imaging Spectrometer;
- MIPAS — Michelson Interferometer for Passive Atmospheric Sounding;
- MWR — Microwave Radiometer;
- RA-2 — Radar Altimeter;
- SCIAMACHY — Scanning Imaging Absorption Spectrometer for Atmospheric Cartography.

Although most of these instruments will provide useful climate data, the following are particularly relevant to this chapter. AATSR is a direct development from the

ATSR on *ERS-1/2* with greater spatial resolution (0.5 km) and spectral resolution (seven wavebands). ASAR will provide bipolar images in five different modes with resolutions varying from 30 m to 1,000 m in global monitoring mode. MERIS will be one of the first hyperspectral instruments to be flown on a satellite offering 300-m spatial resolution in coastal regions and 1,200-m resolution for regional monitoring in fifteen narrow wavebands. RA-2 will provide dual-frequency operation allowing ionospheric correction, while programmable range resolution will allow different surfaces to be tracked. This will allow improved mapping of ice-cap margins and sea ice as well as land elevation and lake levels.

SCIAMACHY, which is an improved version of GOME (Global Ozone Monitoring Experiment) on *ERS-2*, will measure atmospheric trace gas and aerosol concentrations by observing reflected and scattered sunlight, enabling investigation of a wide range of phenomena which have a direct influence on atmospheric conditions – such as forest fires, dust storms, industrial pollution volcanic activity and events associated with the Sun. GOSMOS will monitor upper-atmosphere ozone levels and other trace gases and MIPAS will provide profiles of atmospheric pressures, temperatures and trace gases.

SPOT-5

The next in the *SPOT* series is scheduled for launch in 2001. It will carry an improved pushbroom scanner, the High Resolution Geometry (HRG) imager. Three such instruments will be carried, viewing at nadir and at 19.2° forward and aft, all capable of pointing up to 27° to either side. Spatial resolution is improved to 10 m multispectral (except for band 4 which is 20 m) and 5 m panchromatic, still with sixty swathes. In addition, two of the panchromatic bands can be offset and interlaced to give 2.5-m resolution in the so-called 'Supermode'.

METOP

METOP-1, to be launched in 2005, will be the first of a series of three fully operational polar-orbiting weather satellites which will fit into the existing global meteorological observing system and will provide a service well into the second decade of the century. They will form the space segment of the EUMETSAT Polar System (EPS), which is the European component of a joint European–United States meteorological polar satellite system. *METOP* will provide the morning pass and the US spacecraft will provide the afternoon pass, effectively taking over from the NOAA series of satellites. METOP will be at a height of about 840 km and give a 5-day repeat cycle. The core instruments will be duplicated on all three satellites, but some instruments may evolve for *METOP-3*. The instruments to be carried are:

- AMSU Advanced Microwave Sounding Unit;
- ASCAT Advanced Scatterometer;
- AVHRR Advanced Very High Resolution Radiometer;
- Argos Argos Data Collection and Platform Location System;

- GRAS Global Navigation Satellite System Receiver for Atmospheric Sounding;
- GOME-2 Global Ozone Monitoring Experiment;
- HIRS High Resolution Infrared Radiation Sounder;
- IASI Infrared Atmospheric Sounding Interferometer;
- MHS Microwave Humidity Sounder;
- S&R Search and Rescue;
- SEM Space Environment Monitor.

Although ostensibly a meteorological satellite, *METOP* will provide useful climate data. The main imaging system will be the AVHRR, thus providing continuity of data, essential for long-term monitoring. AMSU-A will measure temperature profiles of global weather conditions in a swath of over 2,000 km mainly exploiting the 50-GHz oxygen band. ASCAT will provide near-surface wind-velocity vectors (accurate to better than $3\,\mathrm{m\,s^{-1}}$) over oceans in all weather conditions over two swathes of 550 km and with a spatial resolution of 25 km. It will also complement other instruments by providing ice and land parameters. GRAS will provide temperature profiles of the stratosphere and upper troposphere with a very high vertical resolution (better than 1.5 km) and an r.m.s. temperature error of better than 1 K. GOME-2 will be a more advanced version of the instrument carried on *ERS-2*. It will measure profiles of ozone and other trace gases in the upper atmosphere in a swathe commandable in various steps up to 1,920 km. HIRS will provide basic twenty-channel temperature and humidity soundings of global atmosphere in cloud-free conditions in a swathe of over 2,000 km. It will also measure surface temperatures, cloud information and total ozone content. IASI will use a Michelson interferometer in the range from the CH_4 (methane) to the CO_2 absorption bands to obtain enhanced soundings of the global atmosphere, while the MHS will measure the humidity profile in the range 89–190 GHz. SEM will sense the flux of charged particles from the solar plasma. The AMSU-A, the AVHRR and the GOME-2 instruments may be updated on *METOP-3*.

Meteosat Second Generation (MSG)

The first *Meteosat* satellite was launched in 1977, since when billions of people throughout Europe, Africa and Asia have benefited from it. Seven *Meteosat* satellites have been launched in the 22 years since then, all with identical specifications, providing continuous data for the meteorological community, but which have also been used by other communities. The frequent, semi-global coverage allows, for example, the monitoring of hazard events such as hurricanes and volcanic eruptions. MSG, will also, like its predecessors, be spin-stabilized and operate from the same position in geostationary orbit, but will have an improved payload. It is scheduled for launch in February 2002 with an anticipated lifetime of 7 years. The instruments it will carry are:

- SEVIRI Spinning Enhanced Visible and Infrared Imager;
- GERB Geostationary Earth Radiation Budget Experiment;

- S&R Search and Rescue;
- MCP Mission Communication Package.

SEVIRI will scan at double the rate of the Spin Scan Radiometer on *Meteosat*, providing full Earth coverage every 15 min, with twelve spectral channels instead of three. The spatial resolution will also be greatly improved, to 1 km (at nadir) in the visible channels and 3 km in the others. Eight of the channels will be in the thermal infrared, enabling the temperatures of cloud, land and sea to be measured, and others will be tuned to ozone, water vapour and CO_2 absorption frequencies. *MSG1* has a planned lifetime of 7 years, and this and its successors will provide Europe's geostationary meteorological capability until at least 2012. GERB will monitor the Earth radiation budget at the top of the atmosphere, allowing calculations of short-wave and long-wave radiation, essential for understanding the Earth's climate balance. It will also be a candidate instrument for *MSG2* (2002) and *MSG3* (2003).

3.5 ESA'S LIVING PLANET PROGRAMME

The future will see more frequent but smaller spacecraft, and missions which are more focused and directed at the fundamental problems of Earth system science. Using the heritage of *ERS-1*, *ERS-2* and *Envisat*, Earth-observation space missions will be much more directly targetted to either the research problems that remain or to service user needs through customized applications. The Living Planet Programme (ESA 1998c, 1999) is a rolling programme designed to underpin European efforts in Earth observation from space. It will focus on using Earth observation for developing our knowledge of the Earth, preserving the Earth and its environment, and managing life on Earth in a more efficient way. It will cover the whole spectrum of users' interests ranging from scientific research through to applications. The research-driven *Earth Explorer* missions will be paralleled by applications-driven *Earth Watch* missions designed to focus on specific applications and service provision. In the long run, the *Earth Watch* missions are expected to become free-standing services outside the European Space Agency. The programme also contains an exploitation and technological development element whose purpose is to underpin market development and ensure cost-efficient implementation of both types of spacecraft. Another departure is that the *Explorer* programme will be funded through a new 'envelope' programme in which member states will subscribe on a rolling basis, rather than in the present *ad hoc* way in which each mission has to seek its own funding, which will allow for a much more coherent scientific planning and targetting. The plan has been drawn up following extensive consultation with the Earth-Observing community, coordinated by the Earth Science Advisory Committee (ESAC), which comprises eminent scientists and representatives of interested bodies such as the European Space Science Committee (ESSC), the European Association of Remote Sensing Laboratories (EARSeL) and the

European Science Foundation (ESF). This is therefore uniquely a 'bottom up' approach, as opposed to the usual agency-driven 'top-down' approach.

Earth Watch

These missions will seek to set up operational systems for monitoring the Earth on a near-continuous basis, and will be self-sustaining in the long term. They will represent a transition from technology development to service provision. Topics currently receiving much attention include disaster management, monitoring desertification, precision agriculture and environmental convention support. Discussions are still taking place as to the ways in which this aspect of the programme will develop.

Earth Explorer

These are a series of space missions tackling critical Earth-science issues and are intended to be research/demonstration missions to advance the understanding of the different Earth system processes and introduce new observational techniques. They are of two types: *core missions* designed to address a set of scientific objectives specified in the programme, and *opportunity missions* which will enable scientists to test the feasibility of concepts and techniques.

Core missions

These missions are intended to be used to conduct research and/or demonstrate the potential of new innovative observational techniques relevant to both the scientific and applications communities. Of the nine core missions identified in ESA SP-1196 (ESA 1998a), from the first Announcement of Opportunity, four 'candidate' missions were selected for in-depth studies as to science and applications and to the feasibility of the proposed ideas. This was carried out by science advisory groups, one for each mission, in parallel with Phase-A studies carried out in industry. The selected missions were:

- Gravity Field and Steady State Ocean Circulation mission (GFSSOC);
- Atmospheric Dynamics mission (now called Aeolus);
- Land Surface Processes and Interaction mission (LSPI);
- Earth Radiation mission (ER).

The 'reports for mission selection' for all four were circulated amongst the European Earth Observing research community and presented at a Consultative Workshop in Granada (Spain) in October 1999 at which the first two listed above were selected for further development with a view to launch in 2005/2006. A second call for proposals took place in late 2000 for proposed launches in 2008/2010.

The GFSSOC mission proposes to determine the Earth's stationary gravity field and its geoid with high accuracy (mm) and at very high spatial resolution (100 km). This information is of importance for studies on the internal structure of the Earth and also for determining the mean state of the ocean circulation and thereby the ocean current variability (determined by means of altimeters). The mission would fly on a low-orbit (250 km) satellite and would use an electrostatic gravity instrument consisting of six accelerometers mounted in three orthogonal directions, and a combined GPS/GLONASS receiver.

Aeolus aims to provide global observations of vertical wind profiles so as to improve the knowledge of the global wind field of importance to climate research and climate and weather predictions. In particular, the mission will fill the gap of knowledge as to the wind field in the Tropics and over the major oceans. The instrument to be used will be a powerful UV laser at 35° incidence angle exploiting scattering from atmospheric aerosols (Mie scattering) and from atmospheric molecules (Rayleigh scattering). It will measure the (near) vertical wind profiles by Doppler techniques in the look direction, but methods are available to assimilate this with other available data into numerical models to produce a horizontal wind component. The mission will acquire about 3,000 wind profiles a day with the potential to determine wind profiles even in partly cloudy conditions.

Opportunity missions

These are intended to correct a serious deficiency of the past whereby it has often been impossible to react quickly to evolving situations. They will involve less consultation and therefore provide a mechanism to respond quickly to a given need. They will be smaller and less complex than the core missions (with a cost limit of €100M as opposed to €400M), and will be implemented over short timescales. These are fast and flexible responses to a single critical scientific issue, whereas the core missions are more complex with a larger scope, being required to tackle a range of fundamental problems of wide community interest. The first opportunity-mission launch is planned for 2002.

In response to the first call (ESA 1998b), twenty-seven full proposals were received, from a large number of outline proposals. These were evaluated by teams of scientists from the Earth-observation research community, and two were selected by ESAC in March 1999 for feasibility and definition study (i.e. combined Phase A/B). Subject to the successful conclusion of this phase, one will go forward for full development and launch in 2003 and the second will be prepared for implementation and launch in 2005. A third mission will be maintained in reserve. This decision was ratified by the Earth Observation Programme Board (EOPB) in May 1999. It should be noted that this is possibly the first ever example of a space mission being chosen by scientific researchers rather than by an agency.

The first opportunity mission will be CRYOSAT: a mission to determine fluctuations in the mass of the Earth's land and marine ice fields. It is planned as a 3-year mission using radar altimetry to measure variations in the continental ice sheets and marine ice cover. One of the goals of CRYOSAT will be to test the predictions of thinning, perennial Arctic sea ice as a result of global warming and to reduce the uncertainty in the contribution to global mean sea-level changes. The system will consist of a single satellite in a high inclination (94°) orbit carrying a microwave K_u band altimeter capable of operating in conventional pulse-limited mode, in synthetic-aperture radar mode and in interferometric mode. The precision of the measurements is estimated to be of the order of a few centimetres of yearly ice-equivalent thickness.

The second mission is the Soil Moisture and Ocean Salinity (SMOS) mission. SMOS will use an L-band two-dimensional interferometric radiometer to provide observations of soil moisture and sea surface-salinity data globally. Water and energy fluxes at the land surface/atmosphere interface depend strongly on soil moisture, and ocean salinity is the key variable in the determination of ocean circulation and the water cycle. Both these observations should help in the understanding of the medium- to long-term components of the climate system.

The third mission chosen as reserve was the Atmosphere Climate Experiment (ACE). This aims to monitor temperature, pressure and ozone profiles in the stratosphere and upper troposphere and the humidity profile in the lower troposphere for assimilation into climate models. Two other missions were included in the short list. These were SWARM, a small multi-satellite mission to observe the dynamics of the Earth's magnetic field, and SWIFT, a stratospheric wind interferometer for transport studies. Other flight opportunities of some of these instruments alongside other international programmes are being investigated, such as on Japan's GCOM missions, and working groups have been established with both NASA and NASDA to set up a framework for international co-operation in mission implementation.. A second call for proposals will take place early in 2001.

3.6 CONCLUSIONS

This chapter has presented a brief summary of some current and future planned Earth-observing programmes which can provide data for input to climate models. It can only present a snapshot of the situation as it is at present. Satellites and systems can be rather ephemeral, as are some of the programmes! Plans and programmes could easily change over the next few years as the political climate, and hence the economic forcing, inevitably changes.

4

Atmospheric data assimilation

A. O'Neill

4.1 INTRODUCTION

Observational data for the Earth's atmosphere are derived from a variety of sources, including satellites. The data are of different kinds (e.g. temperature measurements by radiosonde or radiance measurements by satellites), and they suffer from different errors. To get the maximum benefit from the data, for weather forecasting or research, we need to find a way to combine the heterogeneous mixture of data we receive. Moreover, we should like to do this in a manner that leads to a sequence of analyses (three-dimensional states) that is consistent with our knowledge of how the atmosphere evolves (the equations of motion with a representation of physical and chemical processes). The modern method of achieving this goal is data assimilation, a technique at the heart of present-day weather forecasting, which is increasingly being used to optimize the scientific return from expensively acquired satellite data.

This chapter presents a brief overview of the techniques used in data assimilation, focusing on its application to the atmosphere. For further information about data assimilation see Daley (1991), about its mathematical basis see Tarantola (1987) and about its application to weather forecasting see Talagrand and Courtier (1987), Lorenc *et al.* (1991), Courtier *et al.* (1994).

4.2 WHAT IS DATA ASSIMILATION?

Data assimilation is the technique whereby observational data are combined with output (forecast fields) from a dynamical (i.e. evolving) numerical model to produce an 'optimal' representation of the evolving state of the atmosphere. The word optimal means that information from instruments and models is combined in a manner that allows for the (supposedly known) errors in the observations and in the short-range forecasts made by the model. The model brings dynamical consistency to the observational data and interpolates data, in a dynamically consistent

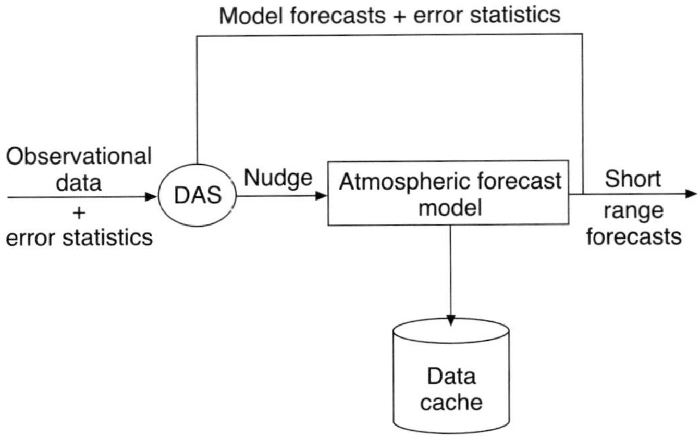

Figure 4.1. Schematic of the data assimilation process. DAS is the data assimilation system, or algorithm, used to combine observational and model data. See Section 4.2 for details.

manner, into data-void regions. The observational data correct the trajectory through state space of the imperfect model, keeping it 'on the rails'.

This synergism is illustrated by Figure 4.1, a schematic of the data-assimilation process. This figure shows that atmospheric data assimilation has many features in common with the broader field of control theory, used for example to design automatic landing systems for aircraft.

The main elements are:

1. A dynamical model makes a short-range forecast of the future state of the atmosphere, $\mathbf{x}_B$ (B for background field).
2. The state $\mathbf{x}_B$ is passed to the data-assimilation algorithm, where two things occur:
   - the observational data $\mathbf{y}$ are quality controlled by comparison with each other and with $\mathbf{x}_B$ – statistical methods are used to reject data 'outliers' (i.e. bad data);
   - the observational data $\mathbf{y}$ are combined with the forecast data $\mathbf{x}_B$ to produce, by allowing for errors in $\mathbf{y}$ and $\mathbf{x}_B$, an optimal estimate of the current state of the atmosphere $\mathbf{x}_A$ (A for analysis).
3. The state $\mathbf{x}_A$ is fed into the numerical model as the initial field to produce the next forecast $\mathbf{x}_B$ and so on.
4. Periodically (e.g. every 12 hours) long-range weather forecasts are run from initial fields $\mathbf{x}_A$.
5. The states $\mathbf{x}_A$ are archived (typically four times per day), to produce global three-dimensional datasets for researchers. Also archived are two additional important datasets:
   - statistics showing how well observational data $\mathbf{y}$ compared with the 'truth' $\mathbf{x}_A$ – this information can be used to correct problems with instruments;

- statistics showing how well model forecasts $\mathbf{x}_B$ compared with the 'truth' $\mathbf{x}_A$ – this information can be used to correct problems with the model.

4.3 WHAT ARE THE BENEFITS OF DATA ASSIMILATION?

Quality control

The model background state $\mathbf{x}_B$ provides a method of data quality control. By comparison with $\mathbf{x}_B$, we can determine whether a datum is likely to be erroneous.

Combination of data

Data assimilation combines and reconciles observational data from different sources (e.g. radiosondes and satellites) with a priori data from a forecast model to get a best estimate of the atmospheric state $\mathbf{x}_A$.

Errors in data and in the model

By comparing observational data $\mathbf{y}$ and forecast data $\mathbf{x}_B$ with our best estimate $\mathbf{x}_A$ of the true state, we build up valuable error statistics which can be used to improve instruments or retrievals and to improve models.

Filling in data-poor regions

The model provides a way to propagate information from data-rich regions of space and time to data-poor regions (e.g. over the oceans).

Designing observational systems and networks

The data-assimilation system can be used to design Observing System Simulation Experiments (OSSEs) to test the value of particular measurements and observing systems. Typically, the model is sampled by the observing system as if it were the real atmosphere. These synthetic measurements are then fed through the assimilation system to see if $\mathbf{x}_A$ comes closer to the actual state of the model.

Maintaining physical and chemical consistency

The model ensures that physical and chemical relationships among variables are maintained, to a good approximation, despite data errors that would lead to inconsistencies.

Estimating unobserved quantities

Through the relationships expressed in the governing equations of the model, the model provides information about quantities that are unobserved or inadequately observed (e.g. temperature observations are used to deduce winds, concentrations of unmeasured chemicals can be inferred from the observed evolution of others).

4.4 HOW IS DATA ASSIMILATION DONE?

How does one combine different estimates of atmospheric variables provided by a set of instruments and a model background field? An optimal estimation method is used to combine an a priori estimate $\mathbf{x}_B$ of the atmospheric state with an estimate based on observations $\mathbf{y}$ of that state. One method is to take a linear combination of these estimates weighted by the inverses of error covariance matrices that measure our confidence in the estimates. For this method to work as stated any biases in $\mathbf{x}_B$ and $\mathbf{y}$ should be removed beforehand. Such a statistical method is widely used in data assimilation for the objective analysis of atmospheric data. When the statistics are accurate and when no approximations are used in applying them, the data-assimilation method is called Optimal Interpolation (OI). In practice, methods are seldom optimal.

4.4.1 A simple statistical algorithm

We consider for simplicity the case of combining estimates of a scalar quantity $f(\mathbf{r})$ (the univariate case) where $\mathbf{r}$ indicates the three-dimensional spatial co-ordinates (x, y, z). Let $f_A(\mathbf{r}_i)$ be the desired analysed value of f at the analysis (model) gridpoint $\mathbf{r}_i$, $f_B(\mathbf{r}_i)$ the background (first-guess or a priori) value of f at $\mathbf{r}_i$, and $f_O(\mathbf{r}_k)$ and $f_B(\mathbf{r}_k)(\mathbf{r}_i)$ the observed and background values at the observation points, the latter obtained by interpolation in space from gridpoint values (i = gridpoint, k = observation point). Then our estimate for $f_A(\mathbf{r}_i)$ can be written:

$$f_A(\mathbf{r}_i) = f_B(\mathbf{r}_i) + \sum_{k=1}^{K} W_{ik}[f_O(\mathbf{r}_k) - f_B(\mathbf{r}_k)] \tag{4.1}$$

where K is the number of observation points and W_{ik} is the weight, to be determined, indicating the weight to be given to mismatches between the background field and the observations. The weights will depend on our confidence in $f_A(\mathbf{r}_k)$ and $f_B(\mathbf{r}_k)$ and on the correlation between quantities at $\mathbf{r}_i$ and $\mathbf{r}_k$. Weights are chosen to minimize the expected analysis error variance, see chapter 4 of Daley (1991). Thus, if the model background value is believed to be less accurate than the observed value, the model value is given the lower weight.

If we call the quantity in square brackets an increment, the procedure expressed by Equation (4.1) amounts to adjusting the model background field by a weighted field of increments, calculated at the observation points and then interpolated onto the model's gridpoints. Figure 4.2 shows schematically a possible horizontal distribution of increments with respect to the model's gridpoints. Figure 4.3 shows schematically the effect on the model's trajectory through state space when this adjustment procedure is applied periodically. The adjusted model state, from which the next forecast is made, falls within the error bounds of the observed state and the model background state, lying closer to the state with the smallest errors. In practice, the adjustment may be spread out over a few hours to avoid a shock to the model, which would generate noisy fields by exciting inertia-gravity

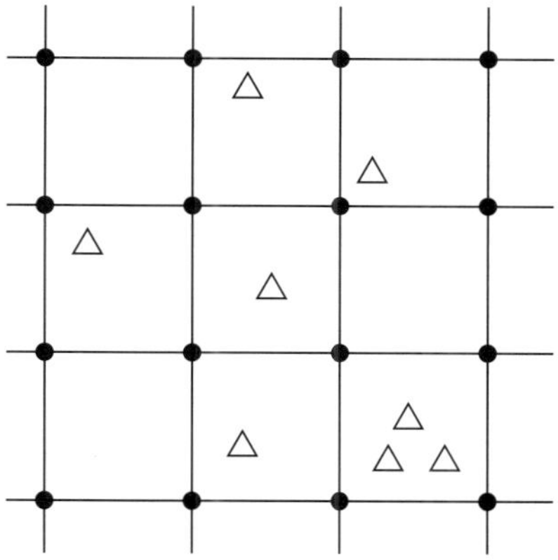

Figure 4.2. Schematic showing the horizontal positions, with respect to a model grid, of increments △. The increment is the difference between the observed value at the indicated position and the estimated model value, obtained by interpolating from the grid to the position of the observation.

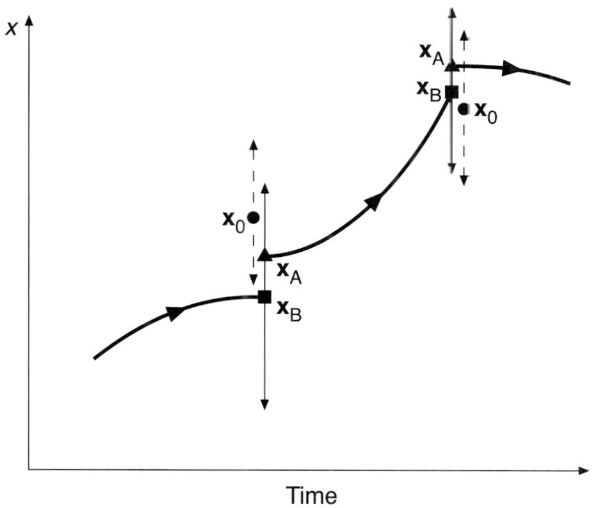

Figure 4.3. Schematic showing how a model's trajectory through its state space is adjusted to accommodate new observational data. The solid two-ended arrow indicates the expected error in the model forecast; the dashed two-ended arrow indicates the expected error in observation. The adjusted model state falls within the ranges of these errors.

modes. Alternatively, the model may be initialized before a forecast is made, to prevent the excitation of inertia–gravity modes (Machenauer 1977; Baer 1977).

4.4.2 A generalized algorithm: multivariate case

Because atmospheric variables are related to each other (e.g. temperature and winds; ozone and nitrous oxide) we need a multivariate algorithm that can treat all these variables simultaneously. We now write down a general method for combining observations and model background fields. Let $\mathbf{y}$ be the vector of observations available at (or attributed to) a certain time, and $\mathbf{x}_B$ the vector of forecast values of all atmospheric variables at all grid points ($\mathbf{x}_B$ has dimension 10^6 or 10^7!). Let $\mathbf{x}_A$ be the desired analysis that combines the available information. We construct a so-called penalty function that penalizes mismatches between our estimated state vector $\mathbf{x}_A$ and the observations $\mathbf{y}$, and between $\mathbf{x}_A$ and $\mathbf{x}_B$. We penalize ourselves according to our error estimates (error covariances) of $\mathbf{y}$ and $\mathbf{x}_B$; the greater the confidence we have in, say, $\mathbf{y}$, the more we penalize mismatches between our estimate $\mathbf{x}_A$ and $\mathbf{y}$. Thus:

$$J(\mathbf{x}_A) = [\mathbf{y} - \mathbf{K}(\mathbf{x}_A)]^T \mathbf{O}^{-1} [\mathbf{y} - \mathbf{K}(\mathbf{x}_A)] + [\mathbf{x}_B - \mathbf{x}_A]^T \mathbf{B}^{-1} [\mathbf{x}_A - \mathbf{x}_B] \tag{4.2}$$

where $\mathbf{O}$ is the error covariance matrix of the observations telling us, for example, how observation errors are correlated spatially (see Section 4.6) which explains what an error covariance matrix is; $\mathbf{B}$ is the error covariance matrix of the model forecast (background) field, providing us with analogous information; $\mathbf{K}$ is the operator that takes us from model space to observation space. (e.g. the elements of $\mathbf{y}$ refer to observation points and the elements of $\mathbf{x}_A$ refer to model grid points). To calculate the mismatch, we interpolate from model grid points to the observation points. Then $\mathbf{K}$ would be the interpolation operator.

Procedure

To get our optimal estimate of $\mathbf{x}_A$, minimize J with respect to $\mathbf{x}_A$ by moving down the (local) gradient of J until the minimum in J is reached to a prescribed tolerance (Figure 4.4). A Newtonian descent algorithm can be used, or a variant thereof, to get J_{min}. See, for example, Gill *et al.* (1981) for an account of descent algorithms. A complication is that if $\mathbf{K}$ is a non-linear operator there may be more than one J_{min}.

Notes on the method

(1) The procedure depends crucially on our knowledge of the error covariance matrices $\mathbf{O}$ and $\mathbf{B}$. Often these are not well known or may change as the atmosphere evolves. The error covariances are often modelled in terms of algebraic functions (e.g. of the distance between points); the greater the distance between points, the less errors are assumed to be correlated.

(2) This generalized algorithm allows us to assimilate the measured quantity directly as long as that quantity can be estimated from the model state. Thus if $\mathbf{y}$ were a vector of satellite radiances, $\mathbf{K}(\mathbf{x}_A)$ would be the corresponding estimate of radiances derived, using a radiative transfer algorithm, from the model state $\mathbf{x}_A$. Thus we avoid introducing errors by an off-line retrieval process.

(3) Procedures can be designed to accept data in finite windows of time (analysis times) or continuously (better for asynoptic satellite data). Some procedures lead to states $\mathbf{x}_A$ that are not in balance (e.g. geostrophic balance), in which case an initialization step is included in the assimilation cycle. If suitable constraints are built into the penalty function J, an initialization step can be avoided.

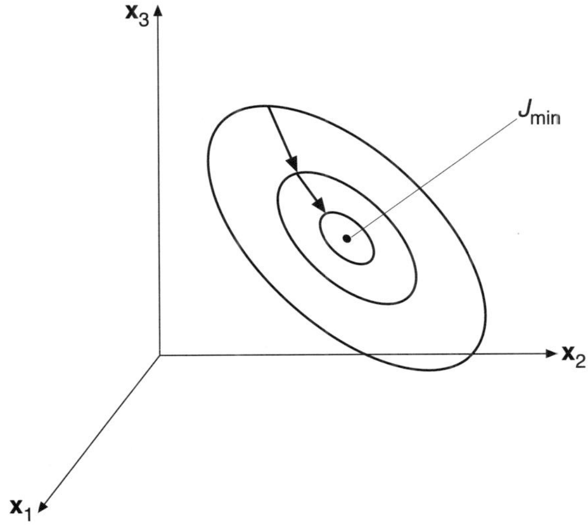

Figure 4.4. Surfaces of constant J in the model's state space at a given time. Arrows show the direction movement as the state vector is adjusted to minimize J.

(4) The method just described takes in data only within a certain (short) time window to produce a three-dimensional analysis $\mathbf{x}_A$. Past data affects this analysis implicitly and information about the future state of the atmosphere is not included. Because $\mathbf{x}_A$ is varied at a given time to find the minimum of J, and hence the optimal analysis, the method is referred to as 3DVAR.

4.4.3 4DVAR

The formulation of the 3DVAR method just described serves as the basis of a more general, and potentially more powerful, formulation of the atmospheric data-assimilation problem. This is four-dimensional variational data assimilation (4DVAR). It demands a close fit to the data (within the error bounds of the data), plus consistency with a dynamical model over an extended period of time. Since the model is deterministic (i.e. given the initial state, future states are pre-ordained), the method involves adjusting this initial state, $\mathbf{x}_{A_0}$, to give a best fit to the data, weighted by error covariances (Figure 4.5). The state $\mathbf{x}_{A_0}$ is then a vector of control variables of the system, which controls the fit to the data. These control variables could include model parameters that we should like to estimate from the demand of consistency with the observational data (e.g. an unknown chemical-reaction rate in a set of photochemical equations).

Procedure

Construct a penalty function as before, but this time (a) summing over time to measure the misfit over an extended period of time, (b) treating the model as perfect, (c) minimizing $J(\mathbf{x}_{A_0})$ with respect to the initial state $\mathbf{x}_{A_0}$.

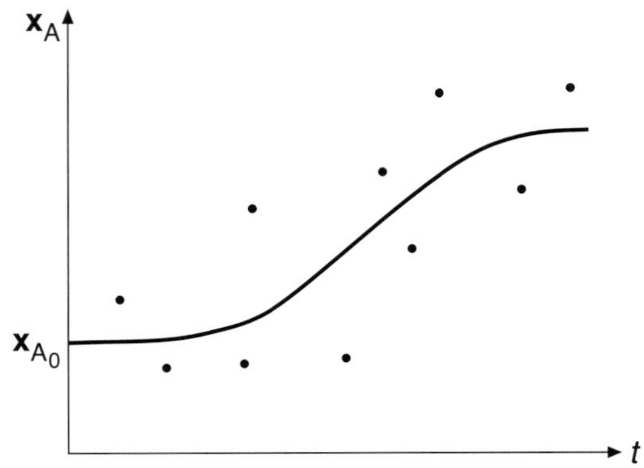

Figure 4.5. Schematic showing the model's trajectory through state space resulting from a given choice of initial conditions (and possibly of other control variables). The dots denote observed values. The initial state is determined to minimize the sum of the error-weighted mismatches between the trajectory and the observational data.

The penalty function is

$$J(\mathbf{x}_{A_0}) = \sum_{n=0}^{N} [\mathbf{y}_n - \mathbf{K}(\mathbf{x}^F)]^T \mathbf{O}^{-1} [\mathbf{y}_n - \mathbf{K}(\mathbf{x}^F)] \tag{4.3}$$

where $\mathbf{O}$ is the error covariance matrix of the observations, $\mathbf{x}^F$ is the model forecast starting from state $\mathbf{x}_A$O (which therefore depends implicitly on $\mathbf{x}_{A_0}$) and other notation is as in Section 4.4.2.

How do we achieve this minimization? Details will not be given here; a readable account can be found in Daley (1991, section 13.2). One method – not at all an obvious approach – is to find the minimum of J by constructing the linearised version of the model (the so-called tangent linear model), taking the adjoint of this tangent linear model and integrating it backwards in time. In order to ensure that the initial state (and consequently all later states in the time window) is consistent with the end state of a previous assimilation cycle, a background term may be added to Equation (4.3) to penalize mismatches between initial and end states (Zupanski 1997).

Rabier *et al.* (1993) found that 4DVAR has some advantages over other data-assimilation methods. It produces more accurate analyses at the end of the data-assimilation period, there is much smaller error growth in subsequent analyses and it is more robust in controlling the spurious excitation of inertia–gravity modes. A particular benefit of the method is that it can be used to derive wind fields from the evolution of trace-chemical fields, since these fields are implicitly related in time in Equation (4.3). Fisher and Lary (1995) have used a variant of the method, using precomputed air-parcel trajectories, to assimilate photochemically active trace chemical data.

4.4.4 Kalman filter

The 4DVAR algorithm is very powerful but it has the drawback that the forecast model is introduced as a strong constraint on the minimization of J – which is appropriate only if the model is perfect. This limitation is avoided by use of the Kalman filter technique. The model is assumed to be imperfect, and its forecast error-covariance matrix is updated as the integration proceeds by using the governing equations of the model and error fields at the previous time step. Descriptions of the application of this method to the atmospheric data assimilation problem can be found in Ghil *et al.* (1991), Dee *et al.* (1985) and Parrish and Cohn (1985). Though theoretically very attractive, this method is very expensive to use because it requires many matrix operations with matrices of dimension $10^6 \times 10^6$!

4.5 CONCLUDING REMARKS

The stored archives of meteorological analyses (three-dimensional state vectors) produced by operational data-assimilation systems have provided a very valuable resource for climate modellers. The analyses enable detailed studies to be made of climate processes, including calculations of momentum and heat budgets. For instance, the European Centre for Medium Range Weather Forecasts (ECMWF) has produced a 15-year series of global analyses from 1979 to 1993 by assimilating historical meteorological data into a data-assimilation scheme that was not changed during the period (Gibson *et al.* 1997). Keeping a fixed assimilation system reduced the risk of introducing changes in the resulting time series that might have been interpreted as a change in climate. ECMWF plans to produce a 40-year reanalysis of meteorological data, dating from 1958. There are similar activities at other operational weather centres.

New satellite programmes should provide a wealth of information on the climate system. The European Space Agency's Envisat satellite, scheduled for launch in 2001, will carry instruments making measurements of the atmosphere, ocean and land. The US National Aeronautic and Space Administration's Earth Observer System, which involves the deployment of several satellites, will also provide a comprehensive suite of climate measurements. Data assimilation will be an essential tool to maximize the scientific return from the expensively acquired data.

4.6 SUPPLEMENT: THE (ERROR) COVARIANCE MATRIX S_x

The concepts of variance (and standard deviation) of a scalar quantity x will be familiar. When we work with a vector quantity $\mathbf{x}$, the analogue of variance is covariance. The covariance matrix $\mathbf{S_x}$ of a vector quantity $\mathbf{x}$ is defined as:

$$\mathbf{S_x} = E\{\mathbf{x}\mathbf{x}^T\}$$

where $\mathbf{S_x}$ is a square matrix, $E\{\ \}$ is the expectation of what is inside $\{\ \}$, $\mathbf{x}^{\mathrm{T}}$ is the transpose of $\mathbf{x}$ (so $\mathbf{xx}^{\mathrm{T}}$ is a square matrix). For a sample with N elements,

$$\mathbf{S_x} = N^{-1} \sum \{(\mathbf{x}_i - \langle\mathbf{x}\rangle)(\mathbf{x}_i - \langle\mathbf{x}\rangle)^{\mathrm{T}}\}$$

where $\langle\mathbf{x}\rangle$ is the mean of the sample. If not stated explicitly, assume the sample mean has been subtracted off. The diagonal elements of $\mathbf{S_x}$ are the variances of the elements of $\mathbf{x}$, i.e.

$$\sigma_i^2 = (\mathbf{s_x})_{ii}$$

where σ_i^2 is the variance of x_i. The off-diagonal elements give a measure of the correlation between vector elements or of their errors (with respect to the sample mean). Error-covariance matrices are used in Equations (4.2) and (4.3).

5

Land-cover datasets derived from Earth-observation systems for global-change science

A. S. Belward

5.1 INTRODUCTION

In his review of climate-change science and policy interactions Shardul Agrawala points to a gradual move from policy as a driver for the advancement of science to a situation where scientific results are themselves influencing policy development (Agrawala 1999). This is a continuing dialogue that can trace its roots back the best part of two centuries. By the beginning of the nineteenth century Jean-Baptiste-Joseph Fourier had already developed his suggestions that the Earth's atmosphere traps heat; citations as early as 1807 can be found, but the first separate publication is that of Fourier (1824). John Tyndall's work measuring the radiative absorption of atmospheric constituents was published in the second half of the century (Tyndall, 1863) and, by the end of the nineteenth century, Svante Arrhenius had published work on the likely effects of coal burning on atmospheric CO_2 (carbon dioxide) concentration and subsequent warming (Arrhenius 1896). At the start of the twenty-first century we are now in a position where the political agenda is adapting to scientific advance. This 'dependence' on science in the political arena considerably raises the expectations on science to be an accurate and reliable information provider. This chapter discusses how well the expectations for quality products can be met by using remote-sensing methods.

5.2 LAND COVER AND THE GLOBAL CLIMATE AGENDA

Only 30 per cent of our planet's surface is 'land' and of this around 24 per cent is barren or covered by snow and ice (Figure 5.1). Hence it is not surprising that the notion of 'The Blue Planet' came into being. Yet information on vegetation growth,

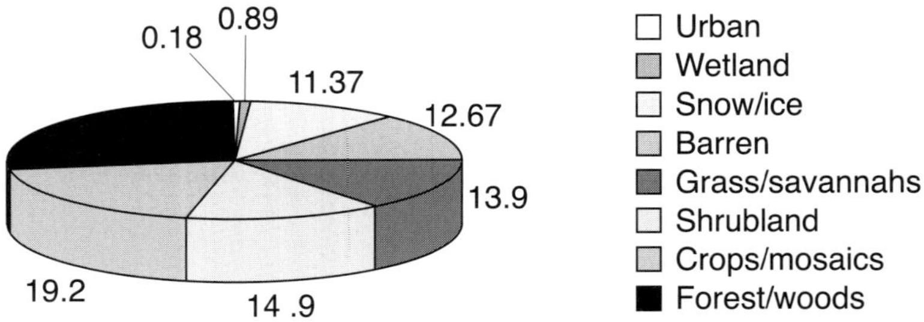

Figure 5.1. Cover types over Earth's terrestrial surface (expressed as percentages) (Loveland *et al.* 1999).

differentiation and distribution remains central to the global-climate agenda. Vegetation accounts for 99 per cent of the Earth's total biomass and is a major controlling factor concerning energy, water and material transport within the Earth system. Not only is it a strong determinant, it also provides us with food, fuel and fibre. As such it is highly susceptible to anthropogenic activities (i.e. the ways in which the land is used). Changes in land use associated with population migration or changes in land-use practice, or even climate, can thus cause change in the vegetation cover. In either case vegetation change may be an indicator of environmental pressure or indeed, given the overall importance to the Earth system, may even cause global change.

The concentrations of atmospheric CO_2 and other greenhouse gases are rising. The Mauna Loa observatory's record shows a 15.2 per cent increase in the mean annual concentration, from 315.83 ppmv (parts per million by volume) of dry air in 1959 to 363.82 ppmv in 1997 (Keeling and Whorf 1998; Keeling *et al.* 1982).

The Intergovernmental Panel on Climate Change (IPCC) categorically asserts that the biosphere is a strong determinant of the chemical composition of the atmosphere (IPCC 1996). Deforestation, conversion of natural vegetation to agricultural land and biomass burning are all part of this process (Crutzen and Andreae 1990, Houghton et al. 1991). Biomass burning is also implicated in the formation of acid rain (Dignon and Penner 1991). Repeated burning can force the transition from one ecosystem to another (Menaut 1983) and ecosystem changes subsequently affect the Earth system's energy, water and material transport. Deforestation, for example, is linked to climate change by altering surface–atmosphere energy exchanges (Nobre *et al.* 1991) and local and regional hydrological regimes because of changing fluxes and storage elements of the hydrological cycle (Salati 1987).

Indeed, the IPCC (1996) believes there is evidence that exploitation of the biosphere for fuel, food and fibre is contributing to increasing atmospheric concentrations of greenhouse gases, especially CO_2. For this reason the IPCC includes human activities that change the way land is used or affect the amount of biomass in existing biomass stocks in national calculations of greenhouse-gas emissions.

5.2.1 Land-use change, forestry and IPCC reporting

The radiant energy of the Sun is converted into chemical energy of plant tissue through the process of photosynthesis:

$$CO_2 + 4H_2O \xrightarrow{\text{light+chlorophyll}} (CH_2O) + 3H_2O + O_2$$

Likewise it is essential to remember that plants also respire, hence CO_2 is returned to the atmosphere through the process of respiration:

$$(CH_2O) + O_2 \xrightarrow{\text{protoplasm}} CO_2 + H_2O$$

Heterotrophic respiration also occurs in the decay of organic matter in litter and soils. Respiration by all living organisms and the burning of carbonaceous fuels has the capacity to use the entire atmospheric oxygen supply in a matter of a few thousand years, if there were no source to replace that used. Under ideal conditions though, the rate of photosynthesis in green plant tissue is around thirty times that of respiration (in the same tissues) thus playing a vital role in regulating the Earth's atmosphere's CO_2 and O_2 (oxygen) content. In the absence of major anthropogenic disturbances the respiration and photosynthesis fluxes balance on the global scale. Even on a local scale individual plant communities sooner or later reach the stage where growth rates (relying on carbon assimilation through photosynthesis) considerably slow down. Photosynthesis and respiration continue, but are now more in balance. This is one of the reasons why the IPCC guidelines for national greenhouse-gas inventories do not consider mature forests. These are assumed to be in a 'steady state'. Having said this, forests can change from net sink to net source of atmospheric CO_2 depending on local conditions, and if the entire forest ecosystem is considered (i.e. including the soil, litter layer and associated micro-organisms). And, of course, dramatic changes in cover as a result of human intervention will affect both carbon flux and carbon reservoirs.

The IPCC offer an illustrative numerical example of carbon fluxes associated with land-use change over a 20-year period (IPCC 1996). This is represented in diagrammatic form in Plate 1 and extended to a 40-year period to illustrate the effect afforestation and reforestation could have on the carbon cycle.

The diagram shows the conversion of primary forest to pasture, abandonment of the pasture, recovery of abandoned pasture as secondary forest, and finally the establishment of mature secondary forest. The entire process occurs over a 40-year period (1970–2009) and is limited to a 20-ha plot. In 1970 we see a 20-ha block of mature primary forest. This is in a steady state. In 1971 the first hectare of forest is cleared. Assuming there are 200 tC ha^{-1} (tonnes of carbon per hectare) of forest biomass and all are destroyed in the clearing process this would represent a net flux of 200 tC to the atmosphere. In 1972 a second hectare is cleared (again giving a gross flux to the atmosphere of 200 tC). However, the first hectare cleared is now established as a pasture. Assuming there are 5 tC ha^{-1} in the pasture biomass this would represent a net flux to the atmosphere of only 195 tC. In 1973 a third hectare of forest is cleared (200 tC to atmosphere), the second pasture is established (5 tC to biosphere), but the first pasture is now in a steady state; net flux to atmosphere is again 195 tC. We would see exactly the same pattern in 1974. In the fifth year, however, the first pasture is abandoned and secondary forest starts to establish itself. Assuming secondary forest grows linearly at a rate of 10 tC ha^{-1} per year for 15 years (after which it assumes a steady state once more) this would give a net flux to the atmosphere of 185 tC. So while the gross annual flux to the atmosphere continues to be 200 tC the net flux drops by 10 tC each year so that in 1989 net flux to the atmosphere is only 45 tC. In 1990 the last hectare of primary forest is cleared. The net flux to the atmosphere will remain at 45 tC because by now the first pasture has become a hectare of secondary forest in steady state. Assuming that the process of deforestation does not recommence the next 19 years will see no new flux to the atmosphere and a steadily declining flux to the biosphere until the entire 20-ha block returns to secondary forest in a steady state.

Of course, the above is a highly simplified picture of the effects of land-cover change on carbon flux, but it does provide us with a sound conceptual model. It clearly shows how changes in land use have both immediate and long-term effects on carbon flux and reservoirs and points to the need for long-term observation of land cover and land-cover change.

5.2.2 Land-cover requirements of the International Geosphere–Biosphere Programme

The International Geosphere–Biosphere Programme (IGBP) examines the broader implications of land cover. Land cover representing baseline conditions is an important input to global dynamic models and the quality and validity of such data define the reliability of simulated future scenarios. As a great deal of atmospheric variability is directly related to its interaction with the surface vegetation (Verstraete and Dickinson 1986; Avissar and Verstraete 1990), it is essential to represent this complex interaction with the greatest levels of accuracy possible. The core projects of IGBP have clearly made this case (Townshend *et al.* 1994).

BAHC (Biospheric Aspects of the Hydrological Cycle), GCTE (Global Change and Terrestrial Ecosystems), IGAC (International Global Atmospheric Chemistry Project) and latterly LUCC (Land Use Cover Change) all identify the need for improved baseline land-cover datasets for their studies of global change. Existing global land-cover databases such as Baily's world ecoregions (Baily and Hogg 1986), major world ecosystems ranked by carbon in live vegetation (Olson *et al.* 1985) and Matthews' vegetation, land use and seasonal albedo datasets (Matthews 1983) are only available at coarse resolutions (50–100 km). Their thematic accuracy too is unknown, and the classifications used to create them are inappropriate for some core-project needs. Without exception they were derived from disparate primary-data sources and aggregated to provide the global picture. This raises issues of internal consistency and thus overall accuracy. Most of the thematic datasets also have a climate element as part of the classification scheme and therefore often have an undesirable mixture of potential versus actual land-cover. Moreover, in common with all existing global land-cover maps they are dated (Townshend 1992). The paucity of global thematic-map products is testimony to the difficulties in mapping on these scales. All of the above criticisms are undeniable, but by the same token the authors of these maps deserve our gratitude. Before their products there were no data sets describing global vegetation in any consistent way. Nevertheless, the false idea that the maps we want are readily accessible, up-to-date, accurate and inexpensive to produce continues to endure (Estes and Mooneyhan 1994).

The land-cover categories identified by the IGBP are related to the needs of gas-exchange studies, vegetation attributes for modelling net primary production, burn emissions and gas exchange, wetlands cover and water regimes, changes in vegetation and land cover over time, biological attributes, physical attributes and landscape characteristics.

Any land-cover dataset created for IGBP core projects should be derived from a single data source obtained within a fixed and sequential time period. It should cover every part of the Earth's surface and each class must be exclusive. IGBP requirements for land-cover data specify that it must discriminate between perennial and annual above-ground biomass, as this is important for seasonal climate and carbon-balance modelling. It is also a major vegetation determinant of the surface roughness length parameter required by climate models for energy- and momentum-transfer equations. Evergreen and deciduous canopy types should be separated as this is a critical variable in carbon-cycle dynamics of vegetation and affects seasonal albedo and energy-transfer characteristics of the land surface. Leaf longevity is an important factor as this indicates whether a plant must completely regrow its canopy, or a portion of it, annually with inferred consequences for carbon partitioning, leaf-litter fall dynamics and soil carbon. It should also separate needle leaf, broadleaf and grasses as gas-exchange characteristics are affected by leaf type.

5.3 EARTH OBSERVATION AS AN INFORMATION SOURCE

So the question arises as to how Earth observation can help. The spatial and temporal dimensions of spaceborne Earth observations are such that they have the potential for global observation, often daily and over a range of wavelengths suited to the study of terrestrial surfaces. Though a recent innovation in terms of the timescales relevant to climate change, Earth-observing satellites can now provide observations going back to 1972 (and even beyond with the declassification of early military-satellite imagery). A daily global record is also available from 1981 to the present from the global area coverage archives of AVHRR (Advanced Very High Resolution Radiometers) data (Belward 1996).

5.3.1 Land-cover mapping

Land-cover classes such as those specified by the IGBP can be derived from Earth observation using established methods of digital-image classification. These usually rely on clustering algorithms designed to group pixels with the same spectral and/or temporal–spectral characteristics. On the global scale, coarse resolution data (typically from the AVHRR) are usually used to create general classifications, with some kind of geospatial sampling method employed to provide accuracy assessments and to produce more accurate statistics on area changes (Richards *et al.* 1999).

Another area of mapping relevant to global land-cover change relies on the use of thermal and mid-infrared data to map vegetation fires and burn scars associated with vegetation dynamics at the community, biome and ecosystem levels of organization. Interest in global fire and burn-scar mapping arises because of the impact fires have in both the short and long term on land cover and the atmosphere. The next generations of sensors (e.g. *Meteosat* Second Generation [MSG]) specifically consider fire monitoring in their science plans (Cihlar et al. 1999). Much of our current knowledge of global fire dynamics comes from AVHRR. The spectral channels of AVHRR are well placed for the detection of hot targets (Matson and Dozier 1981; Muirhead and Cracknell 1984), and the NOAA (National Oceanic and Atmospheric Administration) satellites' orbits, in combination with AVHRR's wide field of view, provide daily, global Earth observation. Global pictures of fire activity and burned-area estimates are now emerging (Barbosa *et al.* 1998; Dwyer *et al.* 1999; Stroppiana *et al.* 1999). Near real-time information on global patterns of fire activity are also being created through an international co-operative programme, the World Fire Web (WFW) (Pinnock and Grégoire 1999).

Improvements in computer technology, mass-storage devices and faster processing algorithms have also led to a new and exciting possibility for global environmental studies, namely the creation of very large datasets suitable for mapping from Synthetic Aperture Radar (SAR) data (Rosenqvist *et al.* 1999). The all-weather capabilities of SAR make the acquisition of time-sequential data over the persistently cloudy parts of the globe a real possibility. Given the capacity for 'guaranteed' data acquisition it has proved possible to time multiple acquisitions to coincide with particular aspects of phenology. The resulting multi-temporal data sets are providing a new view of the complex tropical-forest ecosystems and for the first time maps of flooded-forest ecosystems are being created (De Grandi *et al.* 1999).

5.3.2 Biophysical parameters

In addition to mapping, Earth-observation techniques are increasingly being used for extraction of quantitative parameters describing changes in vegetation dynamics (e.g. the use of vegetation indices as surrogate measures of plant growth and differentiation).

Optical remote-sensing techniques for studying vegetation dynamics generally rely on the contrasting spectral response of green vegetation and background material in the red and near-infrared wavelengths (Tucker and Sellers 1986). Combinations of these two wavebands as vegetation indices, especially the Normalized Difference Vegetation Index (NDVI) (Rouse *et al.* 1974), have been extensively used as surrogate measurements of various attributes of a vegetation canopy (i.e. total biomass and leaf-area index).

Though the NDVI has been widely used, it tends to produce a saturation response to very high green-leaf densities, and at low-vegetation cover soil background reflectance can exert a considerable influence. Other problems in using these indices for quantitative measurement are the effect of variable viewing conditions, including both geometry and atmosphere. In common with all natural surfaces (except very calm water bodies) the reflectance field of vegetation is anisotropic. Leaf orientation, self-shading of canopies, vegetation clumping and gaps and even roughness of the underlying soil all control the overall shape of the Bidirectional Reflectance Distribution Function (BRDF). Atmospheric conditions are also highly variable and, depending on the spectral region, scattering and absorption by different atmospheric components along the radiation path can affect radiance values. And, of course, the atmospheric components themselves have particular angular reflectance patterns. Thus the reflectance measurements over vegetation derived from the satellite sensor measurements can change as a function of viewing geometry and atmosphere rather than as a result of some change in the vegetation itself. Verstraete and Pinty (1996) point out that an ideal vegetation index should have the following characteristics:

- it should show maximum sensitivity to the presence of, and changes in, green vegetation;
- it should not be sensitive to atmospheric absorption and scattering;
- it should be independent of spatial and temporal variation in Sun-target-sensor geometry;
- it should be insensitive to soil colour and brightness effects.

These requirements are partly contradictory, but a concentrated research effort can yield optimal spectral indices to determine biophysical variables such as the Fraction of Absorbed Photosynthetically Active Radiation (FAPAR) (Gobron *et al.* 1999). Unlike NDVI, FAPAR has direct biological significance and can be directly used in biological models to help in the calculation of important ecological variables such as net primary productivity.

New approaches, such as that of Gobron *et al.* (1999) take into account both the atmospheric and directional problems in computing new indices. Indeed 'angular signatures' are now being measured and exploited not just to correct the single-angle approaches inherent in vegetation-index construction but also as a source of information in their own right (Martonchik *et al.* 1998). The angular approach is capable of providing additional information to that provided by the spectral solution. By using anisotropic-reflectance characteristics it is possible not just to resolve sources of 'error' in the single-angle indices but also to glean new information (Diner *et al.* 1999). Indeed, the multi-angular approach is expected to achieve a significant reduction of the errors in the estimation of geophysical variables from remotely sensed observations. Running *et al.* (1994) suggest that errors of less than 10 per cent may be achievable in place of the current 20–40 per cent through the use of multi-angle observations and radiative-transfer models.

Perhaps the last word on the advantages of employing the angular domain should be left to Diner *et al.* (1999) who state 'By providing additional information that constrains the physical nature of the scene being observed, whether it be an identification of aerosol type, cloud height and form, or surface cover, there is an attendant increase in confidence in the uniqueness of the retrieved signal and considerably reduced ambiguity and error in the retrieved geophysical or biophysical parameters.'

5.3.3 Need for calibration and validation

Confidence in the accuracy and quality of products derived from Earth-observation systems starts with confidence in the quality of the original data emanating from the sensors. This calls for sensor-specific calibration and validation, and secondly geophysical-parameter and derived-product validation. Calibration is defined by the Committee on Earth Observation Satellites (CEOS) as 'the process of quantitatively defining the system response to known, controlled signal inputs', and validation as 'the process of assessing by independent means the quality of the data products derived from the system outputs' (CEOS 1995).

Sensor-calibration coefficients are needed to convert the raw digital counts collected at the satellite into physical quantities of radiance or reflectance; a detailed discussion of pre-launch and post-launch calibration will be found in Chapter 6. Radiance or reflectance values in turn may then be used to derive geophysical parameters (i.e. albedo, land-surface temperature or vegetation indices). As the sensitivity of sensors may change with time, absolute calibration and temporal radiometric stability must be determined before judgements can be made concerning geophysical quantities derived from Earth-observation data (Slater 1997).

Though always prerequisite, the level to which calibration and validation are required is to some extent user dependent. Short-term meteorological applications may be satisfied by 'relatively' coarse calibration (users assume that there are no major changes from one day to another in sensor response or sensor stability). Long-term climatology observations require observing systems capable of detecting small changes over a long period of time, calling for far more stringent calibration and validation requirements (Allen *et al.* 1994).

Using data from a series of satellites carrying nominally identical sensors and/or different sensors measuring similar parameters also calls for calibration and validation. Even when users are presented with well-calibrated datasets, they must still check that different algorithms generating the same high-level product produce consistent results and that, where appropriate, accuracy statements accompany these results. The creation of the first land-cover dataset to satisfy IGBP science requirements had to have a validation component if it was to be useful to the IGBP core projects (see Chapter 11).

5.4 CONCLUSIONS

Global-change science needs information from disparate and diverse sources. Earth observation can play a role in this. Indeed, providing consistent, quality views of our planet's surface adapted to global-change studies is a key application for

Earth-observation systems. We could say that remote sensing plays a small part in global-change science, yet global-change science plays a large part in remote sensing.

Growing expectations on science by the policy makers and implementers is demanding better results, documentation and implementation by the scientific information providers. Advanced algorithms, improved awareness of the needs for calibration and validation, increasing commitment to provide them and ever-improving computer power mean that these expectations can increasingly be met.

6

Calibration of the Advanced Very High Resolution Radiometer

C. R. Nagaraja Rao and J. T. Sullivan

6.1 INTRODUCTION

The Advanced Very High Resolution Radiometer (AVHRR) flown on the Polar-orbiting Operational Environmental Satellite system (POES) spacecraft of the US National Oceanic and Atmospheric Administration (NOAA) is an invaluable source of long-term records of geophysical parameters such as sea-surface temperature, columnar aerosols over oceans, Earth-radiation budget and the Normalized Difference Vegetation Index (NDVI), to name a few. The data are unique in that they provide perhaps the only long-term satellite data record which can be used as a test bed for climate-change studies, and for establishing the feasibility or effectiveness of various techniques that are used in climate-change studies. AVHRR measurements also find application in the areas of fire and drought monitoring.

AVHRR is a cross-track scanning filter radiometer with a swathe width of the order of 2,500 km. It has an instantaneous field of view of approximately 1.1 km at nadir; consecutive scans are contiguous at nadir. The Earth is sampled six times per second and the whole Earth is imaged several times each day; any given Earth scene is viewed at the same angle once in 9 days. Three versions of AVHRR – with four, five, and six spectral channels – have been flown to date. The main features of the three versions of AVHRR are shown in Table 6.1, along with typical applications. We show in Figure 6.1(a) the nominal response functions of the six-channel AVHRR, along with sea-level atmospheric transmission and in Figure 6.1(b) the spectral distribution of the extraterrestrial solar irradiance (e.g. Neckel and Labs 1984), and of a 300-K black body which is representative of typical Earth scenes.

We shall discuss the impact of atmospheric transmission (absorption) on the upward radiation measured in the various channels later. Further details of the instrument are to be found in the technical publications of ITT Aerospace/Communications (Fort Wayne, Indiana) – the instrument maker – and in Cracknell (1997). For programmatic reasons, the NOAA/NASA (National Aeronautics and Space

Table 6.1. Basic features of the Advanced Very High Resolution Radiometer.

Channel	Wavelength version 1 (μm)	Wavelength version 2 (μm)	Wavelength version 3 (μm)	Detector	Applications
1	0.58–0.68	0.58–0.68	0.58–0.68	Silicon photodiode	Aerosols; cloud properties; Normalized Vegetation Index (NDVI)
2	0.72–1.10	0.72–1.10	0.72–1.10	Silicon photodiode	Aerosols; cloud properties; Normalized Vegetation Index (NDVI)
3	3.55–3.93	3.55–3.93	–	Indium antimonide	Sea surface temperature; fire monitoring
3A	–	–	1.58–1.64	Indium-gallium-arsenide	Aerosols; cloud–snow discrimination
3B	–	–	3.55–3.93	Indium antimonide	Sea surface temperature; fire monitoring
4	10.5–11.5	10.3–11.3	10.3–11.3	Mercury-cadmium-telluride	Sea surface temperature
5	–	11.5–12.5	11.5–12.5	Mercury-cadmium-telluride	Sea surface temperature

Administration) AVHRR Pathfinder Program (see Section 6.2) was focused on the rehabilitation of data obtained with AVHRRs on afternoon satellites (i.e. satellites which cross the equator from south to north in the afternoon as opposed to morning satellites which cross the equator from north to south in the morning). We shall discuss in detail the calibration of AVHRRs on afternoon satellites – namely, *NOAA-7*, *N0AA-9*, *NOAA-11* and *NOAA-14*. However, the principles of calibration are the same for AVHRRs on both the afternoon and morning satellites, with the difference that the impact of progressively later equator-crossing times of afternoon satellites because of orbital drift must be taken into account in the inter-calibration of various sensors. An AVHRR chronology is given in Table 6.2 for later reference.

To place the importance of satellite-sensor calibration in proper perspective, we shall briefly discuss the evolution of a meteorological satellite sensor. The basic stages of this evolution are shown in Figure 6.2. It is the user community that initiates the process with a statement of requirements for a geophysical product, with desired accuracy and precision. The statement of requirements may or may not be accompanied by suggestions of spectral intervals to be used in the remote sensing of the geophysical parameter. A conceptual design is then developed taking into consideration product-algorithm issues, if any. A key component of the conceptual design is the identification of spectral intervals which carry a strong signature of the desired geophysical product, or of the set of atmospheric and surface parameters that should be measured to derive the desired geophysical product using state-of-the-art algorithms. An initial set of sensor specifications is established based on the conceptual design. A decision has to be made as to whether established

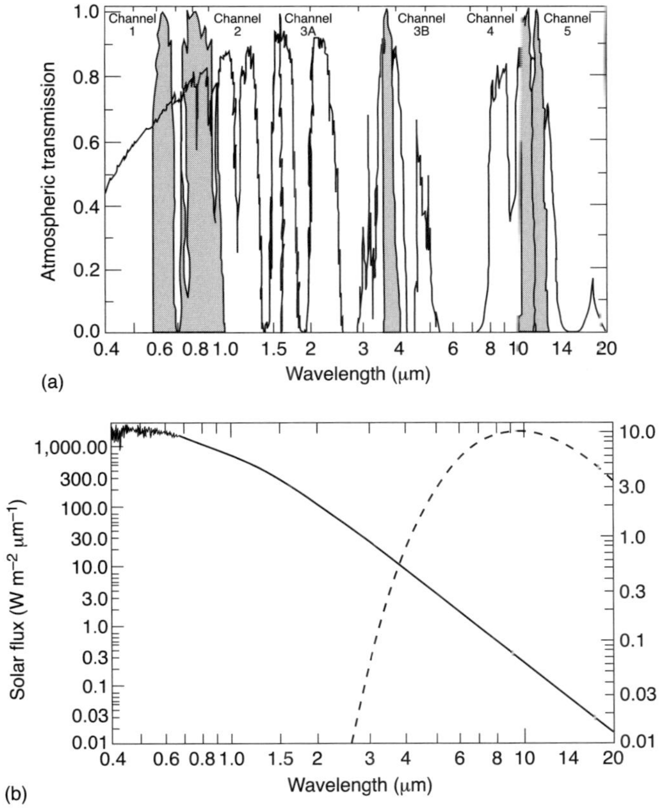

Figure 6.1. (a) Typical AVHRR normalized response functions for the 6-channel AVHRR, and sea level atmospheric transmission and (b) spectral distribution of solar irradiance at the top of the atmosphere (continuous line), and of the radiance of a 300-K black body, typical of Earth scenes (broken line).

instrumentation technology or experimental technology will be used to fabricate the sensor. If a decision is made to use experimental technology, appropriate risk-reduction studies are undertaken. This is the stage when the desired product accuracy and precision, and their translation to the sensor domain play a key role. Estimates of attainable absolute radiometric accuracy, using either on-board calibrators or vicarious post-launch calibration, or a combination of both are made. These estimates are used to determine whether the desired product accuracy can be attained, taking into account product-algorithm-specific effects. The input of the user community is sought to determine the impact, if any, of technology limitations on attainable radiometric calibration accuracy on the usefulness of the desired geophysical product. Through an iterative process between the user community and the instrument scientists and engineers, an optimum set of sensor specifications is developed.

Table 6.2. AVHRR chronology.

Spacecraft	Number of Channels	Launch Date	Operational Dates
NOAA-7(P)	5	23 June 1981	19 Aug. 1981–7 June 1986
NOAA-8(A)	4	28 March 1983	20 June 1983–12 June 1984
			1 July 1985–31 Oct. 1985
NOAA-9(P)	5	12 Dec. 1984	25 Feb 1985–7 Nov. 1988
NOAA-10(A)	4	17 Sept. 1986	17 Nov. 1986–16 Sept. 1991
NOAA-11(P)	5	24 Sept. 1988	8 Nov. 1988–11 April 1995
NOAA-12(A)	5	14 May 1991	14 May 1991–14 Dec. 1998
NOAA-13(P)	5	9 Aug. 1993	9 Aug. 1993–21 Aug. 1993
NOAA-14(P)	5	30 Dec. 1994	11 April 1995–present
NOAA-15(A)	6	13 May 1998	26 Oct. 1998–present
NOAA-16(P)	6	21 Sept. 2000	Being commissioned

Note: (A): morning satellites; (P) afternoon satellites.

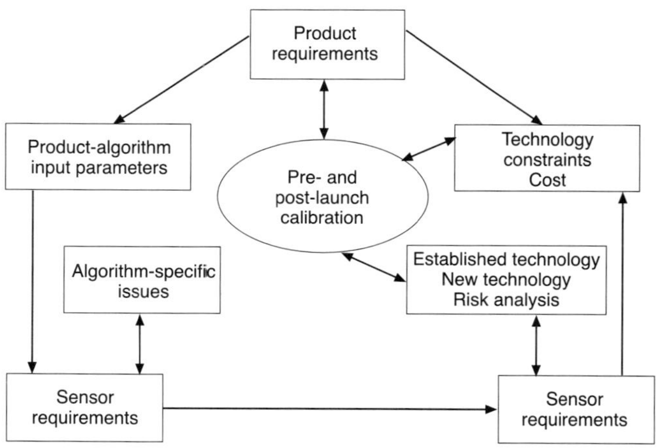

Figure 6.2. Evolution of a meteorological satellite sensor.

6.2 AVHRR PATHFINDER PROGRAM

NOAA and NASA in the United States jointly established the Pathfinder Program in the early 1990s with the objectives of (a) ensuring and optimizing the quality and continuity of environmental products derived from measurements made by meteorological satellite sensors and (b) evaluating the potential and limitations of various techniques available for acquisition, analysis, interpretation and management of large environmental datasets to set the stage for processing the very large volumes of data that would be generated by different sensors on the Earth System Enterprise (ESE) spacecraft. Separate Pathfinder programmes were established to address long term environmental records generated with AVHRR, Geostationary Operational Environmental Satellites' (GOES) imager and sounder, *Landsat*, and the microwave sensors flown on the polar orbiters of the US Department of Defense (DMSP). The AVHRR Pathfinder Program was made up of four different working groups to address (1) calibration, (2) AVHRR-derived atmospheric data, (3) AVHRR-derived land data and (4) AVHRR-derived ocean data. The atmosphere, land and ocean working groups would incorporate the findings and recommendations of the calibration working group in the rehabilitation and reprocessing of their respective datasets. The main objectives of the calibration working group were (a) to establish effective means for the post-launch calibration of the visible and near-infrared channels of AVHRRs on the *NOAA-7*, *-9* and *-11* spacecraft in the absence of on-board calibrators and (b) to develop corrections for the non-linear response of the thermal infrared channels. The initial choice of AVHRRs on the *NOAA-7*, *-9* and *-11* spacecraft was governed by the fact that they together covered the Pathfinder period 1981-1991, during which comprehensive field campaigns such as the First International Satellite Land Surface Climatology Field Experiment (FIFE) and the First International Satellite Cloud Climatology Regional Experiment (FIRE), which used AVHRR data, were conducted. The need for proper calibration of AVHRR was keenly felt in the analysis and interpretation of AVHRR data obtained during these campaigns. The work reported here is essentially a summary of the AVHRR Pathfinder Calibration Working Group activities, and extensions thereof to AVHRRs on *NOAA-14* and to AVHRRs on subsequent polar orbiters.

6.3 CALIBRATION OF THE ADVANCED VERY HIGH RESOLUTION RADIOMETER

As mentioned earlier, the visible and near-infrared channels (channels 1, 2 and 3A, see Table 6.1) of the AVHRR have no on-board calibration devices. These channels are known to degrade in orbit. Initially this is due to the contamination of the scan mirror by launch-vehicle exhaust and by the outgassing of various components of the optical train. Subsequently it is due to the deleterious effects of continued exposure to the harsh space environment (e.g. weathering because of ultraviolet radiation, pitting of the scan mirror by micrometeors, cosmic rays and extreme temperature shifts). We shall first describe very briefly the pre-launch calibration of the visible and near-infrared channels (channels 1, 2 and 3A), the main purpose of which is to ensure that the instrument is performing according to specifications in the laboratory environment. This will be followed by a very detailed discussion of the various techniques presently in use for the vicarious post-launch calibration of these channels (Section 6.4). Results obtained by different investigators to characterize the in-orbit performance of the AVHRRs on *NOAA-7*, *-9*, *-11* and *-14* will be presented and discussed. A technique developed at NOAA to use the AVHRR as a calibration reference to characterize the in-orbit performance of other meteorological satellite sensors in the visible and near-infrared will also be described.

As opposed to the visible and near-infrared channels, AVHRR thermal infrared channels (channels 3 (or 3B), 4 and 5, see Table 6.1) have on-board calibration, capable of yielding a two-point linear calibration by looking at an on-board black body and space. However, the mercury–cadmium–telluride (Hg-Cd-Te) detectors used in the infrared channels exhibit non-linear response, and appropriate non-linearity corrections have to be developed so that the scene brightness temperatures derived from measurements of the upwelling infrared radiation at the top of the atmosphere are accurate (Brown *et al.* 1985; 1993; Weinreb *et al.* 1990; Steyn-Ross and Steyn-Ross 1992). We shall describe the technique developed by the AVHRR Pathfinder Calibration Working Group (Walton *et al.* 1998) to account for these sensor non-linearities in the derivation of the Earth-scene brightness temperatures. It should be noted that the non-linearity corrections are based on the pre-launch calibration of the thermal infrared channels in the laboratory; it is thus assumed that the pre-launch calibration has been performed in an optimum manner, and that the corrections derived from pre-launch calibration are valid in the post-launch environment.

6.3.1 Pre-launch calibration of the visible and near-infrared channels

Calibration is defined as the process of determining the response of an instrument to known input signals. We shall describe here the pre-launch calibration of the visible and near-infrared channels of the six-channel AVHRR which is essentially the same as the procedure used to calibrate earlier versions of the instrument. The instruments are calibrated at the facilities of the instrument maker, ITT Aerospace/Communications (Fort Wayne, IN, USA), following a protocol that has evolved over the past two decades. The experimental set-up is shown schematically in Figure 6.3. A 102-cm (40″) integrating sphere, coated on the inside with white pressed 'Haylon', is used as the source of illumination. The inside of the sphere is equipped with seventeen 45-W lamps and three 150-W lamps mounted in a ring pattern. The radiance emerging from the integrating sphere exit port of diameter 35.6 cm (14″) can be varied over three orders of magnitude by illuminating the sphere with various combinations of the lamps. A separate 11.4 cm (4.5″) sphere source, equipped with a 30-W lamp, and an aperture wheel is used as the source when lower levels of illumination are desired. The level of illumination is varied in twenty-five steps for channel 1, sixteen steps for channel 2 and eleven steps for channel 3A. At each level of illumination of the integrating sphere, 3,600 measurements of the signal issuing from AVHRR, expressed in digital counts on a 10-bit scale, are determined, with the standard deviation serving as an indicator of the precision of the measurements. The integrating sphere is periodically calibrated for its spectral output and linearity with an Optronics Laboratory Model OLA410 integrating-sphere standard, the calibration of which in turn is traceable to radiometric standards maintained at the National Institute of Standards and Technology (NIST) in the USA. The uncertainty in the calibration of the 102 cm (40″) integrating-sphere source is estimated to be of the order of 2 to 4 per cent.

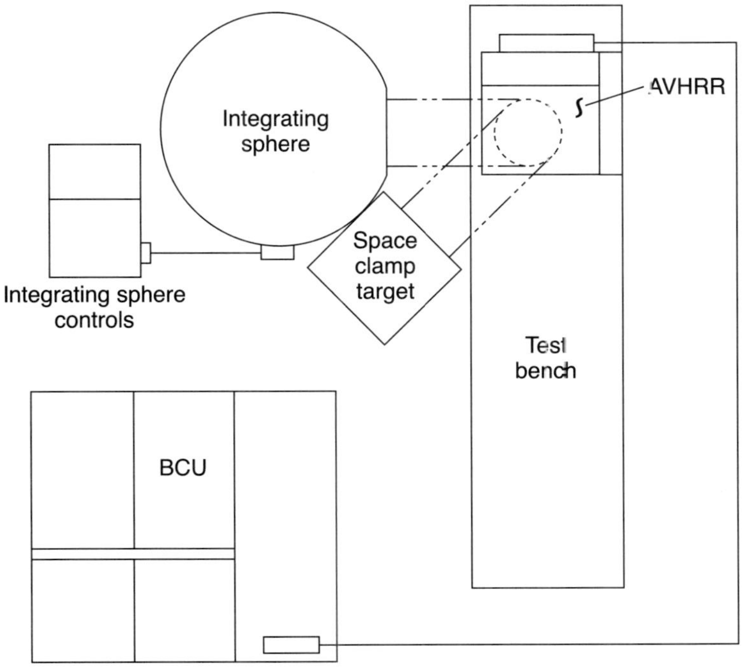

Figure 6.3. Schematic of the experimental set-up for the pre-launch calibration of AVHRR channels 1, 2, and 3A (BCU: Bench Control Unit); the space clamp target furnishes the zero radiance (albedo) reference to the instrument.

Another sensor feature that has to be known accurately is the normalized response function R_λ. This is defined as the ratio of the radiometric response of AVHRR at any given wavelength λ to the peak response of the instrument at wavelength λ_p within the spectral channel of interest. It is determined by comparing the response of the instrument and a reference detector when they both observe a radiation source through a monochromator. The reader is referred to Rao (1987) for further details of the pre-launch calibration of AVHRR channels 1 and 2.

We define the basic sensor characteristics which will be used quite often in our discussion. The effective width ω (μm) of an AVHRR channel is given by:

$$\omega = \int_{\lambda_1}^{\lambda_2} R_\lambda \, d\lambda \tag{6.1}$$

where R_λ is the normalized response function of the channel at wavelength λ, and λ_1 and λ_2 are the short-wavelength and long-wavelength limits of the AVHRR channel. The extra-terrestrial solar irradiance F within the passband of a given channel is given by:

$$F = \int_{\lambda_1}^{\lambda_2} F_\lambda R_\lambda \, d\lambda \tag{6.2}$$

where F_λ is the solar irradiance ($\mathrm{W\,m^{-2}\,\mu m^{-1}}$) at wavelength λ at the top of the atmosphere at the mean Earth–Sun distance (1 astronomical unit). The effective wavelength λ_e is given by:

$$\lambda_e = (\int_{\lambda_1}^{\lambda_2} \lambda R_\lambda F_\lambda \, d\lambda)/F \tag{6.3}$$

The reflectance factor (or AVHRR albedo or scaled radiance), expressed in per cent, when the instrument looks at a source of radiation (integrating sphere or Earth scene) is given by:

$$A\% = (100\pi I\omega)/F \tag{6.4}$$

where I is the average radiance from the source over the bandwidth of the channel. Thus, in the pre-launch calibration the response of the instrument in 10-bit counts C_{10} is measured for various levels of illumination of the sphere. A linear regression is established between the reflectance factor of the integrating-sphere source and AVHRR response in 10-bit counts:

$$A = mC_{10} + c \tag{6.5}$$

In the radiance representation, the regression relationship is:

$$I = A' + B'C_{10} \tag{6.6}$$

A' and B' are obtained by multiplying A and B by $(F/100\pi\omega)$; A' is expressed in units of $\mathrm{W\,m^{-2}\,sr^{-1}\,\mu m^{-1}}$ and B' in units of $\mathrm{W\,m^{-2}\,sr^{-1}\,\mu m^{-1}\,count^{-1}}$. We list in Table 6.3 the values of ω, F and λ_e for different AVHRRs.

A dual-gain feature was incorporated into the six-channel AVHRR (channels 1, 2 and 3A) to enhance the resolution of the instrument at low values of the reflectance factor by allocating half of the dynamic range of the instrument to cover either 0–25 per cent reflectance factor (channels 1 and 2), or 0–12.5 per cent reflectance factor (channel 3A). Details are given in Table 6.4. The crossover point (in counts) between the two gain ranges is maintained at a constant value.

We show the results of the pre-launch calibration in Figure 6.4(a) for channel 1 of AVHRR on *NOAA-14* which is typical of the pre-launch calibration of channels 1 and 2 of the four- and five-channel AVHRRs, and in Figure 6.4(b) for channel 3A of the six-channel AVHRR scheduled for flight on *NOAA-L* (*NOAA-16* after launch) which illustrates the dual-gain feature.

6.3.2 Pre-launch and post-launch calibration of the thermal infrared channels

6.3.2.1 Background

The three AVHRR thermal infrared channels respond to thermal radiation emitted from objects that have temperatures in the 185–335-K range. Data from these channels are used to investigate a wide variety of natural processes, such as:

(a) global sea-surface temperature (SST);
(b) outgoing long-wave radiation (OLR);
(c) cloud detection and rainfall from clouds;

Table 6.3. Equivalent widths, effective wavelengths and in-band solar irradiance values for the AVHRRs on the *NOAA-7*, *-9*, *-11*, *-14*, and *-L* spacecraft.

	ω (μm)	F (W m^{-2})	λ_e (μm)
NOAA-7			
Channel 1	0.108	177.5	0.630
Channel 2	0.249	261.9	0.830
NOAA-9			
Channel 1	0.117	191.3	0.635
Channel 2	0.239	251.8	0.833
NOAA-11			
Channel 1	0.113	184.4	0.635
Channel 2	0.229	241.1	0.832
NOAA-14			
Channel 1	0.129	207.1	0.641
Channel 2	0.244	251.0	0.842
NOAA-L			
Channel 1	0.081	133.2	0.632
Channel 2	0.235	243.1	0.843
Channel 3A	0.045	10.9	1.605

Table 6.4. Dual-gain features of the six-channel AVHRR.

Channel	Albedo range (%)	Count range
1	0–25	0–500
	26–100	500–1,000
2	0–25	0–500
	26–100	500–1,000
3A	0–12.5	0–500
	12.6–100	500–1,000

(d) fire and ice location;
(e) global vegetation patterns.

Many projects that monitor natural processes are concerned with year-to-year change. This requires that the channel data must be relatively accurate (consistent) from year to year. High absolute accuracy is also welcome, but often is not required. The way to guarantee relative or absolute accuracy is by calibration and AVHRR thermal channels are calibrated effectively, both in orbit and before launch.

The AVHRR receives radiant energy from 2,048 different surface or atmosphere scenes in each scan line. Radiant energy from six separate spectral intervals or channels is measured, three thermal-infrared channels, two near-infrared and one

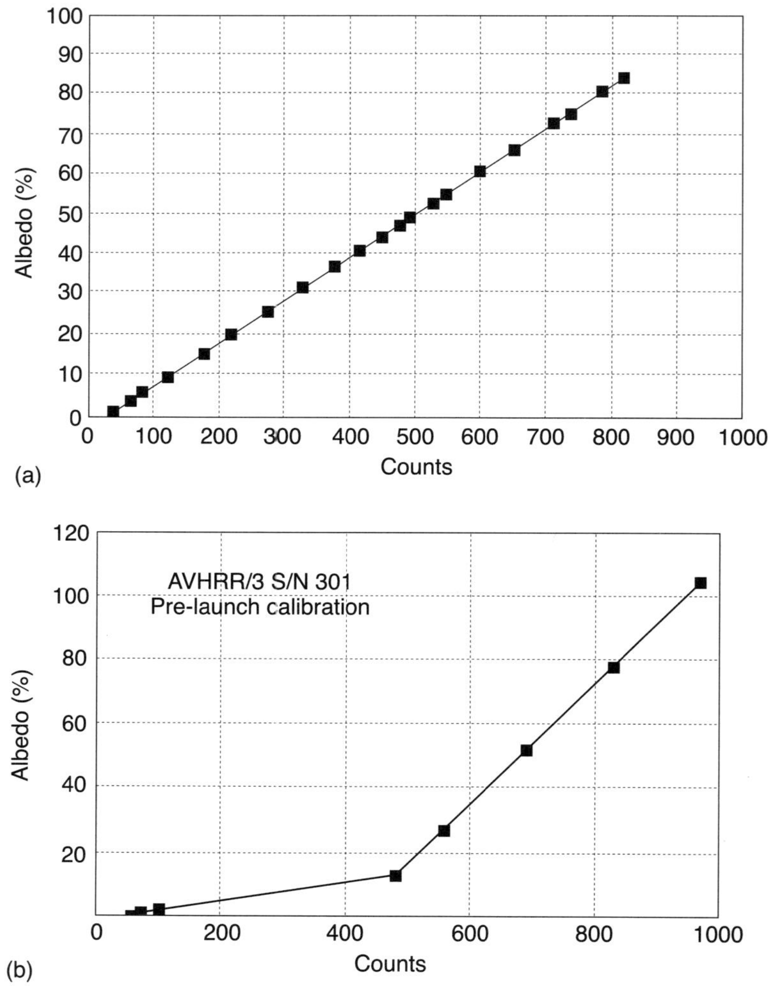

Figure 6.4. Pre-launch calibration of (a) *NOAA-14* AVHRR channel 1 and (b) *NOAA-L* AVHRR channel 1.

visible, as shown in Figure 6.1. The AVHRR also receives radiant energy in every scan line from two calibration targets, cold space and an internal warm black body attached to the baseplate of the AVHRR. Information from the Earth scene, from the two in-orbit calibration targets and from pre-launch calibration, is used to compute a radiance value for each pixel, for each of the three thermal channels.

6.3.2.2 *Calibration overview*

We outline here the basic calibration process, the on-board calibration sequence and the pre-launch calibration runs conducted by ITT/Aerospace Communications, the instrument maker. Basic AVHRR calibration is simple in principle. The radiometer views a black body at constant temperature T for an interval of time and outputs count values. As long as the measured black-body temperature is constant, the output of the AVHRR in counts should also be stable. When this relationship is verified over a wide range of black body temperatures, ITT certifies that the AVHRR is a usable radiometer. This is qualitative; for each channel a quantitative AVHRR radiance is computed by weighting the Planck function for a black body at temperature T by the channel's spectral response function, which is also measured before launch. For each channel, the final result of pre-launch calibration is a set of (AVHRR count, AVHRR radiance) pairs.

The on-board calibration uses the basic calibration twice for each calibration target. Each scan line the radiometer outputs counts values when it views cold space and when it views the warm internal black-body target. In addition to these radiometric measurements, the warm black-body temperature is measured by four platinum resistance thermometers (PRTs) embedded in it. The warm black-body temperature is then converted to AVHRR radiance. Cold space is assigned a non-zero radiance value; the rationale behind this is explained in Section 6.3.2.4(b). This provides two accurately determined (count, radiance) points. For the first approximation of the Earth-scene radiance, the two calibration points are connected by a straight line. A linear estimate of the Earth-scene radiance is generated by substituting the Earth-scene count value into the straight-line equation.

The in-orbit linear radiance estimate suffices for many uses but it is only a first approximation to the pre-launch calibration results produced by the instrument maker. The primary difference between in-orbit and pre-launch calibration is that an in-orbit surface/atmosphere scene at unknown temperature is replaced in pre-launch calibration by a black-body target, commonly referred to as the external calibration target, which is held at a precisely measured temperature. AVHRR counts are measured for different settings of this black-body's temperature. When black-body temperatures are converted to AVHRR radiances, the radiance versus count curve is slightly non linear; the data can be fitted very well by a quadratic curve. A second pre-launch result is that the radiance versus count curve also depends on AVHRR operating temperature itself, as measured by the PRTs in its baseplate. These two results are provided to AVHRR data producers and users so that they can generate more accurate thermal-channel radiances.

6.3.2.3 Black-body temperature to AVHRR radiance conversion

The topic of conversion of black-body temperature to AVHRR radiance is explained here in detail, both for the importance and for the confusion often associated with it. The reverse process, the radiance-to-temperature conversion, needed by both data producers and data users, will also be discussed. Each AVHRR thermal channel responds only to energy from a limited spectral region, as shown by Figure 6.1. The response functions are measured before launch (e.g. the responses of *NOAA-15* thermal channels were measured at about 200 different wavelengths). This information is first converted to the wave-number domain at the NOAA National

Environmental Satellite Data and Information Service (NESDIS) and AVHRR-radiance computations using wave numbers are subsequently carried out. The AVHRR radiance N_{BB} sensed in each thermal channel from a black-body at temperature T_{BB} is defined by:

$$N_{BB} = \sum B(\nu_j, T_{BB})\phi(\nu_j)\,\Delta\nu_j / \sum \phi(\nu_j)\,\Delta\nu_j \tag{6.7}$$

where $B(\nu, T_{BB})$ is the Planck function at wave number ν and temperature T_{BB} and $\phi(\nu)$ is the spectral response function for the channel. The summation goes from wavenumber ν_1 to ν_j, the cut-off wave numbers for the response function; $\Delta\nu_j$ is the width of the jth spectral interval, and there are J spectral intervals. In practice, a look-up table relating AVHRR radiance to black-body temperature is generated for each channel, which specifies the radiance at every tenth of a degree (K) between 180 K and 335 K; the tables are referred to as energy tables.

For each thermal channel, an accurate approximation to the results produced by Equation (6.7) is given by a single expression, a modified Planck function that uses a centroid wave number ν_C and an 'effective' temperature T^*. The AVHRR radiance N_{BB} from a black body at temperature T_{BB} is approximated using the following two-step process:

$$\begin{aligned} T^* &= a + bT_{BB} \\ N_{BB} &= C_1\nu_C^3/[\exp(C_2\nu_C/T^*) - 1] \end{aligned} \tag{6.8}$$

The numerical values for C_1 and C_2 are chosen so that N_{BB} has units $\text{mW m}^{-2}\,\text{cm sr}^{-1}$ when T_{BB} is measured in degrees K ($C_1 = 1.1910659 \times 10^{-5}\,\text{mW m}^{-2}\,\text{sr}^{-1}\,\text{cm}^4$, $C_2 = 1.438833\,\text{cm K}$).

Data users who compute AVHRR radiance N_{BB} from the Earth-scene AVHRR count value and need to calculate the equivalent black-body temperature T_{BB} use the inverse of Equation (8):

$$\begin{aligned} T^* &= C_2\nu_C/\log[1 + (C_1\nu_C^3)/N_{BB}] \\ T_{BB} &= (T^* - a)/b \end{aligned} \tag{6.9}$$

The approximation is very accurate. To assess its accuracy, we can choose a temperature value T_{BB} and substitute T_{BB} into Equation (6.7) to produce the value of N_{BB}. We can then use Equation (6.9) on N_{BB} to invert the process, resulting in the approximate value t_{BB}. Typically, the difference between T_{BB} and t_{BB} will be found to be less than 0.003 K.

6.3.2.4 Calibration details

(a) Pre-launch calibration

Pre-launch calibration of the thermal infrared channels is carried out in a thermal vacuum chamber at the facilities of the instrument maker, ITT Aerospace/Communications. ITT uses three black bodies in the pre-launch calibration set-up. Cold space is simulated by a black body at a temperature near 90 K; its energy output

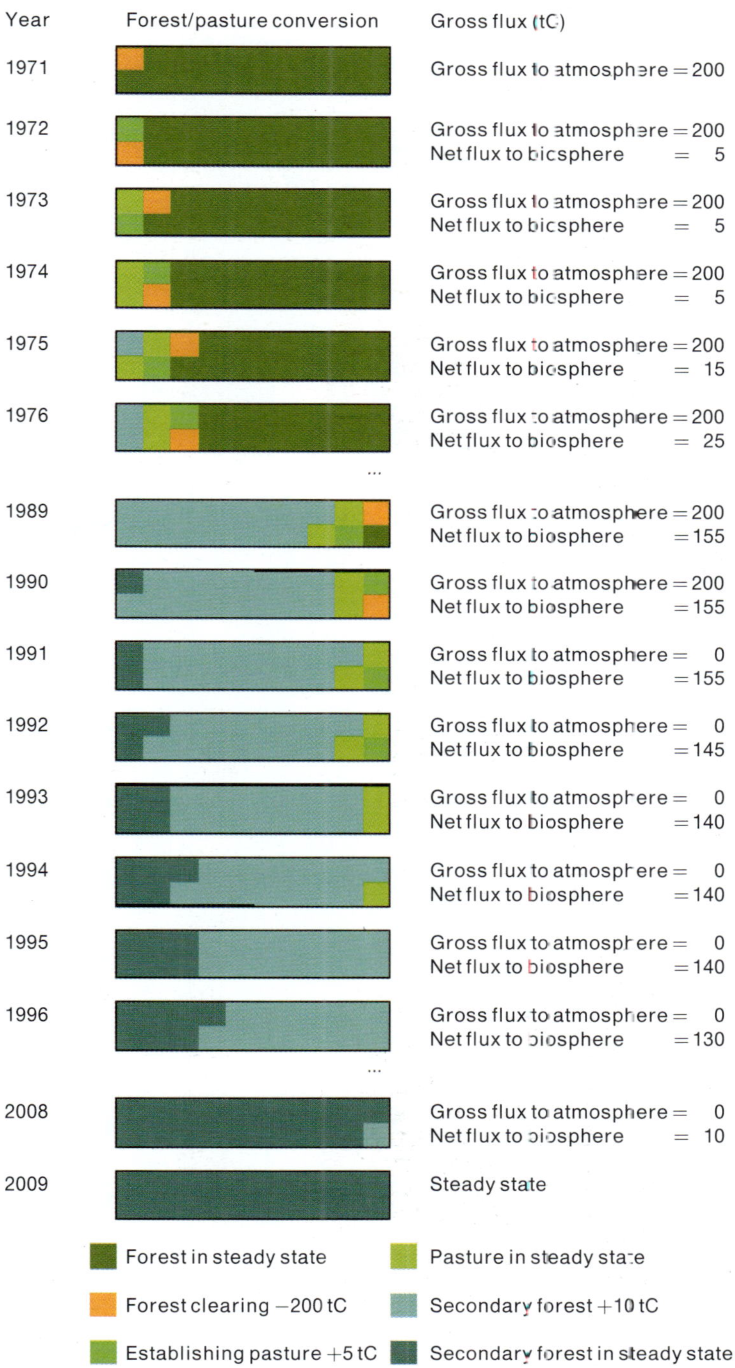

Year
Forest/pasture conversion
Gross flux (tC)
1971
Gross flux to atmosphere = 200
1972
Gross flux to atmosphere = 200
Net flux to biosphere = 5
1973
Gross flux to atmosphere = 200
Net flux to biosphere = 5
1974
Gross flux to atmosphere = 200
Net flux to biosphere = 5
1975
Gross flux to atmosphere = 200
Net flux to biosphere = 15
1976
Gross flux to atmosphere = 200
Net flux to biosphere = 25
...
1989
Gross flux to atmosphere = 200
Net flux to biosphere = 155
1990
Gross flux to atmosphere = 200
Net flux to biosphere = 155
1991
Gross flux to atmosphere = 0
Net flux to biosphere = 155
1992
Gross flux to atmosphere = 0
Net flux to biosphere = 145
1993
Gross flux to atmosphere = 0
Net flux to biosphere = 140
1994
Gross flux to atmosphere = 0
Net flux to biosphere = 140
1995
Gross flux to atmosphere = 0
Net flux to biosphere = 140
1996
Gross flux to atmosphere = 0
Net flux to biosphere = 130
...
2008
Gross flux to atmosphere = 0
Net flux to biosphere = 10
2009
Steady state
Forest in steady state
Forest clearing −200 tC
Establishing pasture +5 tC
Pasture in steady state
Secondary forest +10 tC
Secondary forest in steady state

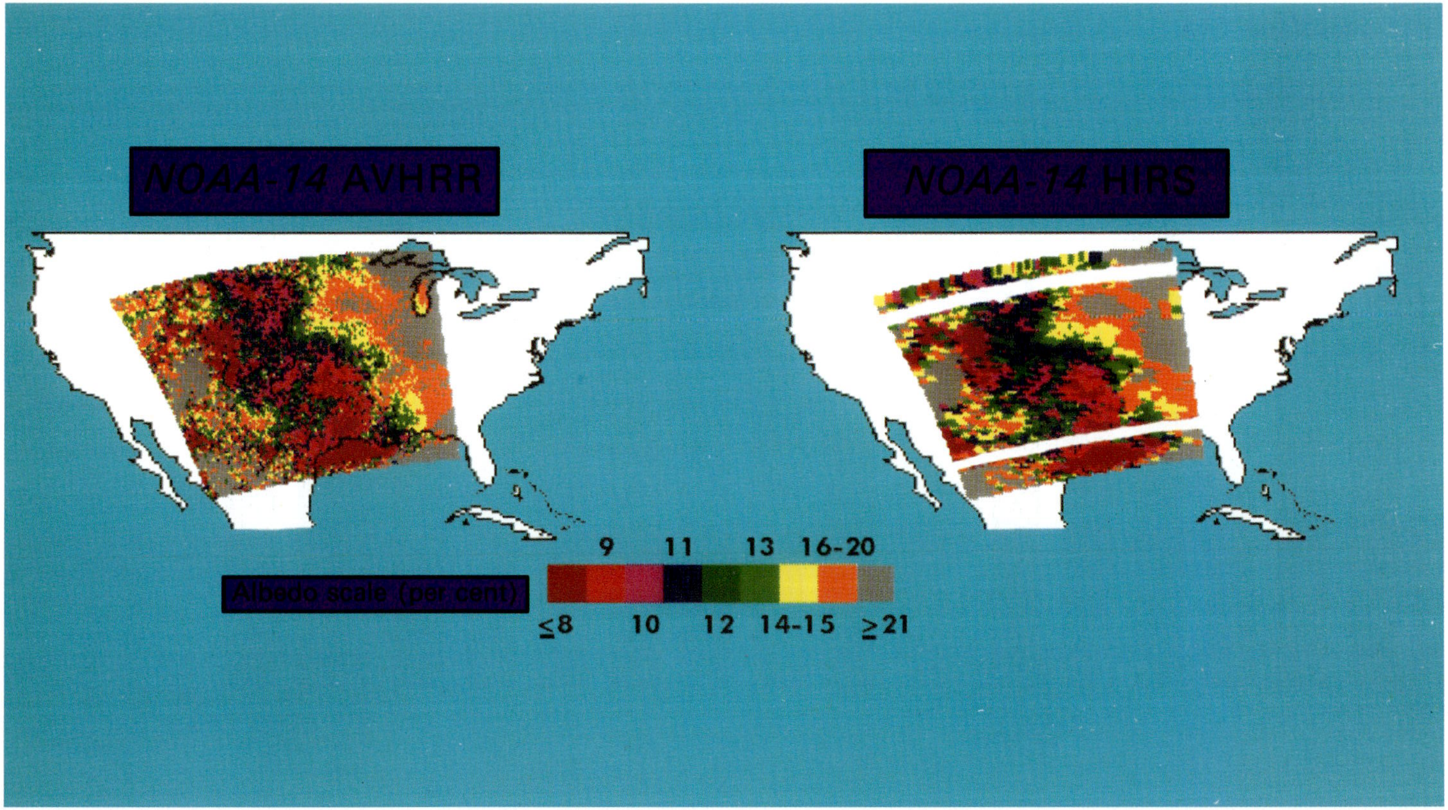

Plate 2. *NOAA-14* AVHRR and HIRS visible albedos measured over the contiguous United States on October 5, 1999; Earth scene data are not available for the white swaths in the HIRS image.

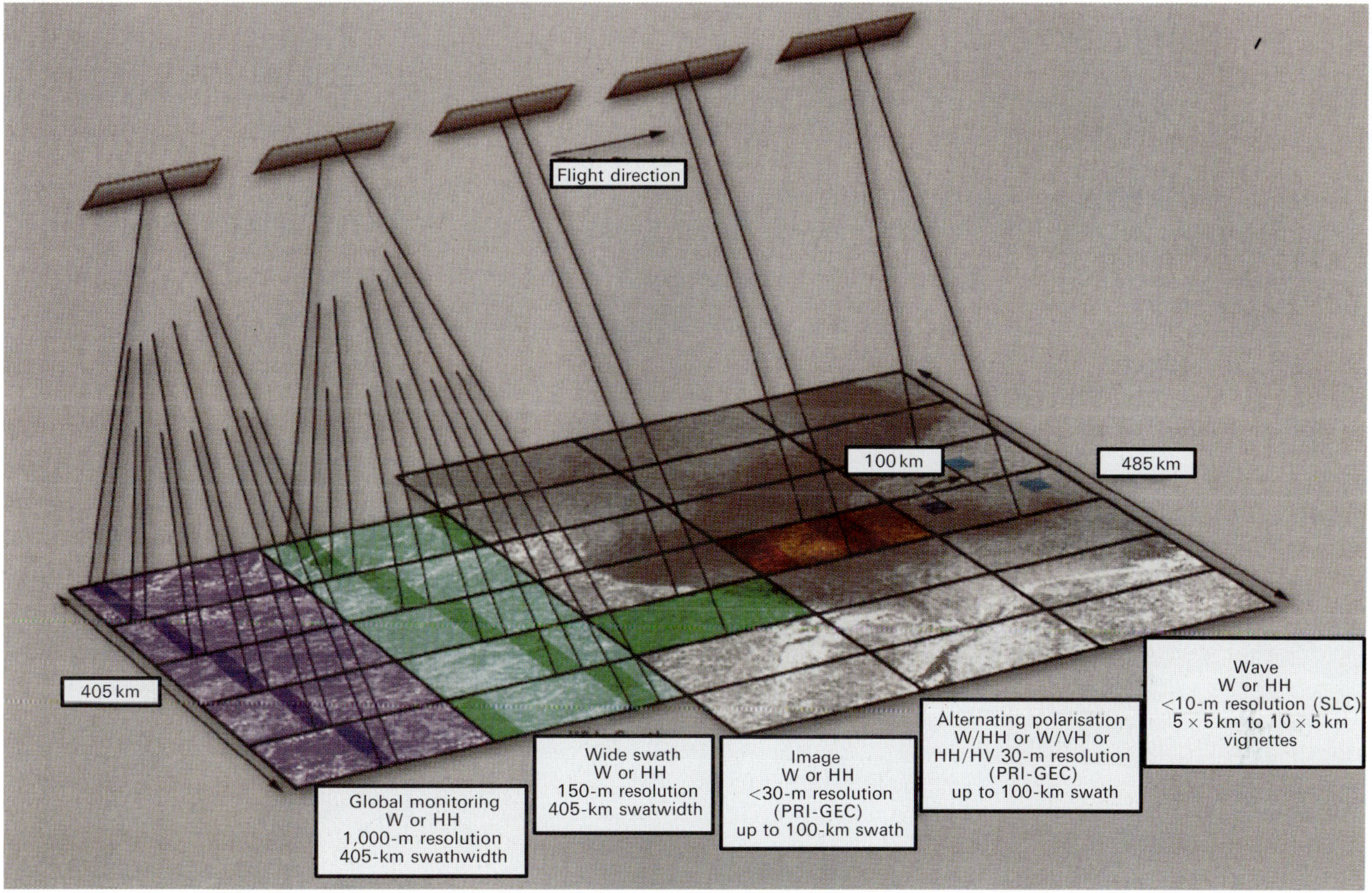

Plate 3. ASAR operating modes.

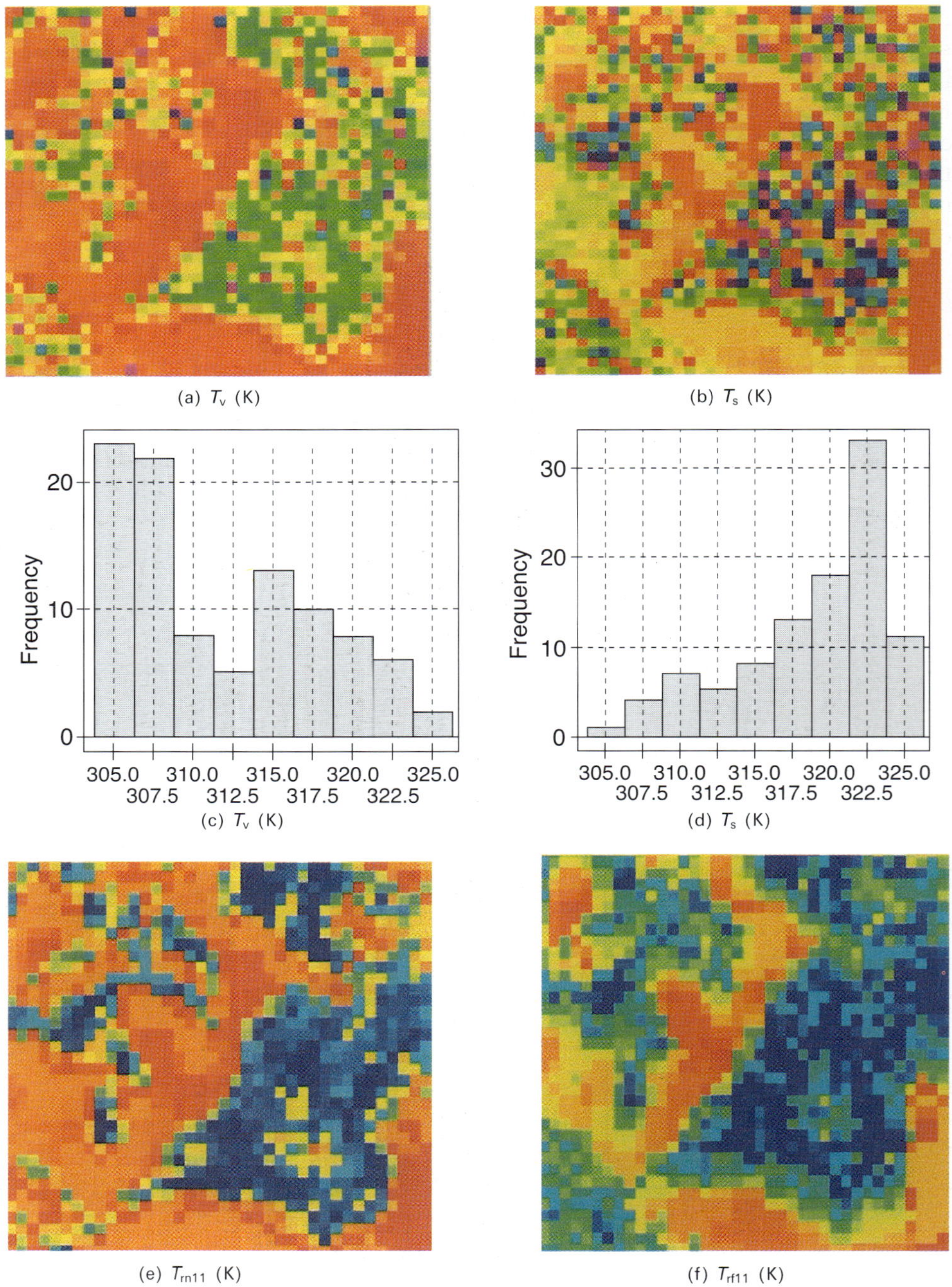

Plate 4. Component surface temperatures, retrieved by ATSR/TIR, 19 August 1991; (a) temperature of vegetation, T_v (K); (b) temperature of soil, T_s (K); (c) histogram of the temperatures of vegetation, T_v (K); (d) histogram of the soil temperatures; (e) radiometric temperature at 11 μm, nadir, T_{rn11} (K); (f) radiometric temperature at 11 μm, forward.

in the thermal infrared wavelength region is too small to affect the AVHRR-count output. The same internal black body target carried in orbit is also used in pre-launch; it is kept at temperatures in the 15–25°C (288 K–298 K) range. The third black body is a substitute for the Earth scene and is not available in orbit; it is an external target whose temperature is precisely measured. The temperatures of both the internal and external black bodies are measured by PRTs embedded in them. The internal black-body temperature is considered to be a good indicator of AVHRR operating temperature itself.

In a single calibration run conducted in a thermal vacuum chamber, the internal black-body temperature is fixed (e.g. at 20°C), and the external black-body temperature is changed from 180 K to 335 K in sixteen discrete steps. At each step the black body is brought into thermal equilibrium with the environment, and 1 min (360 scan lines) of AVHRR count data is collected, with a minimum of ten AVHRR signals (counts) collected from each scan line for each of the three black bodies. The radiance of the external black body measured by the AVHRR is computed by substituting the 360-scanline average temperature into Equation (6.7). The corresponding AVHRR signal (count value) is taken as the average of the 3,600 count values obtained from viewing the external black-body target. The AVHRR radiance is then plotted against AVHRR counts for each of the seventeen external black-body temperature settings. We observe that for channel 3B (or 3), AVHRR radiance is linear in AVHRR count; however, for channels 4 and 5, the resulting curve is slightly non-linear. This behaviour, which is characteristic of Hg-Cd-Te sensors used in the photoconductive mode, is illustrated in Figure 6.5(a) where we show the linear regression of the radiance in channel 4 of *NOAA-11* AVHRR on the output counts, and in Figure 6.5(b) where we show the resulting residuals as a function of the external calibration-target temperature. The shape of the residual curve is indicative of the non-linear response of the channel. Thus, a quadratic curve in count value is a good representation of the radiance.

When the internal black-body temperature is changed (e.g. 15°C), and the pre-launch calibration sequence is repeated, the radiance versus count curves for channels 4 and 5 are again non-linear but differ from the curves produced by the 20°C run. In orbit, an AVHRR calibration that closely resembles the laboratory one is performed. However, only enough data are available from the in-orbit calibration to fit radiance as a straight line in AVHRR count value. The residual information, the difference between pre-launch measurements and the in-orbit linear estimate, must be provided to data producers. This difference is called the non-linear radiance correction; how NESDIS provides users with this information is explained in the following paragraphs.

(b) In-orbit calibration

In orbit, AVHRR transmits counts from the space look, the on-board warm black body (Internal Calibration Target or ICT), the Earth-scene look and the PRTs to ground receiving stations. PRT counts are converted to PRT temperatures using equations determined in the pre-launch calibration and the average of the four

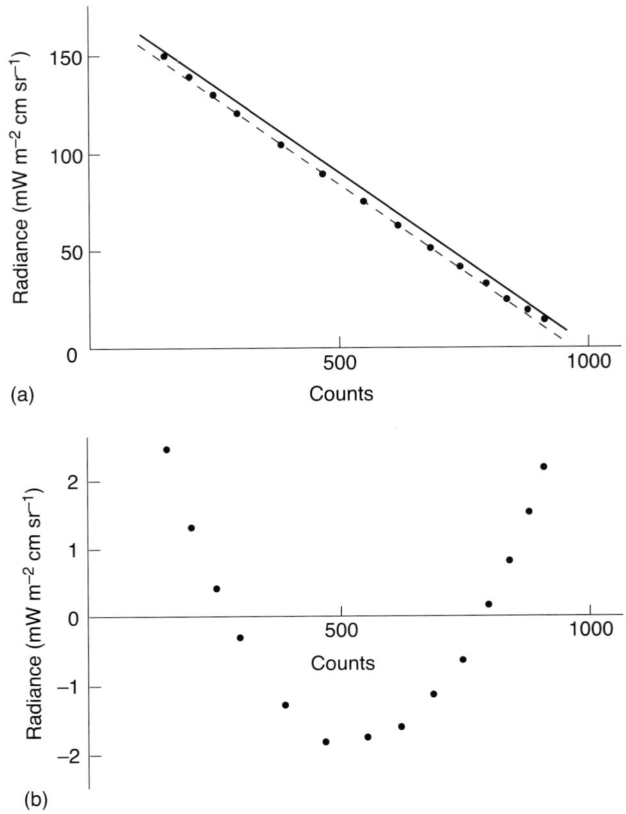

Figure 6.5. (a) Linear regression of the radiance in channel 4 of *NOAA-11* AVHRR on output counts (baseplate temperature 15°C), and (b) the variation of the radiance residual with counts when the instrument is looking at the external calibration target; the solid line in (a), joining the end points of the calibration run, renders the non-linearity more obvious.

PRT temperatures is taken as the warm black-body temperature, T_{BB}. T_{BB} is then converted to AVHRR radiance N_{BB}, using equation (6.8). The averages of space counts and warm black-body counts from five consecutive scan lines, C_S and C_{BB}, are used for all five lines. Using an Earth-scene count C_E from one of the 2,048 scenes in a scan line, the linear radiance estimate N_{LIN} for the scene is computed from the equation:

$$N_{LIN} = N_S + \frac{(N_{BB} - N_S)}{(C_{BB} - C_S)}(C_E - C_S) \tag{6.10}$$

N_S is a non-zero radiance of space value that NESDIS uses to incorporate pre-launch information into the data-producing process in an accurate and unambiguous way. Both N_{BB} and N_S use information derived from pre-launch calibration.

Physically, the radiance of space value N_S in AVHRR channels 4 and 5 is effectively zero. Mathematically, N_S can be any value as long as it remains

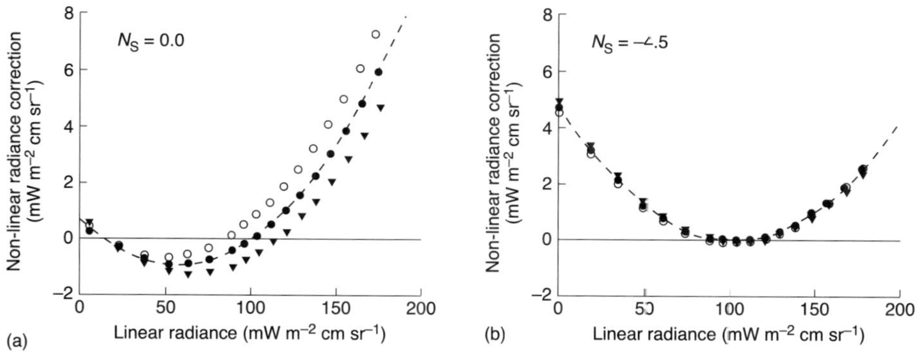

Figure 6.6. Non-linear radiance corrections plotted against linear radiance for internal black-body temperature of 10°C (circle), 20°C (filled circle), and 30°C (triangle) using radiance of space values (a) $N_S = 0.0$ and (b) $N_S = -4.5$. The dashed curves are least squares quadratic fits to the 20°C curve. *NOAA-15* AVHRR pre-launch data were used.

constant in data processing during the satellite's lifetime. The criterion that NESDIS used to choose the value of N_S was that

> non-linear radiance corrections should be independent of the internal black-body temperature.

The usefulness of satisfying this criterion is illustrated in Figure 6.6, which shows non-linear correction curves obtained from *NOAA-15* channel-4 pre-launch data, for calibration runs at 10°C, 20°C and 30°C settings of the internal black-body temperature. The non-linear correction curves were generated by computing linear radiances using Equation (6.10) on the pre-launch data and subtracting these linear estimates from the measured non-linear curves. Figure 6.6(a) shows the correction curves obtained from setting $N_S = 0$; the curves are noticeably separated. Figure 6.6(b) shows correction curves using $N_S = -4.50\,\mathrm{mW\,m^{-2}\,cm\,sr^{-1}}$; corrections for the three calibration runs effectively fall on a single correction curve. (Using a larger negative value of N_S causes the correction curves to separate again.) The dashed lines in the figure show the correction curve for the 20°C run modelled as a quadratic in linear-radiance value.

To summarize, for the correct value of N_S, there is a single non-linear correction curve independent of the internal black-body temperature and the curve is a quadratic in N_{LIN}, the linear-radiance estimate.

Data producers using operational data now need to use only one correction curve to incorporate non-linear-radiance-correction information derived from pre-launch data. The final Earth-scene radiance estimate N_E is computed by adding the non-linear correction N_{COR} to the linear estimate N_{LIN}:

$$N_E = N_{LIN} + N_{COR} \tag{6.11}$$

In operational practice there are more details than are given here. Users of NESDIS

1b operational data compute scene radiance N_E from scene count C_E from the equation:

$$N_E = a_0 + a_1 C_E + a_2 C_E^2 \tag{6.12}$$

where the coefficients a_0, a_1 and a_2 are found in the 1b data stream. The interested reader is referred to the calibration section, section 7, on the *NOAA-K*, *-L* and *-M* satellite documentation website (http://www2.ncdc.noaa.gov/docs/klm/index.htm).

6.4 POST-LAUNCH CALIBRATION OF THE VISIBLE AND NEAR-INFRARED CHANNELS

In order to illustrate the importance of the post-launch calibration of the visible and near-infrared channels of AVHRR, we show in Figure 6.7 the variation in time (Figure 6.7(a)) and with solar-zenith angle (Figure 6.7(b)) of the signal (counts) from channel 1 of *NOAA-9* AVHRR for the period 1985–1988 over a stable site in Sudan (see Section 6.5). The downward trend in time of the signal (Figure 6.7(a)) is due to: (a) the degradation of the sensor in orbit, and (b) effects of orbital drift – a characteristic of afternoon spacecraft. The progressively later equator-crossing times as a result of orbital drift result in correspondingly larger solar-zenith angles over the site, thereby leading to lower levels of illumination. These two effects are apparent in the data shown in Figure 6.7(b). The signal decreases from one year to the next at the same solar-zenith angle over the site because of the degradation of the sensor. In addition, the solar-zenith angles at which measurements can be made over the site move towards progressively larger values from one year to the next, thereby resulting in lower levels of illumination at the site. It is thus necessary to correct for instrument degradation, and to account for the dependence of AVHRR signal on the solar-zenith angle to ensure the accuracy of AVHRR-derived geophysical products. We shall therefore describe in this section some of the techniques presently in use for the vicarious post-launch calibration of the visible and near-infrared channels of the AVHRR. The term vicarious is used to denote any calibration procedure that takes the place of conventional pre-launch or on-board calibration.

6.4.1 Determination of relative in-orbit degradation rates using radiometrically stable calibration sites

Bright desert sites have been used as radiometrically stable calibration targets to determine the relative degradation rates of the AVHRR since the early 1990s (e.g. Staylor 1990; Kaufman and Holben 1993; Wu and Zhang 1994; Rao and Chen 1995, 1999). Radiometric stability implies that the top-of-the-atmosphere albedo is constant in time, for all practical purposes, and that seasonal variations in the same, if any, are repeated from one year to the next. In addition, spatial homogeneity on a scale much larger than the pixel size ($\approx$1.1 km $\times$ 1.1 km) is a stringent requirement. The top-of-the-atmosphere albedo (in per cent) is given by

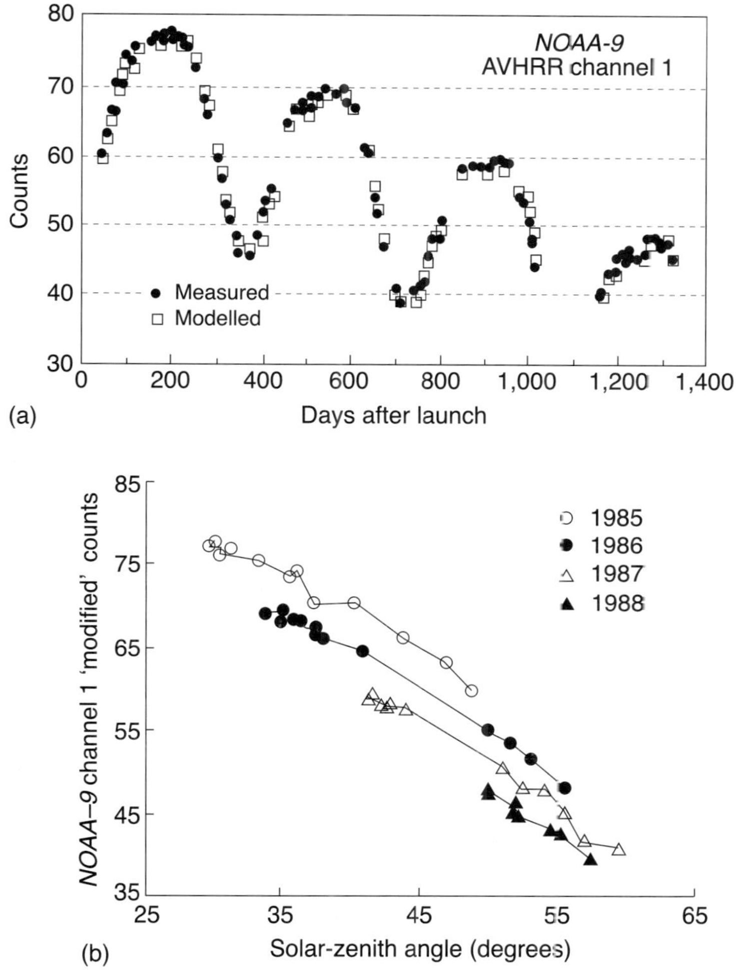

Figure 6.7. Variation of the *NOAA-9* AVHRR channel 1 signal (a) in time and (b) with the solar-zenith angle.

$(100\pi I\omega/F\cos\theta_0)$; here, I is the upward radiance measured by the AVHRR in a given channel which has been corrected for the seasonal variations in the incident solar radiation, F is the extra-terrestrial solar irradiance in the channel at mean Earth–Sun distance and θ_0 is the solar-zenith angle. Use of bright, desert calibration sites offers the advantage that the effects of overlying atmosphere are minimized since most of the upward radiation originates at the surface. However, the technique requires multi-year, multi-seasonal data to account for seasonal variations, if any, in the top-of-the-atmosphere albedo. In addition, independent, absolute calibration of the instrument at several points in time during its operational life is needed to render the relative degradations absolute so that the variation in time of the radiometric calibration 'slope' can be determined. The radiometrically stable ice/snow fields of Greenland and Antarctica have also been used to determine the relative degradation of the AVHRR in orbit (e.g. Loeb 1997; Tahnk and Coakley 2001).

6.4.2 Absolute radiometric calibration using congruent-path aircraft and satellite measurements of the upwelling radiation

The technique of using congruent-path measurements of diffusely reflected radiation with a well-calibrated spectroradiometer on-board a high flying aircraft (e.g. ER-2 aircraft flying at an altitude of $\approx$20 km), and with the satellite radiometer over a stable, highly reflecting desert surface yields absolute radiometric calibration of the satellite radiometer; here congruency implies simultaneous aircraft and satellite radiation measurements at the same view angle. The need for the congruency of the airborne and satellite measurements imposes very stringent requirements on the accurate determination and prediction of the satellite-target viewing geometry so that the spectroradiometer on the aircraft can be properly coaligned with the AVHRR during satellite overpass. The radiance measurements by the instrument on the aircraft are then directly related to the signal from the satellite radiometer, and an absolute calibration of the AVHRR established. Use of a high-flying aircraft such as the ER-2 minimizes the impact of the atmospheric layer between the aircraft and satellite; however, appropriate model-based corrections for the effect of Chappuis-band absorption by ozone, a considerable part of which lies in the atmospheric layer above the ER-2 aircraft, are made in the interpretation of measurements in AVHRR channel 1.

This technique has been used to calibrate the visible and near-infrared channels of AVHRRs on the *NOAA-9* and *-11* spacecraft with measurements made over the White Sands desert located in New Mexico (USA) (Smith *et al.* 1988; Abel *et al.* 1993). The AVHRR on *NOAA-14* has been calibrated using congruent-path radiance measurements made over Rail Road Playa, a site located in Nevada, USA (P. Abel, pers. commun.). The major constraints on using this technique are the cost and logistics of performing aircraft measurements. In addition, the on-board spectroradiometer has to be very accurately calibrated; the impact of the differences in footprints of the aircraft spectroradiometer and of the satellite radiometer should be assessed, and error budgets associated with the same should be determined. The procedure is shown schematically in Figure 6.8 (Smith *et al.* 1988).

6.4.3 Absolute calibration based on radiative model simulations of the upwelling radiation at the top of the atmosphere

The basic principles of calibrating broad-band sensors like the AVHRR using model simulations of the upwelling radiation sensed by the instrument have been discussed fairly extensively in the literature (e.g. Slater *et al.* 1996; Teillet *et al.* 1990; Mitchell *et al.* 1992). The upwelling radiation $I_\lambda(\theta_0, \theta; (\phi - \phi_0))$ is dependent on (a) the solar and satellite-zenith angles θ_0, θ and the relative azimuth angle $(\phi - \phi_0)$ between the direc-

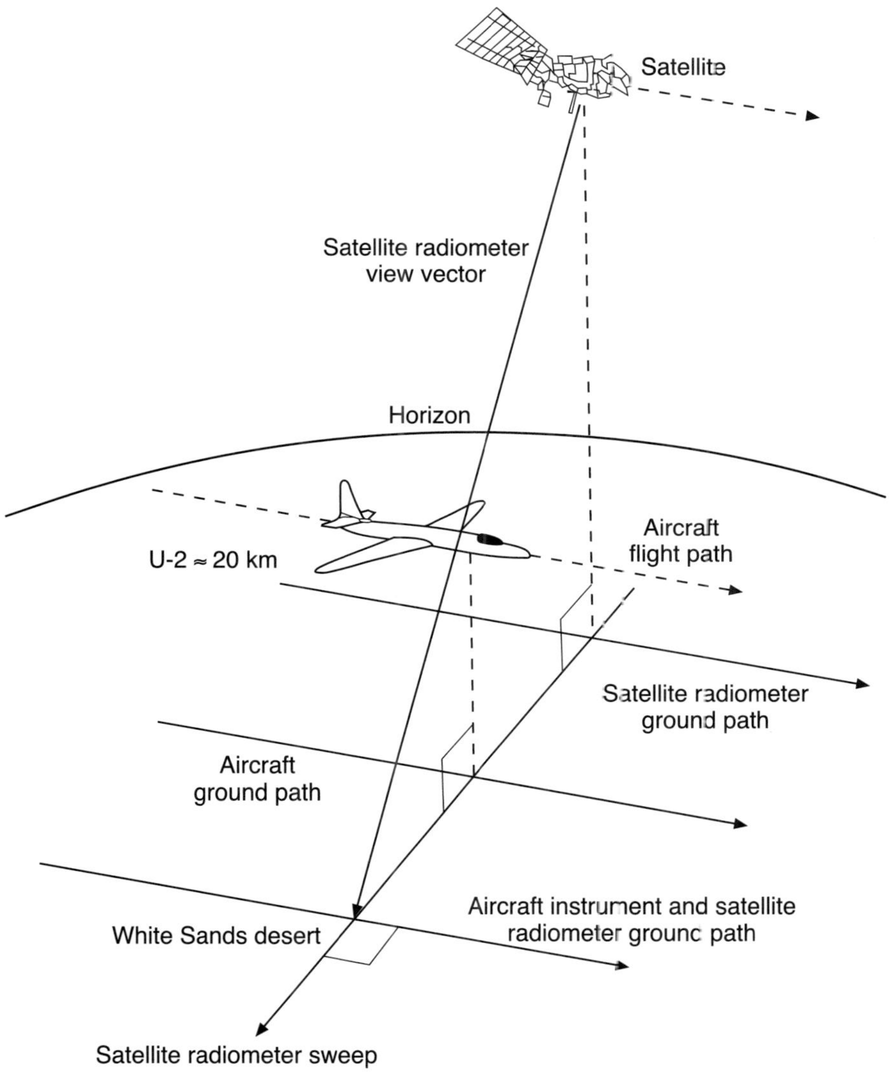

Figure 6.8. Schematic representation of calibration of satellite radiometer using aircraft/satellite congruent-path radiance measurements (Smith *et al.* 1988).

tions of illumination and observation; (b) absorption and scattering of radiation by the gaseous and particulate constituents of the atmosphere and (c) the bidirectional reflection properties of the underlying surface. The principle of this method is to require that the measured satellite signal should yield the computed value of the radiance or the albedo at the top of the atmosphere. Thus, inclusion of the measured values of at least some of the model input parameters – surface reflection properties, atmospheric aerosols, ozone and precipitable water – is desirable since it would help render the model-based calibration more accurate, because uncertainties in the atmo-

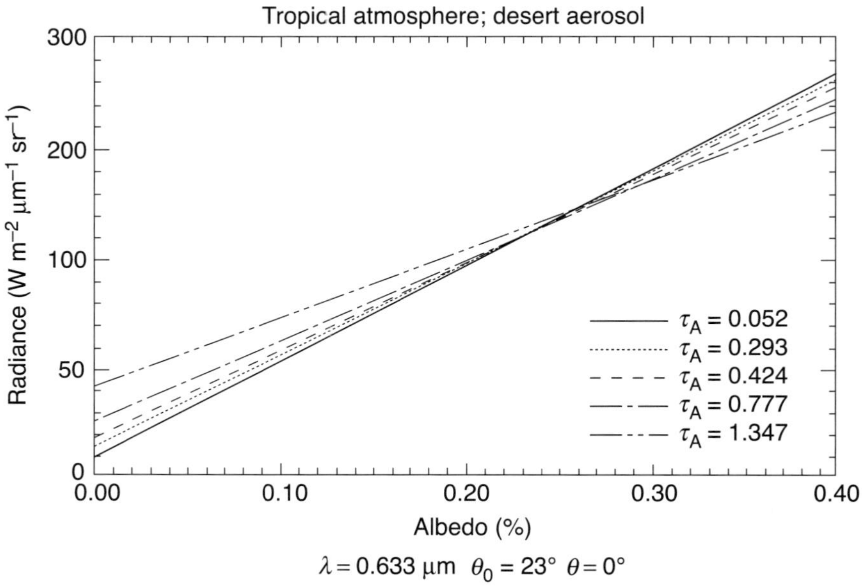

Figure 6.9. Model simulations of the diffusely reflected radiation at the top of the atmosphere in the *NOAA-14* AVHRR channel 1; τ_A is the aerosol optical thickness at 0.5 μm.

spheric and surface parameters contribute to the overall uncertainty in the attainable radiometric-calibration accuracy (Slater *et al.* 1996). The accuracy reported for the calibration of AVHRR visible and near-infrared channels using model simulations ranges from 3 to 19 per cent (Slater *et al.* 1996; Vermote and Kaufman 1995; Mitchell *et al.* 1992; Teillet *et al.* 1990). To illustrate the competing and compensating effects of the atmosphere and surface on the emergent radiation, we show in Figure 6.9 the simulated values of the radiance at the top of the atmosphere for different values of the Lambert surface reflectance, and of atmospheric turbidity labelled in terms of the aerosol optical thickness τ_A at the wavelength of 0.5 μm; the larger the value of τ_A the greater the atmospheric turbidity. We used the LOWTRAN/MODTRAN radiative transfer code (Berk *et al.* 1989) to compute the upward radiation. When the surface reflectance is very low, as for oceans, the upward radiance is very sensitive to the amount of aerosols in the atmospheric column; this is the basis for the retrieval of columnar aerosols over oceans (Rao *et al.* 1989). On the other hand, when the surface is very bright, the radiance decreases with increasing atmospheric turbidity. For certain values of surface reflectance, of the order of 20 to 30 per cent depending on wavelength, the upward radiation is relatively insensitive to atmospheric turbidity. This serves as a criterion to choose a calibration site to determine relative degradation rates. The attainable calibration accuracy is determined by the degree of correspondence between the model atmosphere and the real atmosphere.

6.4.4 Calibration based on intercomparison of sensors

Calibration of the visible and near-infrared channels of the AVHRR on any given spacecraft by comparison with other calibrated satellite sensors, such as the Along-Track Scanning Radiometer, has been advocated. It is important that the impact of (a) differences in the spectral characteristics of the two sensors which determine how the surface and atmospheric properties affect the upward radiation at the top of the atmosphere, (b) differences in footprints and (c) differences in the illuminating and observational geometries is properly understood to develop error budgets for the calibration obtained using this technique. We shall describe this technique in greater detail in Section 6.6 on using the AVHRR as a calibration reference.

6.5 CALIBRATION OF THE AVHRRs ON THE *NOAA-7*, *-9* AND *-11* SPACECRAFT

6.5.1 Relative degradation in orbit

After a survey of the various vicarious techniques presently in use to calibrate the AVHRR, the NOAA/NASA AVHRR Pathfinder Calibration Working Group recommended (Rao *et al.* 1993) that a radiometrically stable desert calibration site should be employed to determine the relative degradation in orbit of channels 1 and 2 of AVHRRs on *NOAA-7*, *-9* and *-11* spacecraft, and that the same should be rendered absolute, using available absolute calibrations obtained with congruent-path aircraft/satellite measurements of the diffusely reflected radiation at the top of the atmosphere. Accordingly we used a site located in Sudan (21–23°N, 28–29°E) in the vast Libyan desert, the radiometric stability of which had been established by researchers elsewhere (e.g. Staylor 1990; Cosnefroy *et al.* 1993; Henry *et al.* 1993). We give details of the method below.

The governing equations, when observations are confined to small range of angles about the local nadir, are:

$$Y = AX^B \exp(-kd) \tag{6.13}$$

where $Y = \mu(C_{10} - C_0)R^2$; C_{10} = AVHRR Earth-scene signal in 10-bit digital counts; C_0 = AVHRR offset or dark counts; k = daily degradation rate on the assumption that the instrument degrades exponentially in time; d = number of days after launch; R^2 = the Earth–Sun distance correction factor which accounts for the seasonal variation in solar illumination at the top of the atmosphere; $X = (\mu\mu_0)/(\mu + \mu_0)$ where μ and μ_0 are respectively the cosines of the satellite- and solar-zenith angles; A, B are regression parameters and X is the bidirectional reflection function associated with the desert surface for near-nadir observations when azimuthal dependence of reflection can be neglected, and is generally referred to as the Minnaert reflection function (Barkstrom 1973; Staylor 1990). The values of A, B, and k were obtained by a repetitive non-linear least-squares process; the value of k approaches convergence after seven iterations (runs). We have shown this process for channel 1 of the *NOAA-7* AVHRR in Figure 6.10.

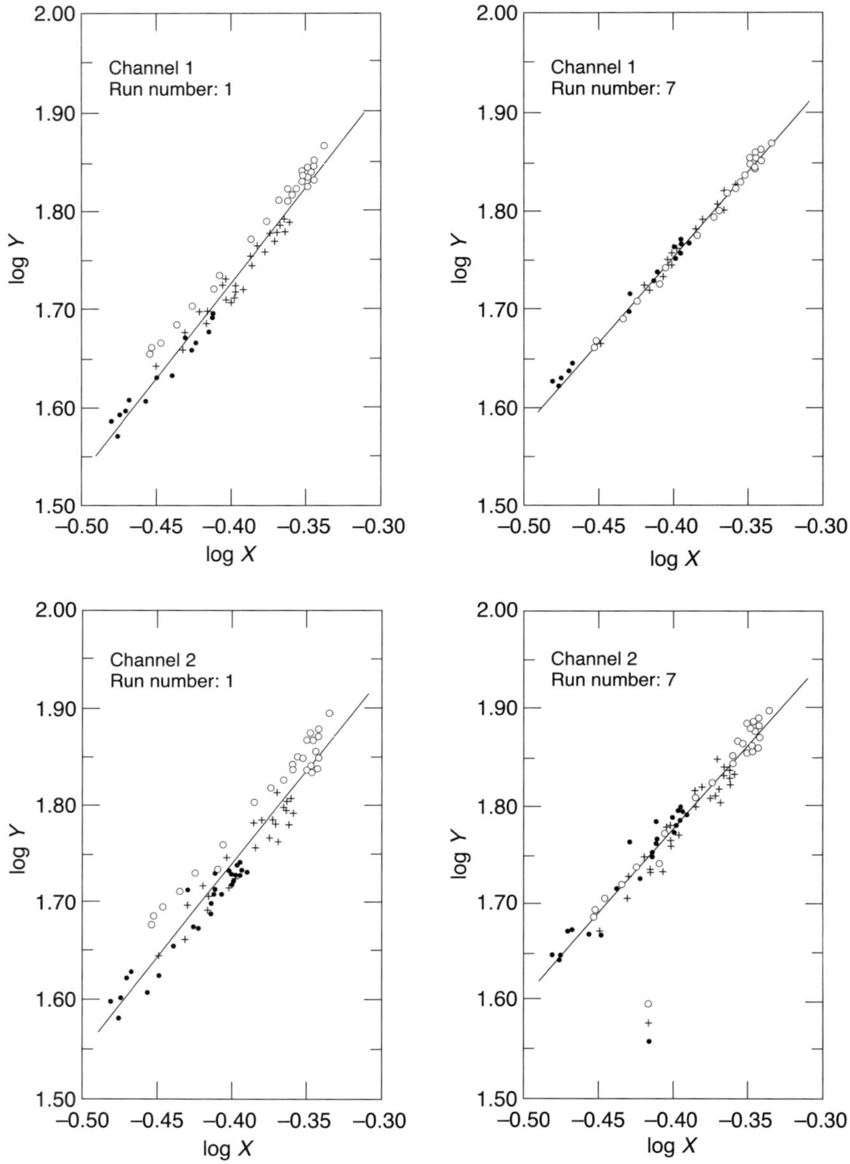

Figure 6.10. Regression of AVHRR counts in channels 1 and 2 of the *NOAA-7* AVHRR on the reflection function; the scatter in the data is minimized after seven iterations (runs).

We used the near-nadir AVHRR Global Area Coverage (GAC) data, taken from the B3 dataset of the International Satellite Cloud Climatology Project (ISCCP) (e.g. Schiffer and Rossow 1983; Brest and Rossow 1992) obtained under clear-sky conditions. The quality-control criteria used to ensure that only clear-sky data were included in our investigations are listed below (greater details can be found

Table 6.5. Regression parameters and daily degradation rates for AVHRRs on *NOAA-7*, *-9* and *-11* spacecraft.

	Channel 1				Channel 2			
	A	*B*	*r*	*k*	*A*	*B*	*r*	*k*
NOAA-7	285.5	1.753	0.997	0.000 101	295.1	1.734	0.982	0.000 120
NOAA-9	307.9	1.756	0.996	0.000 166	300.9	1.699	0.976	0.000 098
NOAA-11	311.8	1.785	0.994	0.000 033	323.9	1.723	0.982	0.000 056

r = correlation coefficient.

in Rao and Chen 1993); they are essentially the same as those that were used in Staylor (1990):

(a) the standard deviation of the measured counts C_{10} from the various ISCCP B3 GAC pixels (≈ 35) over the site should not exceed 3;
(b) the dispersion, defined as the ratio of the standard deviation to the mean of the effective counts ($C_{10} - C_0$) should not exceed 0.03 in channels 1 and 2;
(c) the dispersion in the brightness temperature measured in channel 4(≈ 10.8 μm) should be less than or equal to 0.004.

Also data obtained when the satellite-zenith angle was greater than 14° were discarded as were the data obtained when the solar-zenith angle was greater than 60°. Thus, the total number of data points (average of AVHRR signals over the desert site on any given day) over a 10-year period were reduced from 562 to 253 for the three AVHRRs. We show in Table 6.5 the regression parameters and the degradation rates per day for channels 1 and 2 of AVHRRs on *NOAA-7*, *-9* and *-11* spacecraft. It can be seen that the values of B are very close to one another. This may be interpreted as an indication of the validity of the Minnaert reflection function we have used.

The relative degradation rates reported by different investigators, using techniques similar to or different from those we have used here, are shown in Table 6.6; the spread in the values is an indication of the difficulty of the post-launch characterization of meteorological satellite sensors in the absence of on-board calibrators. The close agreement between the values reported by Staylor (1990) for channel 1 of AVHRRs on *NOAA-7* and *-9* and those found in Rao and Chen (1995) is due to the fact that we used in our work the same target, and the same technique; however, Staylor (1990) used AVHRR channel-1 data from the Heat Budget Parameter (Gruber and Winston 1978) compilation of AVHRR data which has coarser spatial resolution than the modified AVHRR GAC data we used for the same site from the stage B3 dataset of the ISCCP. Kaufman and Holben (1993) used four sites (25.5°N, 14°E; 25°N, 12°E; 29°N, 25°E; 26.5°N, 26°E) located in the vast Libyan desert, and considered AVHRR data for August

Table 6.6. Comparison of the annual degradation rates (%) reported by different investigators.

	NOAA-7		*NOAA-9*		*NOAA-11*	
	Channel 1	Channel 2	Channel 1	Channel 2	Channel 1	Channel 2
Rao and Chen (1995)	3.6	4.3	5.9	3.5	1.2	2.0
Staylor (1990)	3.5	–	6.0	–	–	–
Kaufman and Holben (1993)	4.2	5.6	5.7	3.1	1.3	2.9
Wu and Zhang (1994)	–	–	5.8	4.6	–	–
Che and Price (1992)	4.4	4.6	5.0	6.7	6.7	3.9
Brest and Rossow (1992)	0	–	4.6	–	–	–

and September since the sites are very dry and devoid of vegetation during these months. On the assumption that the sites were radiometrically stable, they derived the relative degradation rates using multi-year data. Wu and Zhang (1994) worked with clear-sky data over thirty-eight spatially homogeneous sites in the Gobi desert, China, each about 20 km × 20 km. Based on a functional relationship between observed top-of-the-atmosphere radiation and viewing geometry, and using data obtained a few years apart, they established the relative degradation in orbit of the two channels of AVHRR on *NOAA-9*. Che and Price (1992) compiled the gains, expressed in units of counts $W^{-1}\,m^2\,\mu m\,sr^{-1}$ for either of the two channels of AVHRR on a given spacecraft, reported by different investigators and grouped the data by month. In the event, only one calibration result was reported in a particular year; it was assigned to the month of July. A simple regression relationship was established between the gain values, and the elapsed time in orbit, expressed in months after launch. Using the regression relationship, the 'slope', expressed in units of $W\,m^{-2}\,\mu m^{-1}\,sr^{-1}\,count^{-1}$, was calculated as the reciprocal of the gain. It should, however, be noted that some of these regression relations have very low values of correlation coefficient.

According to the authors' contribution to Rao *et al.* (1993), the trend analysis by Brest and Rossow (1992), who confined their attention to AVHRR channel 1, was performed with a statistical procedure using a number of large-area targets which cover a significant portion of the globe. The targets cover a wide range of brightness values, and a variety of surface spectral-reflectance properties. The wide variety of surface types ensures the measurements cover a large part of the dynamic range (0–100 per cent reflectance factor) of the instrument. A fundamental assumption behind this analysis is that the global aggregate of regional variations of surface visible reflectance and of the overlying atmosphere is not changing with time, and is not affected by changes in the solar-zenith angle because of AVHRR orbital drift. Since very accurate data on the reflectivity of various surfaces were and still are sparse, this analysis provides a calibration relative to a somewhat poorly defined standard. It was also assumed in this analysis that the visible channel of AVHRR on *NOAA-7* was very stable in orbit, and could therefore be used as a normalization standard to ensure the continuity in time of AVHHR calibration performed for the ISCCP.

The authors have since revised AVHRR calibration (Brest *et al.* 1997) based on the key assumption that the mean properties of the Earth are more nearly constant than the calibration of AVHRR on timescales of the order of a decade. Their revised estimates for the degradation in orbit of channel 1 of *NOAA-7* and *-9* AVHRRs are around 1.3 and 6.0 per cent per year, respectively. While their revised degradation rate for *NOAA-9* AVHRR channel 1 is very close to what we have observed (Table 6.6), they also report an increase in the sensitivity of *NOAA-11* AVHRR channel 1 with time.

6.5.2 Absolute calibration of the AVHRRs on *NOAA-7*, *-9* and *-11*

As mentioned earlier, the AVHRR on *NOAA-9* was perhaps the most extensively studied instrument since its operational life covered the FIRE experiment. In addition, absolute calibration of the two channels had been performed on three different occasions during its operational life, using congruent-path aircraft/satellite radiance measurements (e.g. Smith *et al.* 1988). Assuming the aircraft-based absolute calibration of 1986 to be the most accurate (G. L. Smith, pers. commun.), and using the relative degradation rates we had determined over the Libyan desert calibration site, the 'slope' S_0 $(d = 0)$ on the day of launch was determined by requiring that it should be related to the slope S_d, the 'slope' d days after launch, by the relationship $S_0 = S_d \exp(-kd)$ since the 'slope' increases with time as the AVHRR signal decreases because of instrument degradation. The AVHRR Pathfinder Calibration Working Group further required that the computed trend in slope should yield for November 1987 a value very close to the aircraft-based slope for the same month. We show in Figure 6.11 the 'slope' trends given by different investigators for channel 1 of the *NOAA-9* AVHRR. It is apparent that the NOAA calibration trend, anchored to the aircraft-based calibration of 1986 yields a 'slope' value very close to the aircraft-based calibration of November 1987, thereby meeting the requirement of the AVHRR Pathfinder Calibration Working Group. The formulae for the calculation of the 'slope' either in the radiance or albedo representation for the *NOAA-9* AVHRR are given below:

Radiance representation:

$$\begin{aligned} &\text{Channel 1} \qquad S_d = 0.5406 \exp[1.66 \times 10^{-4} d] \\ &\text{Channel 2} \qquad S_d = 0.3808 \exp[0.98 \times 10^{-4} d] \end{aligned} \tag{6.14}$$

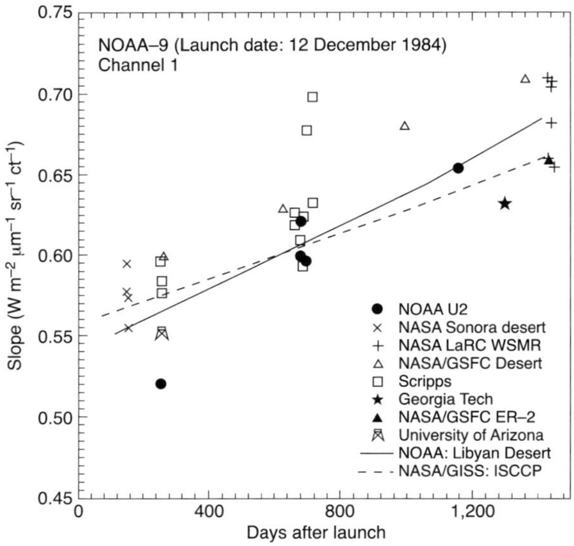

Figure 6.11. Comparison of the post-launch calibration results obtained by different investigators for channel 1 of the *NOAA-9* AVHRR; the sources for the results listed under the headings of NASA Sonora desert, NASA LaRC WSMR, NASA GSFC desert, Scripps, Georgia Tech, NASA GSFC ER-2, University of Arizona and NASA/GISS ISCCP are found in Whitlock *et al.* (1990).

Albedo representation:

$$\left.\begin{aligned} &\text{Channel 1} \qquad S_d = 0.1039\exp[1.66\times 10^{-4}d] \\ &\text{Channel 2} \qquad S_d = 0.1136\exp[0.98\times 10^{-4}d] \end{aligned}\right\} \tag{6.15}$$

where d is the elapsed time in orbit, expressed in days after launch.

We used the *NOAA-9* AVHRR as a calibration reference to render the relative degradation of AVHRRs on the *NOAA-7* and *-11* spacecraft absolute. The method is based on the use of matched datasets consisting of the measured counts, satellite- and solar-zenith angles and the days of the month when measurements of the upwelling radiation over the Libyan desert site were made. The criteria used to generate the matched datasets for the *NOAA-7*/*NOAA-9* combination, and for the *NOAA-9*/*NOAA-11* combination were:

(a) the solar- and satellite-zenith angles for measurements made by the AVHRRs on *NOAA-7* and *NOAA-11* should be within 1° of the corresponding angles for measurements made by the *NOAA-9* AVHRR;
(b) the two measurements should have been made in the same calendar month, with the days of the month being as close to one another as practicable to allow for seasonal variations, if any, of the radiometric characteristics of the calibration site. Sometimes measurements made by AVHRRs on either *NOAA-7* or

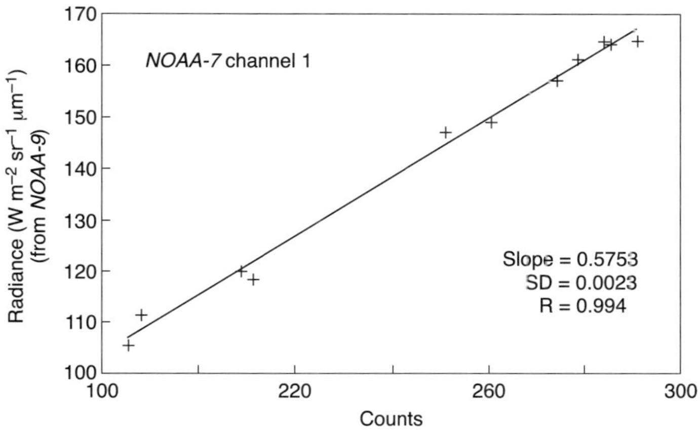

Figure 6.12. Regression of *NOAA-9* AVHRR radiance on *NOAA-7* counts corrected for degradation in orbit.

NOAA-11 in any given month in the *i*th year in orbit of the relevant spacecraft could be matched with measurements made in the same calendar month in the *j*th year in orbit of *NOAA-9*.

These selection criteria resulted in eleven sets of matched data for the *NOAA-7*/*NOAA-9* combination, and ten sets of matched data for the *NOAA-9*/*NOAA-11* combination. The method used to establish interlinkages among the three AVHRRs is:

(a) knowing the relative degradation rates for *NOAA-7* and *NOAA-11* AVHRRs, the scene counts $(C_{10} - C_0)R^2$ were corrected for instrument degradation in orbit; the elapsed time in orbit d was calculated from the known dates of launch and of observation;
(b) the matched *NOAA-9* AVHRR radiance was calculated using Equation (6.14);
(c) the *NOAA-9* AVHRR radiances were then regressed on the corrected *NOAA-7* or *NOAA-11* counts in either channel.

The multiplier m in the regression relationship of the form $y = mx + c$ yields the calibration coefficient ('slope') for the relevant instrument on the day of launch. We show in Figure 6.12 the regression of *NOAA-9* AVHRR channel-1 radiance on *NOAA-7* AVHRR signal. Using the combination of the 'slope' on the day of launch thus determined, and the relative degradation rates, simple formulae were derived to calculate the 'slope' as a function of elapsed time in orbit. We show in Figure 6.13 the top-of-the-atmosphere albedo time series over the Libyan desert site for the period covered by the AVHRRs on *NOAA-7*, *-9* and *-11*. It is apparent that the spurious trends in time when the pre-launch calibrations were used, and the discontinuity during the transition from one spacecraft to the next, are eliminated when we use the post-launch calibrations. The mean and (standard deviation) of the top-of-the-atmosphere albedo (in per cent) in channels 1 and 2 are, respectively, 37.8

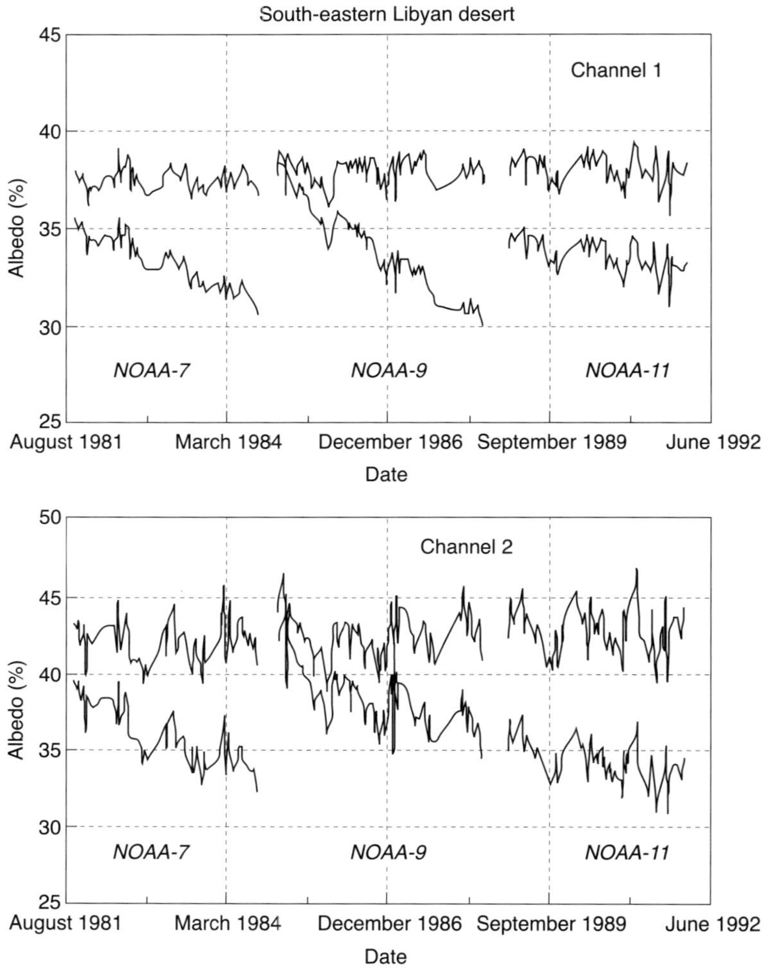

Figure 6.13. Top-of-the-atmosphere albedo time series over the Libyan desert site; the series which is nearly parallel to the x-axis in both panels was generated with post-launch calibration coefficients which accounted for instrument degradation in orbit; the series showing a downward trend in time was generated using the pre-launch calibration coefficients.

(0.7) and 42.6 (1.5) based on measurements by all three AVHRRs. It should be noted that only 15 per cent of the available data for the site were used in the development of the calibration formulae for the three sensors, and we thus feel that using the same to rehabilitate the entire time series is reasonable.

Abel *et al.* (1993) have measured the upwelling radiance from a high-flying ER-2 aircraft over the White Sands area of New Mexico, USA, since late 1988 during satellite overpasses to evaluate the post-launch performance of the AVHRR on *NOAA-11*. We show in Figure 6.14 the aircraft-based calibrations along with the calibration trend we predict, using the method outlined above, for the *NOAA-11*

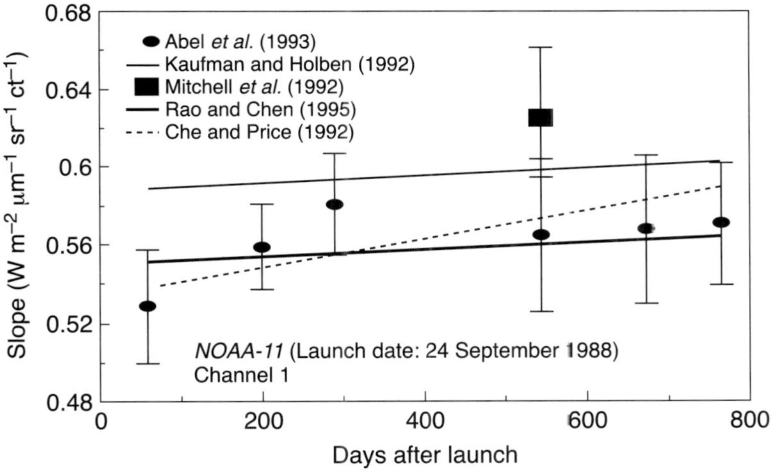

Figure 6.14. Comparison of the post-launch calibration of *NOAA-11* AVHRR channel 1 by different investigators.

Table 6.7. Formulae for the calculation of radiance.

NOAA-7	
Channel 1	$0.5753\exp[1.01\times 10^{-4}d](C_{10}-36)$
Channel 2	$0.3914\exp[1.20\times 10^{-4}d](C_{10}-37)$
NOAA-9	
Channel 1	$0.5406\exp[1.66\times 10^{-4}d](C_{10}-37)$
Channel 2	$0.3808\exp[0.98\times 10^{-4}d](C_{10}-39.6)$
NOAA-11	
Channel 1	$0.5496\exp[0.33\times 10^{-4}d](C_{10}-40)$
Channel 2	$0.3808\exp[0.56\times 10^{-4}d](C_{10}-40)$
NOAA-14	
Channel 1	$(0.566+0.000069\times d)(C_{10}-41)$
Channel 2	$(0.440+0.000044\times d)(C_{10}-41)$

channel 1. The noticeable feature is that the slopes determined using the formulae in Tables 6.7 and 6.8 are wsubjectithin one standard deviation of the mean values of the majority of the aircraft-based absolute calibrations. A revision of the results in Mitchell *et al.* (1992), based on changes in some of the model parameters (R. Mitchell, pers. commun.), brings their calibration closer to the results reported in Rao and Chen (1995).

We show in Figure 6.15 the degradation of channels 1 and 2 of AVHRRs on *NOAA-7*, *-9* and *-11* spacecraft and the variation of the slope (radiance representation) with time. The slope increases in time because of the degradation of the instrument.

Table 6.8. Formulae for the calculation of albedo.

NOAA-7	
Channel 1	$0.1100 \exp[1.01 \times 10^{-4} d](C_{10} - 36)$
Channel 2	$0.1169 \exp[1.20 \times 10^{-4} d](C_{10} - 37)$
NOAA-9	
Channel 1	$0.1039 \exp[1.66 \times 10^{-4} d](C_{10} - 37)$
Channel 2	$0.1136 \exp[0.98 \times 10^{-4} d](C_{10} - 39.6)$
NOAA-11	
Channel 1	$0.1060 \exp[0.33 \times 10^{-4} d](C_{10} - 40)$
Channel 2	$0.1098 \exp[0.56 \times 10^{-4} d](C_{10} - 40)$
NOAA-14	
Channel 1	$(0.1110 + 0.000\,0135 \times d)(C_{10} - 41)$
Channel 2	$(0.1340 + 0.000\,0133 \times d)(C_{10} - 41)$

6.5.3 Calibration of the AVHRR on *NOAA-14*

We transferred the calibration of the *NOAA-9* AVHRR to the *NOAA-14* AVHRR by requiring that the signal issuing from either channel of the latter should yield the appropriate albedos for the two channels determined from the 10-year record of measurements (Figure 6.13; see also Section 6.5.2). Thus, the slopes of the two channels of *NOAA-13* AVHRR were determined as a function of elapsed time in orbit. Based on measurements over a 3-year period, simple formulae have been developed to predict the behaviour of the instrument (Rao and Chen 1999); the variation in time of the slope for the two channels of the *NOAA-14* AVHRR published by different investigators is shown in Figure 6.16. The visible and near-infrared channels of the six-channel AVHRR, presently on *NOAA-15*, and scheduled for flight on future NOAA polar orbiters, will be calibrated using similar procedures. However, the 1.6 μm channel (channel 3A) will be calibrated using a 3-year top-of-the-atmosphere albedo record for the Sudan site established with measurements made in the 1.6 μm channel of the Along-Track Scanning Radiometer-2 (ATSR-2) flown on the European Remote Sensing satellite *ERS-2* as a calibration reference; the ATSR-2 has an on-board calibration device for the visible and near-infrared channels.

We list the annual degradation rates for channels 1 and 2 of AVHRRs on *NOAA-7*, *-9*, *-11* and *-14* spacecraft in Table 6.9. It is to be noted that the degradation rates are different for the four AVHRRs even though they are all based on the same design, and are subject to the same set of specifications; presently, there is no satisfactory explanation for this behaviour.

The global applicability of AVHRR calibration using a desert site with a top-of-the atmosphere albedo around 40 per cent is questioned at times in view of the fact that the Earth scene albedo varies from very low values of the order of a few per cent over oceans to very high values of the order of 70 per cent over the snow fields of

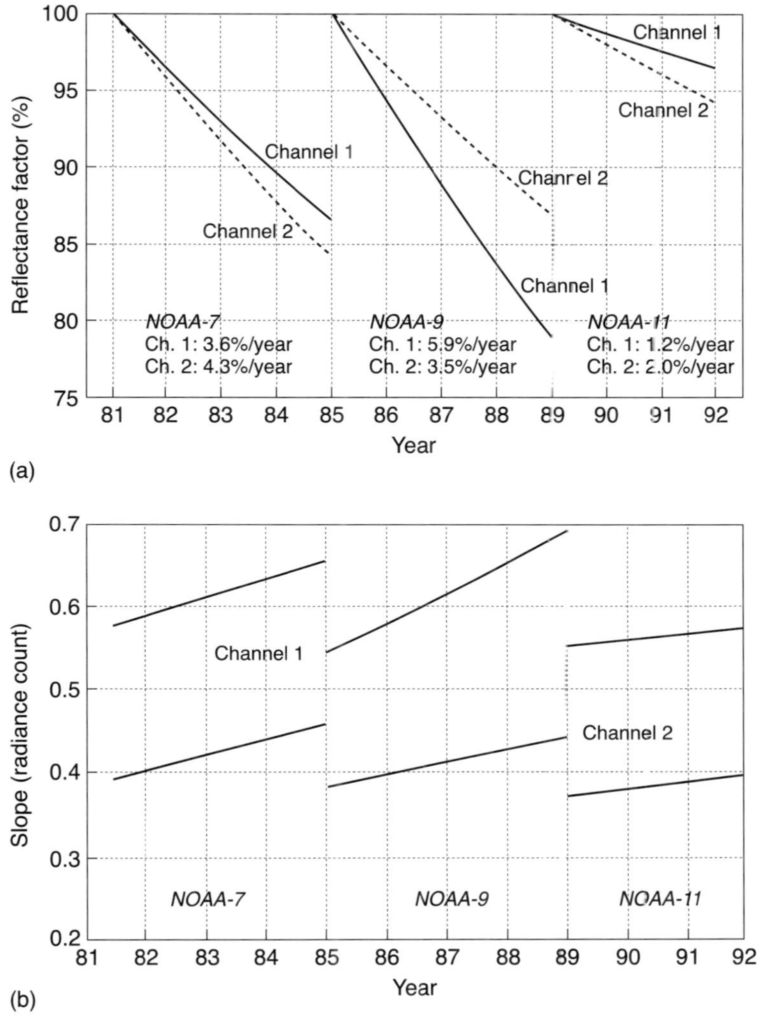

Figure 6.15. The degradation of channels 1 and 2 of AVHRRs on *NOAA-7*, *-9* and *-11* spacecraft (a), and the variation of the slopes (radiance representation) with time (b).

Antarctica and Greenland. However, Loeb (1997) has derived degradation rates for channels 1 and 2 of several AVHRRs, using the ice/snow fields of Greenland and Antarctica, as radiometrically stable calibration sites, and his results are within 3 to 5 (per cent) (relative) of our results. Similarly, when the columnar abundance of the aerosols over the dark ocean is derived using measurements in the visible channel of the AVHRR, the correspondence between the satellite derived aerosol optical thickness and ship-board sun photometer measurements of the same improves noticeably when the desert-based AVHRR calibrations are incorporated in the aerosol-retrieval algorithm. These observations can be viewed as evidence of the

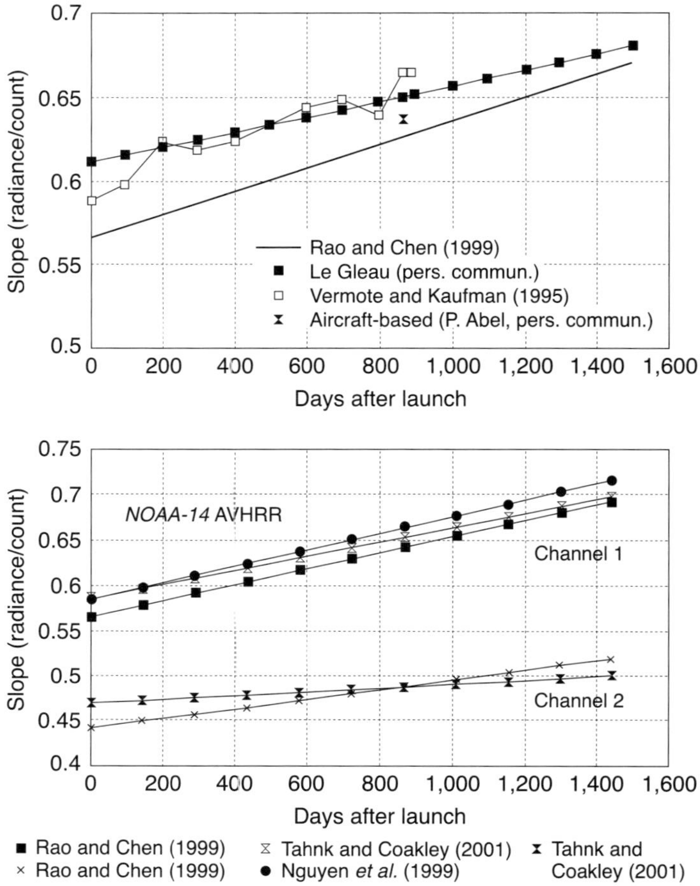

Figure 6.16. Comparison of calibration results obtained by different investigators for channels 1 and 2 of the *NOAA-14* AVHRR.

global applicability of desert-based calibration of the visible and near-infrared channels of AVHRR.

6.5.4 Propagation of calibration uncertainties in products

Generally the errors in the radiometric calibration of the visible and near-infrared channels of the AVHRR are propagated in a non-linear manner with AVHRR-derived products, with the impact of the uncertainties being very large at the lower values of the product. We shall illustrate this with the NDVI which is given by:

$$N \text{ (for NDVI)} = (\alpha_2 - \alpha_1)/(\alpha_2 + \alpha_1) \tag{6.17}$$

where α_1 and α_2 are respectively the albedos measured in AVHRR channels 1 and 2 . It can be shown that an uncertainty in albedo of either channel, $\Delta\alpha_i/\alpha_i$, which is

Table 6.9. Annual AVHRR degradation rates (%).

Spacecraft	Channel 1	Channel 2
NOAA-7	3.6	4.3
NOAA-9	5.9	3.5
NOAA-11	1.2	2.0
NOAA-14	4.4	3.6

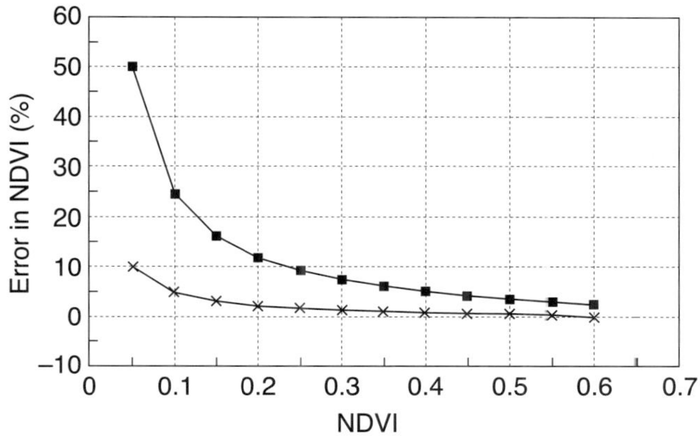

Figure 6.17. Relative uncertainty in NDVI as a function of NDVI for a 5 per cent uncertainty (upper curve) and a 1 per cent uncertainty (lower curve) in channel 2 calibration.

directly proportional to the relative uncertainty in calibration, results in a relative uncertainty $(\Delta N/N)$ in NDVI, given by:

$$(\Delta N/N) = [(1 - N^2)/2N](\Delta\alpha_i/\alpha_i) \tag{6.18}$$

in the first approximation. Here α_i is the albedo in either channel ($i = 1$ or 2). The uncertainties in calibration are thus propagated in a non-linear fashion with the result that the impact is larger at low values of N. The product uncertainty will be proportional to the algebraic sum of the uncertainties in the two albedos. We show in Figure 6.17 the relative uncertainty in NDVI as a function of NDVI for a 5 per cent relative uncertainty in the calibration of AVHRR channel 2. Similar non-linear propagation of uncertainties in the calibration of AVHRR channel-1 is also noticed in NOAA's aerosol product which is derived from AVHRR channel-1 radiance, and is shown in Figure 6.18.

The uncertainty in the calibration of the visible and near-infrared channels using a radiometrically stable desert target can perhaps be traced to the small but finite variations in the top-of-the-atmosphere albedo of the Libyan desert site. The albedo variations could be due to changes in atmospheric ozone which would affect the upward radiation in channel 1, to changes in precipitable water which would affect the upward radiation in channel 2, to changes in the surface reflectance, or to

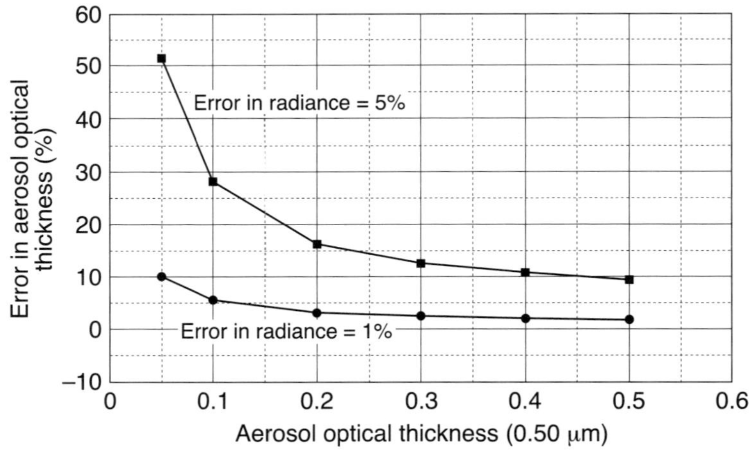

Figure 6.18. Relative uncertainty in the aerosol product for two values of the uncertainty in the calibration of channel 1.

changes in the columnar aerosols which would affect all three channels (1, 2 and 3A). Since we cannot with confidence assign this variation in the top-of-the-atmosphere albedo to any one of the above atmospheric and surface phenomena, we adopted a statistical approach to study the impact of random variations in the AVHRR signal on the calibration of channels 1 and 2. We varied the AVHRR signals in the two channels by ± 5 per cent in a random manner to simulate the cumulative effects of variation in the atmosphere and surface features, and determined the degradation rates. Using the data for channels 1 and 2 of the AVHRR on the *NOAA-9* spacecraft, we found that the relative changes in the degradation of channel 1 were of the same order of magnitude as changes in the signal, and were somewhat larger in channel 2.

6.6 THE AVHRR AS A CALIBRATION REFERENCE

6.6.1 General

Generation of fused or blended records of geophysical products from measurements made by the same instrument on successive members of a spacecraft series (e.g. the AVHRRs on different NOAA polar orbiters), or from measurements made by different sensors on the same platform or sensors on different platforms, is assuming increasing importance in the context of the generation of long-term records of geophysical products for use in climate studies, and in the context of filling lacunae in the existing records. Also, merged geophysical records are an essential requirement for the successful implementation of activities planned under the global climate, ocean and terrestrial observing systems (G3OS), the World Climate Research Program (WCRP) and the International Geosphere–Biosphere

Programme (IGBP). Since the geophysical records generated with the AVHRR will be important to these activities, and since those generated with other sensors will be compared and very likely merged with the AVHRR data, we considered it worthwhile to examine whether the visible and near-infrared channels of the AVHRR could be used as a calibration reference. We have confined our attention to the visible and near-infrared regions of the spectrum, since an appreciable number of sensors in these spectral regions do not have onboard calibration devices.

6.6.2 Method

The calibration of any candidate sensor using the AVHRR as a calibration reference is performed as follows:

(a) The inter-relationship between the radiances/albedos measured by the candidate sensor and the appropriate channel of the AVHRR is established using model simulations of the upward radiation in the passbands of the two instruments for a very representative set of atmospheric and surface conditions, using the LOWTRAN/MODTRAN radiative transfer code (Berk *et al.* 1989). Using this inter-relationship, the albedo/radiance seen by the candidate sensor is determined over sites whose radiometric stability has been established with the AVHRR, and whose albedo has been determined using the AVHRR.
(b) The relative degradation of the candidate sensor is established using radiometrically stable calibration site(s).
(c) Corresponding to *d* days after launch, the slope S_a of the candidate sensor is determined by requiring that it should yield the albedo over a given site, as was determined from the interrelationship established in (a).

The upward radiation has been simulated for the following surface conditions: (1) surface reflection independent of wavelength, (2) surface reflection slowly varying with wavelength and (3) surface reflection strongly dependent on wavelength. We have assumed a Lambertian surface. Atmospheric model parameters used in the simulations of the upward radiation sensed by the candidate sensor and the AVHRR are:

Atmospheric model:	Tropical and mid-latitude summer	
	Solar-zenith angle	30°, 45° and 60°
	Satellite viewing angle	0°, 15°, 30° and 45°
	Relative azimuth angle	0°, 90° and 180°
	Aerosol	Rural
	Surface Meteorological Range	5, 23, and 300 km

As the surface meteorological range decreases, atmospheric turbidity (aerosols) increases. We show in Figure 6.19 the wavelength dependence of surface reflectance that has been used in the model simulations. We use vegetation-reflectance spectra (e.g. Geitelson and Merzlyak 1994) corresponding to three different values of chlorophyll, labelled Chl in the diagram. We see that the reflectance by vegetation shows

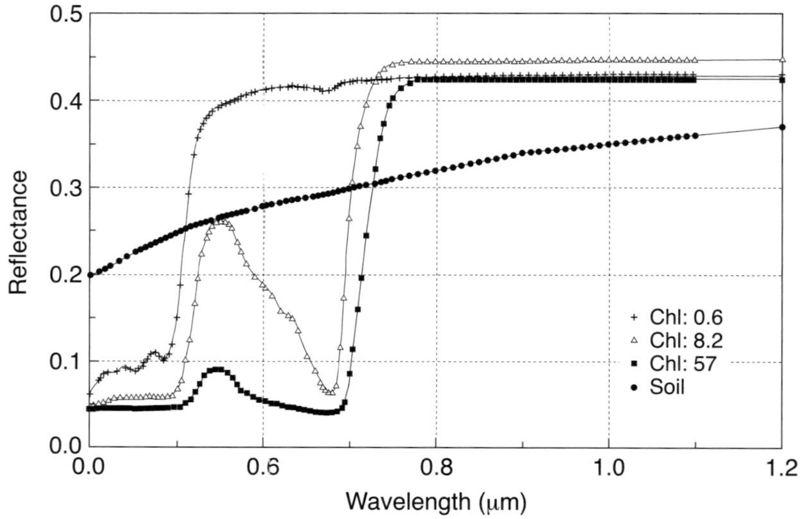

Figure 6.19. Wavelength dependence of surface reflectance of soil and vegetation; vegetation reflectance spectra are labeled in terms of the chlorophyll content (Chl) expressed in units of (mmol cm^{-2}); the larger the chlorophyll amount, the greener the vegetation.

pronounced wavelength dependence for intermediate values (Chl: 8.2 mmol cm^{-2}) of chlorophyll.

6.6.3 Applications

6.6.3.1 High Resolution Infrared Radiation Sounder

We have used the method described above to study the in-orbit behaviour of the visible channel of the the High Resolution Infrared Radiation Sounder (HIRS) flown on *NOAA-14*. The HIRS is a twenty-channel filter-wheel radiometer covering the region of the spectrum from the visible to the thermal infrared, and is designed for temperature and humidity soundings of the atmosphere, utilizing measurements made in the thermal infrared. The visible channel of the HIRS (channel 20, 0.67–0.71 μm) has no on-board calibration device. Since measurements made in the HIRS visible channel are finding increasing quantitative applications in the development of atmospheric corrections, cloud detection, and related activities, HIRS has to be calibrated using vicarious techniques. Since both the AVHRR and the HIRS are part of the polar-orbiter payload, and are essentially coregistered, we used the Sudan calibration site to determine the relative degradation in orbit of the HIRS visible channel, following procedures described elsewhere. We found that the *NOAA-14* HIRS visible channel is extremely stable with an annual degradation of the order of 0.71 (±0.21) per cent. We determined the albedo of the desert site in the visible channel of the HIRS using the simulated interrelationship between the HIRS and AVHRR visible albedos, as percentages, shown in Figure 6.20.

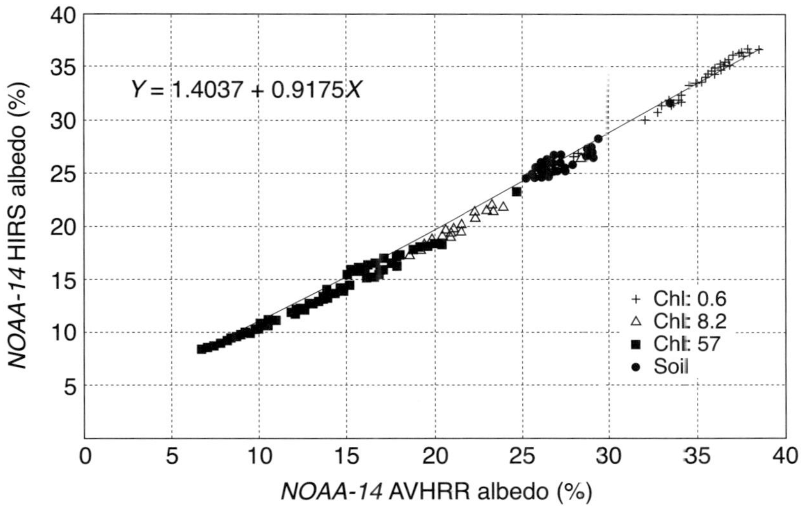

Figure 6.20. Simulated interrelationship between the visible channel albedos of the *NOAA-14* AVHRR and HIRS.

$$\text{HIRS albedo} = 1.4037 + 0.9175 \times \text{AVHRR albedo} \tag{6.19}$$

From this relationship, we find that the HIRS visible albedo over the Sudan calibration site is 31.6 per cent. Combining the albedo value with the degradation rates, and following the procedure given in the last section, we obtain the following relationship for the variation in the 'slope' with the elapsed time in orbit.

Albedo representation:

$$S_d(\text{per cent albedo/count}) = 0.0228 + 0.000\,000\,459d \tag{6.20}$$

Radiance representation:

$$S_d(\text{radiance/count}) = 0.151 + 0.000\,003\,04d \tag{6.21}$$

We show in Plate 2 the top-of-the-atmosphere albedo measured in the visible channels of the *NOAA-14* AVHRR and HIRS over the contiguous USA. Measurements were made on 5 October 1999, and HIRS albedos were calculated using the formula for the slope given in Equation (6.20). The high degree of correspondence between the two is noteworthy, after due allowance is made for the differences in footprints of the two sensors.

6.6.3.2 *Near-infrared channel of the NOAA-14 AVHRR*

We wanted to examine whether the albedos/radiances measured in AVHRR channels 1 (≈0.58–0.68 μm) and 2 (≈0.72–1.1 μm) could be interrelated; the desirability of examining whether such a relationship exists has been keenly expressed by the user community. We show in Figure 6.21 the computed values of the albedos in the two channels. There is hardly any correlation between the albedos in channels 1

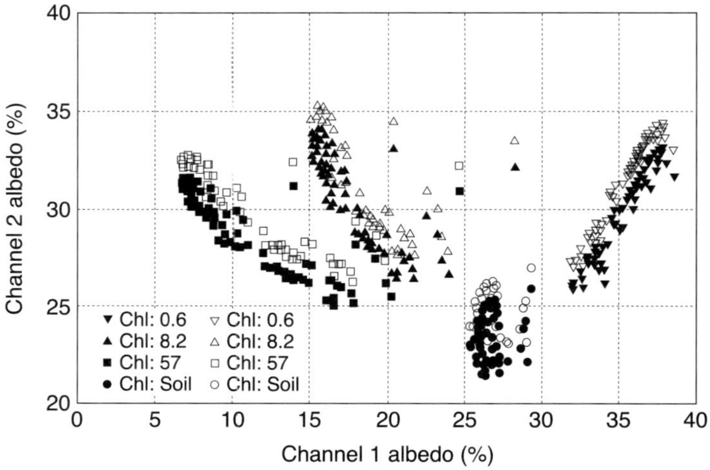

Figure 6.21. Interrelationship between the simulated albedos in the visible (channel 1) and near-infrared (channel 2) channels of the *NOAA-14* AVHRR; the solid symbols refer to the Tropical Atmosphere model, and the open symbols to the Mid-Latitude Summer atmosphere model.

and 2 over the greater range of AVHRR channel-1 albedo which is of great interest to the scientific community. One of the principal reasons for this lack of correlation between the two albedos is that the spectral region covered by channel-2 includes the narrow region of the spectrum where the reflectance by vegetation rises very rapidly, and the absorption bands of water vapour occur around 0.74 μm. However, we do notice some correspondence between the two at either very low or fairly high values of the AVHRR channel-1 albedo. Observations of the top-of-the-atmosphere albedo in channels 1 and 2 of the *NOAA-14* AVHRR support these conclusions. We show in Figure 6.22 the observed albedos in the two channels over the greater part of the contiguous USA for July and December 1995. In July, the surface is covered in the main by natural vegetation and crops under cultivation. The channel-2 albedo is generally greater than the channel-1 albedo and in July there is very little correlation ($R^2 \approx 0.85$) between the two (Figure 6.22(a)). On the other hand, in December (Figure 6.22(b)) most of the vegetation has disappeared and surface reflection is relatively more homogeneous so that the correlation between the two albedos is considerably better ($R^2 \approx 0.85$).

We have also examined the interrelationship between the AVHRR channel-1 albedo and the visible channel of the very high resolution radiometer on *INSAT*, the visible channel of the imager on NOAA's *GOES* and some of the visible and near-infrared channels of the Moderate-Resolution Imaging Spectroradiometer (MODIS).

Based on our work to date, we have established several criteria that govern the degree of correspondence between the visible albedos in the visible channel of the AVHRR and of the candidate sensor, or any pair of reference/candidate sensors in general:

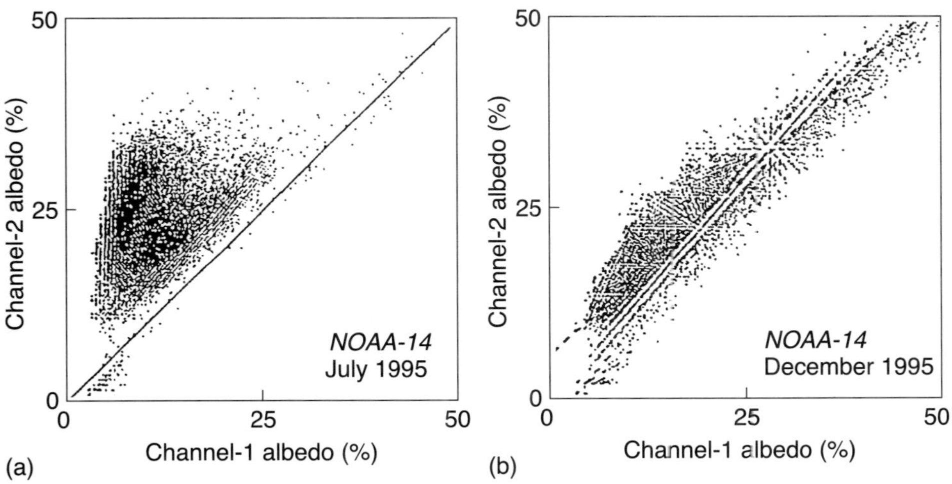

Figure 6.22. *NOAA-14* AVHRR top-of-the-atmosphere albedo in channels 1 and 2 for the months of (a) July and (b) December 1995.

(1) the normalized response curves of the two sensors should have considerable overlap;
(2) gaseous and particulate absorption and scattering of radiation should not be widely different for the two sensors;
(3) the effective wavelengths of the two sensors should be close to each other; and
(4) the wavelength dependence of surface reflection in the spectral regions covered by the two sensors should be similar.

The sensor characteristics of interest are listed in Table 6.10. An examination of the entries in the table indicates that while most of candidate sensors have spectral characteristics which are comparable with those of the *NOAA-14* AVHRR channel 1, the characteristics of the *NOAA-14* channel 2 are distinct in that the effective wavelengths are far apart, and there is no overlap in the spectral regions covered by the two. As mentioned earlier, the lack of correlation between the albedos measured in the two channels is thus due to this incompatibility of the spectral characteristics.

6.7 CONCLUDING REMARKS

The importance of proper in-orbit calibration and characterization of satellite sensors, to ensure the accuracy, continuity and viability of the derived geophysical products, can hardly be exaggerated. There are well-understood techniques for the post-launch calibration of satellite sensors in the thermal infrared region of the spectrum. However, the calibration of sensors in the visible and near-infrared

Table 6.10. Sensor characteristics.

Sensor	Spectral interval (μm)	Effective width (μm)	Effective wavelength (μm)	In-band solar irradiance ($W m^{-2}$)
NOAA-14 AVHRR channel 1	0.58–0.68	0.129	0.641	207.1
NOAA-14 AVHRR channel 2	0.72–1.10	0.244	0.842	251.01
NOAA-14 HIRS channel 20	0.67–0.71	0.031	0.694	64.4
GOES-8 Imager visible channel	0.52–0.72	0.194	0.627	314.5
GOES-9 Imager visible channel	0.52–0.72	0.204	0.629	330.6
INSAT 2B VHRR	0.61–0.74	0.141	0.675	210.1

sensors is rather complex, especially when there are no on-board calibrators. Even when a sensor has on-board calibration such as ATSR, it is necessary to monitor its performance after launch. Thus, there is a need for the development of robust vicarious calibration techniques for the in-orbit characterization of these sensors. A conservative estimate of the absolute, radiometric calibration accuracy that can be achieved with vicarious calibration techniques is of the order ±5–7 per cent (relative). The feasibility of using combinations of different vicarious calibration techniques currently in use to improve the attainable, radiometric calibration accuracy is being investigated. Against this background, the design and development of on-board calibrators for satellite sensors in the visible and near-infrared assumes great importance. Plans are under way to draw upon the heritage of sensors such as the ATSR, the Sea-Viewing Wide-Field-of-View Sensor (SeaWiFS), and the promise of MODIS to design better on-board calibrators. It should, however, be noted that the attainable, absolute, radiometric calibration accuracy with on-board calibrators is of the order of 5 per cent at present; it is expected that advances in technology may permit attainable accuracies of the order of 3 per cent in the future.

It is thus necessary to build into the design and development of satellite-derived product algorithms the technology-imposed limits on attainable, radiometric calibration accuracy. Another issue to be addressed is the traceability of satellite-derived radiance to Système Internationale d'Unités (SI units) to facilitate the intercomparison and fusion of geophysical products derived from (a) either the same sensor on successive platforms in a series (e.g. the NDVI and columnar aerosols from AVHRR on the polar orbiters) or (b) from different sensors on different spacecraft (e.g. sea-surface temperature from thermal infrared measurements made by the AVHRR and the ATSR). Thus, a calibration strategy should be developed to address the pre-launch and post-launch characterization of satellite sensors which would lead to the required traceability to SI units. One of the key elements of this strategy would be the use of retrievable manned spectroradiometers, whose calibration can be traced to

SI standards maintained in national-standards laboratories, for the absolute calibration of satellite sensors using congruent-path radiation measurements, either from high-flying aircraft (see Section 6.4) or from space-based platforms such as the International Space Station (ISS). It is essential that manned spectroradiometers are calibrated at the national-standards laboratories before and after each calibration campaign.

7

Long-term global climate datasets from operational satellites

G. Ohring

7.1 PHYSICAL BASIS FOR EXTRACTING INFORMATION ON CLIMATE VARIABLES FROM SATELLITE MEASUREMENTS

We recall the basic laws of the physics of radiation. Planck's law gives the maximum amount of energy a body can emit. Such a body is called a black body, and the energy $E(\lambda)\, d\lambda$ emitted between the wavelengths λ and $\lambda + d\lambda$ depends on λ and on the temperature T. Wien's law, obtained from Planck's law by differentiating with respect to λ and finding the maximum, gives the wavelength of maximum radiation. This law shows that the wavelength of maximum radiation is inversely proportional to temperature (i.e. $\lambda_{max} T =$ constant). Stefan's law, obtained by integrating Planck's law over the entire temperature range, gives the total amount of radiation emitted by a black body at a given temperature T. This total is proportional to the fourth power of the temperature (i.e. $E = \sigma T^4$).

Figure 7.1 shows the Planck curves for three stars with different temperatures. The emitted radiation has been normalized to 1, otherwise the curve for Spica would dwarf the others. The shift of wavelength of maximum emission to longer wavelengths as temperature decreases is clearly shown.

Figure 7.2 illustrates the heat balance of the Earth and its atmosphere. In the diagram the average incoming solar radiation at the top of the atmosphere is taken as 100 units. The atmosphere is relatively fairly transparent to solar radiation. More than half of the incident solar radiation reaches the surface of the Earth, where it is mainly absorbed. Almost a third is reflected back to space by clouds, by the surface and by air molecules and dust in the atmosphere. The atmosphere is relatively opaque to the radiation emitted by the surface of the Earth. The atmosphere absorbs the surface radiation and re-emits it, at a lower temperature, to space and back to the surface.

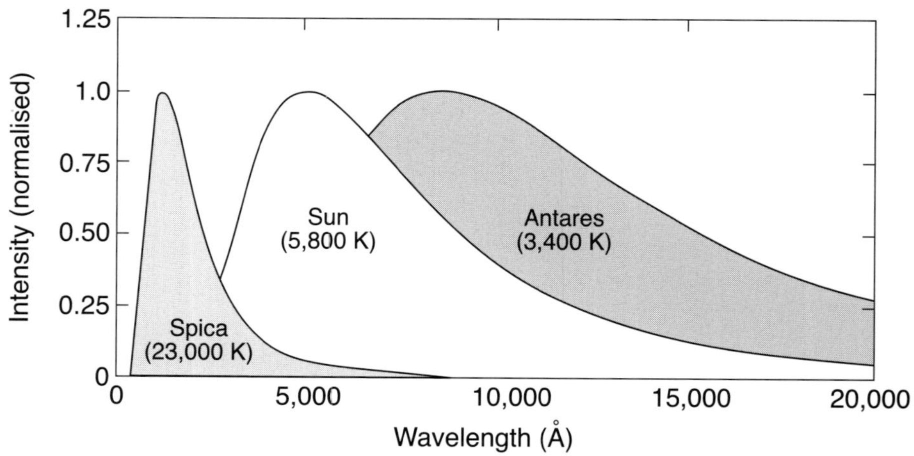

Figure 7.1. Basics of remote sensing: radiation laws.

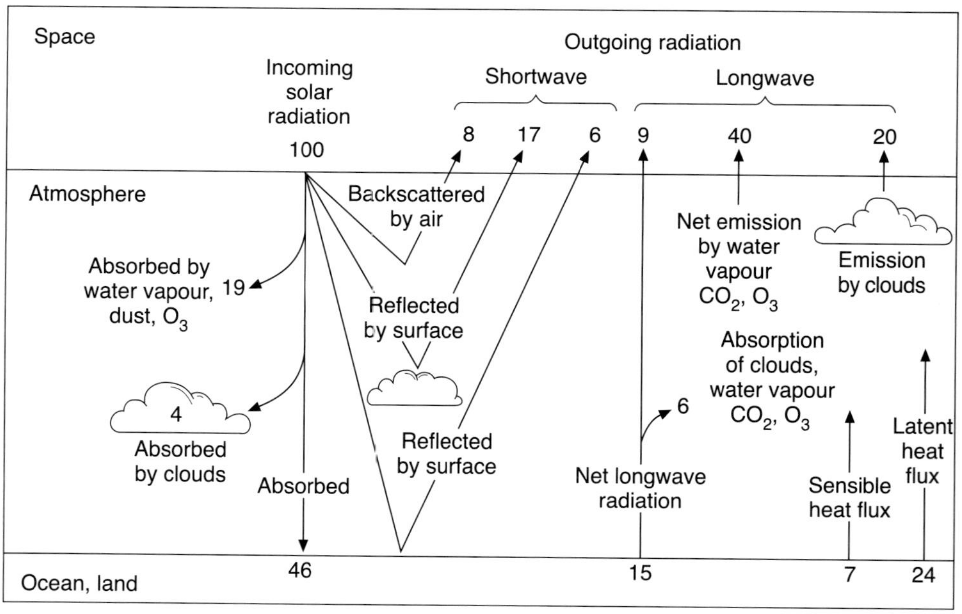

Figure 7.2. Heat balance of the Earth.

We consider now the situation of non-black bodies (i.e. bodies which are not perfect emitters or perfect absorbers of radiation). For any wavelength λ the following quantities are defined:

$$\text{absorptance} = \text{absorbed radiation/incident radiation} = \alpha$$

$$\text{reflectance} = \text{reflected radiation/incident radiation} = \rho$$

$$\text{transmittance} = \text{transmitted radiation/incident radiation} = \tau$$

where $\alpha + \rho + \tau = 1$ and the emittance ε is equal to the absorptance α. Each of these quantities can generally be expected to be a function of wavelength.

The emittance ε of a non-black body is defined as the ratio of the energy emitted by the body between λ and $\lambda + \mathrm{d}\lambda$ at a given temperature T to the energy emitted within that wavelength range by a black body at the same temperature. In other words, for a non-black body the energy emitted between λ and $\lambda + \mathrm{d}\lambda$ is $\varepsilon E(\lambda)\,\mathrm{d}\lambda$. If ε is independent of wavelength, the total energy emitted by the non-black body at temperature T is $\varepsilon\sigma T^4$. The brightness temperature T_{B} of a non-black body is defined as the temperature of a black body that emits the same amount of energy as the non-black body emits at temperature T, i.e.:

$$\sigma T_{\mathrm{B}}^4 = \varepsilon\sigma T^4$$

It is the brightness temperature that can be determined from thermal infrared remotely sensed data if the value of ε for a surface is not known. But, of course, since the ocean radiates as a black body (i.e. $\varepsilon = 1$) we could derive the surface temperature of the ocean from a measurement of the brightness temperature at any wavelength, provided the atmosphere was a perfect window. For the land, however, the value of ε is not necessarily equal to 1, indeed it is quite variable, and so the determination of the surface temperature of the land is more complicated than it is for the sea.

We now consider the variation of the quantities α, ρ and τ with wavelength of the radiation. We consider separately the situation for different types of radiation and summarize some of the general radiative properties of clouds, atmospheric gases, and the Earth's surface for solar radiation, infrared radiation, and microwave radiation:

solar radiation	gases	Rayleigh scattering, absorption, transmission
	clouds	scattering and transmission
	surface	reflection and absorption;
infrared radiation	gases	absorption and transmission
	clouds	absorption and transmission (thick clouds are black)
	surface	absorption (almost black bodies);
microwave radiation	gases	absorption and transmission
	clouds	absorption, scattering and transmission
	surface	absorption, reflection

We have also noted that a good absorber of radiation is an equally good emitter of

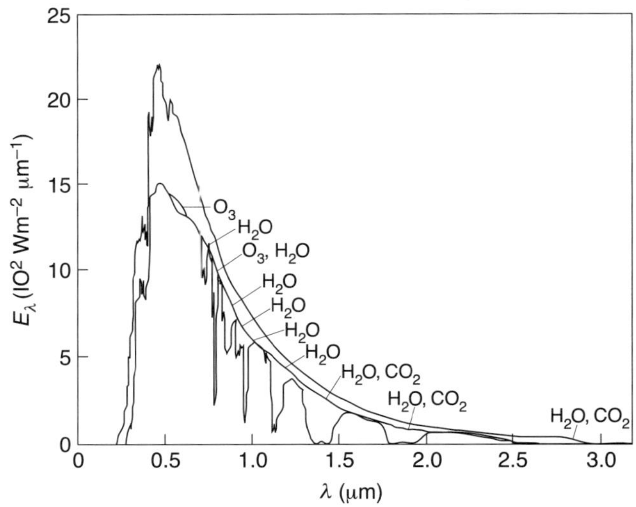

Figure 7.3. Solar spectrum at Earth.

radiation. This fact is known as Kirchhoff's Law; this states that at any wavelength, emittance equals absorptance.

The above properties form the basis of remote sensing of clouds, gases and surface properties in different parts of the spectrum. The electromagnetic spectrum extends from very short-wavelength gamma γ rays to very long-wavelength radio waves.

Figure 7.3 shows the solar spectrum at the top of the atmosphere and at the Earth's surface. The shape of the solar spectrum at the top of the atmosphere is similar to that of a black body at the Sun's temperature of 5,800 K. But the intensities of radiation have been reduced by $1/d^2$, where d is the Earth–Sun distance. The lower curve shows the solar radiation spectrum at the surface of the Earth after transmission through the atmosphere. The radiation reaching the surface is reduced by molecular (Rayleigh) scattering, proportional to $1/\lambda^4$, and therefore particularly effective at the shorter (ultraviolet and blue) wavelengths, and absorption by oxygen, ozone, water vapour and carbon dioxide.

7.2 PHYSICAL BASIS OF RETRIEVALS FROM REFLECTED SOLAR RADIATION

7.2.1 Clouds

One way of detecting clouds objectively is based on the fact that clouds are better reflectors of solar radiation than the surface or the atmosphere. However, it is difficult to detect clouds over other bright surfaces (i.e. snow- or ice-covered surfaces and sandy deserts). This technique obviously only works during the daytime.

7.2.2 Snow and ice

Snow or sea ice can be detected because their reflectances of visible radiation are much higher than those of other natural surfaces, except clouds. One way of separating clouds from snow is based on the fact that clouds move and snow cover is stationary. Over a period of a day for a geostationary satellite, or a week for a polar orbiter, it may be possible to obtain a clear observation of the Earth. Recently, a new infrared spectral channel at 1.6 μm has been added to the AVHRR on the NOAA polar-orbiter series of spacecraft. At this wavelength, clouds are good reflectors and snow and ice are very poor reflectors, making it possible to differentiate snow and ice from clouds by using a combination of visible and 1.6 μm observations.

7.2.3 Aerosols

The ocean has a low reflectivity for solar radiation. Aerosols will enhance the ocean reflectivity because of their scattering ability. Measurements of the increase in reflectivity provide information on aerosol amounts over the oceans. A major problem in this technique is differentiating clouds from aerosol layers. Objective techniques take advantage of the facts that: (1) clouds are usually better reflectors than aerosols and (2) aerosols are usually more spatially uniform than clouds.

7.2.4 Vegetation

Green vegetation has the unique property that its near-infrared reflectivity is much higher than its visible reflectivity. The difference between these two reflectances is the basis for a vegetation index (Figure 7.4). The greater the difference, the greater the vegetation amount. For clouds, snow, ice and deserts this difference is close to zero.

7.2.5 Ozone

Most of the Earth's ozone is concentrated in the stratosphere (at heights of 15–50 km). Ozone absorbs solar radiation at ultraviolet and visible wavelengths. Much of the radiation that is not absorbed on the way down is scattered back by air molecules or clouds, which are concentrated in the troposphere (between the surface and a height of 15 km). Some of this backscattered radiation is absorbed by the ozone layer before reaching the satellite. Since the ozone-absorption coefficients and the physics of Rayleigh scattering are known, and the reflectivity of the cloud can be obtained from a measurement at a wavelength with no ozone absorption, the ozone amount above the reflecting level can be determined. By making measurements at wavelengths with low, moderate and strong absorption, we can determine ozone amounts above low, moderate and high altitudes. This provides information on the vertical profile of ozone concentration.

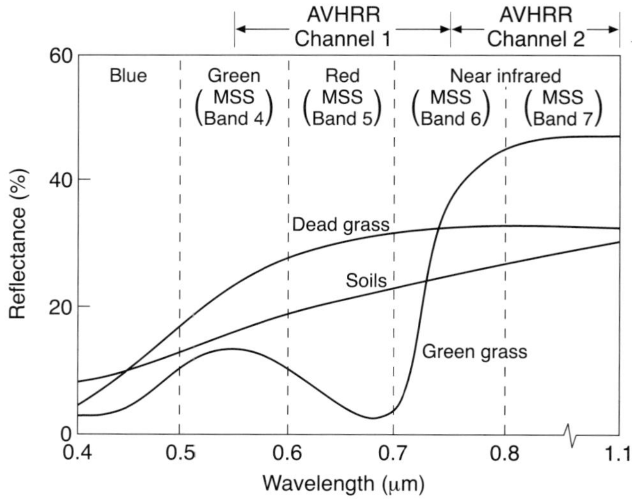

Figure 7.4. Physical basis for measuring vegetation condition (vegetation index).

7.2.6 Theory of ozone profile

We consider the theory of the determination of the vertical ozone profile from Solar Backscatter Ultraviolet (SBUV) measurements. This involves the observation of the spectrum of solar ultraviolet radiation which is backscattered by air molecules (Rayleigh scattering) and clouds. The ozone lies above the scatterers and absorbs the ultraviolet radiation both before and after backscattering. We assume the following:

- the Sun is overhead;
- the instrument is nadir viewing;
- for the stratospheric ozone layer the absorption is greater than the scattering;
- only single scattering, no multiple scattering, is considered.

The absorption leads to a decrease in intensity according to:

$$I(p) = I(0)\exp[-kn(p)] \tag{7.1}$$

where k = absorption coefficient per molecule, $n(p)$ = number of molecules above pressure p. We express thicknesses of the atmospheric layers in terms of differences in pressure rather than in height, so that the radiation scattered vertically upwards from a layer of thickness $\mathrm{d}p$ is proportional to $I(p)$ and $\mathrm{d}p$ and so can be written as:

$$\mathrm{d}I = aI(p)\,\mathrm{d}p \tag{7.2}$$

where a is a constant. Therefore:

$$I_{\text{sat}} = I(0)a\int_0^\infty \exp[-2kn(p)]\,\mathrm{d}p \tag{7.3}$$

Consider a constant ozone mixing ratio $n(p) = n'p$. Therefore:

$$I_{sat} = I(0)a \int_0^\infty p \exp[-2kn'p] \, d \ln p \qquad (7.4)$$

A weighting function that is proportional to the contribution of each atmospheric layer to I_{sat} is given by $W = p \exp[-2kn'p]$ so that:

$$dW/dp = (1 - 2kn'p) \exp(-2kn'p) \qquad (7.5)$$

so that W is a maximum at $p = (2kn')^{-1}$. Thus, by choosing wavelengths with different values of k, the ozone profile can be obtained.

7.3 PHYSICAL BASIS OF RETRIEVALS FROM INFRARED-EMISSION MEASUREMENTS

If both the Sun and the Earth were perfect black bodies the spectral distribution of radiation at the top of the Earth's atmosphere (or as observed by an Earth-orbiting satellite) would be as shown in Figure 7.5. Note that, as a result of Wien's law, the wavelength of maximum radiation is 0.6 μm for the solar radiation and 15 μm for the Earth's radiation. Note also that there is very little overlap between the solar spectrum and the Earth's emission spectrum. Figure 7.6 shows the absorption by the atmosphere. Wavelength ranges for which the absorption is small (i.e. for which the atmosphere is almost transparent), are called atmospheric windows.

The infrared-emission spectrum of the Earth as observed by the *Nimbus-3* IRIS instrument is shown as the irregular curve in Figure 7.7. Also shown are Planck curves for black bodies at various temperatures. The observed emitted energy is often converted to a brightness temperature, which is given by the Planck curves. The wavenumber is $1/\lambda$. In atmospheric windows, we observe brightness temperatures close to that of the Earth's surface (e.g. close to 300 K at 1,000 cm^{-1} [10 μm]).

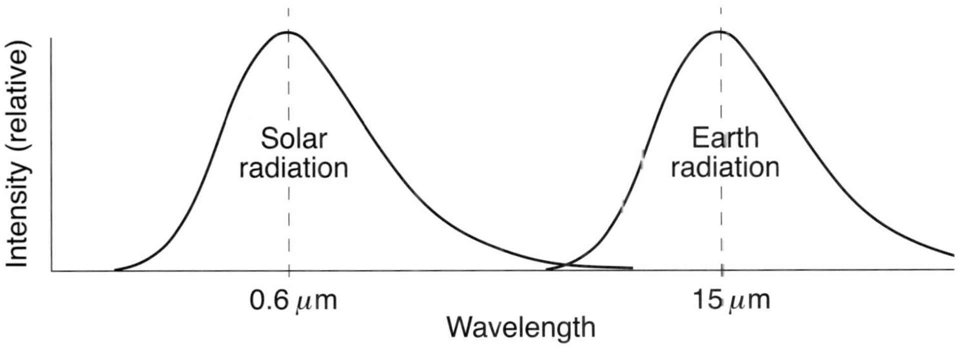

Figure 7.5. Schematic diagram showing black-body solar and Earth radiation at top of atmosphere.

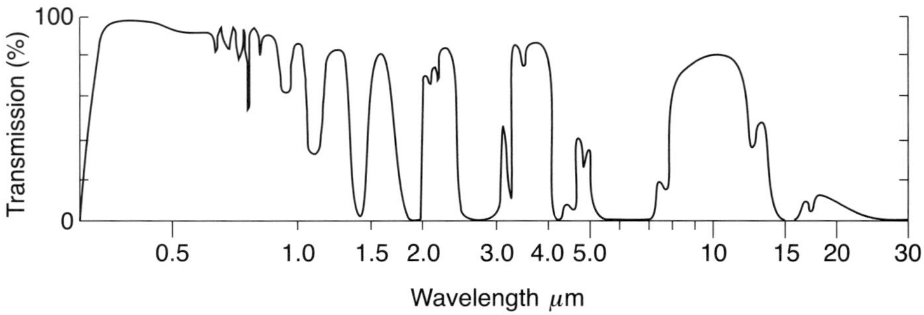

Figure 7.6. Absorption by the atmosphere (%).

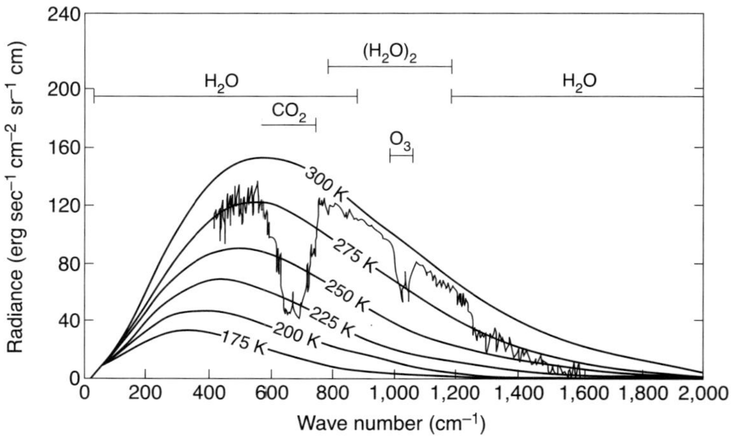

Figure 7.7. Infrared emission spectrum of the Earth.

The Earth's surface and all but thin clouds behave as black bodies. Thus, observations in atmospheric windows form the basis for measurements of surface and cloud-top temperatures and detecting the presence of clouds day and night. But even in the most transparent atmospheric windows, water-vapour absorption may depress the observed brightness temperature by up to 5°C, depending upon the amount of water-vapour present in the atmosphere. By observing at two nearby window wavelengths, it is possible to eliminate the water-vapour effect and solve for the surface temperature. At wavelengths where water vapour and CO_2 (carbon dioxide) absorb the radiation emitted by the surface, the brightness temperatures are lower. Observations in the CO_2 and water-vapour absorption bands form the basis for atmospheric temperature and water-vapour soundings.

The above discussion summarizes some of the general radiative properties of clouds, atmospheric gases and the Earth's surface for solar radiation, infrared radiation and microwave radiation. These properties form the basis of remote sensing of clouds, gases and surface properties in different parts of the spectrum.

7.4 PHYSICAL BASIS OF ATMOSPHERIC SOUNDINGS TO STUDY TEMPERATURE AND WATER VAPOUR

The observed decrease in brightness temperature from the edges to the centre of the CO_2 absorption band (see Figure 7.7) is basically a temperature sounding of the Earth's atmosphere. The temperature decreases with height in the troposphere and increases with height in the stratosphere.

The observation in Figure 7.7 is for a cloudless sky. A cloud at a temperature of 250 K, say, behaving as a black body would effectively screen radiation from the atmosphere below it. The emission spectrum in Figure 7.7 would follow the 250-K isopleth, except for the centre of the CO_2 emission band, which is sensitive to the above-cloud atmosphere. Thus, in the presence of cloud, infrared-emission measurements cannot be used to sound the atmosphere below the clouds.

Radiation emitted by the surface of the Earth is either transmitted to space with little absorption by the atmosphere or absorbed by the atmosphere. By Kirchhoff's law, the radiation absorbed by the atmosphere is re-emitted up and down. At a wavelength with weak absorption, the radiation emitted upward will be transmitted by the atmosphere to space. At a wavelength with moderate absorption, the radiation absorbed by the lowest layer will be re-emitted and absorbed by a higher layer. For a wavelength with very strong absorption, only radiation from the very highest atmospheric layer will escape to space. These three situations are illustrated in Figure 7.8(a).

Thus, to derive an atmospheric temperature profile, observe the spectrum of emitted radiation in the absorption band of an atmospheric case with a fixed mixing ratio (e.g. CO_2 or O_2 [oxygen]). Each wavelength is sensitive to a different layer of the atmosphere, depending on the value of the absorption coefficient. The sensitivity is defined by a weighting function, as derived and illustrated in the next section. Actual weighting functions for the High Resolution Infrared Radiation Sounds (HIRS) are shown in Figure 7.8(b).

7.4.1 Temperature-profile theory

To derive an atmospheric water-vapour profile, observe the emitted radiation in an atmospheric water-vapour band. If the temperature profile is known, the water vapour profile can be retrieved. The satellite-observed radiance in the CO_2 absorption band I_λ is given by:

$$I_\lambda = \int_0^\infty B_\lambda(T)[\mathrm{d}\tau_\lambda(z,\infty)/\mathrm{d}z]\,\mathrm{d}z + B_\lambda(T_\mathrm{s})\tau_\lambda(0,\infty) \qquad (7.6)$$

where $\lambda =$ wavelength, $z =$ height, s refers to the surface, $B_\lambda(T) =$ Planck function, $\tau(z,\infty) =$ transmittance, $T =$ temperature, $k =$ absorption coefficient, $\rho = CO_2$ density and $c = CO_2$ mixing ratio. We assume that $\varepsilon_{\lambda,\mathrm{s}}$ the emissivity at

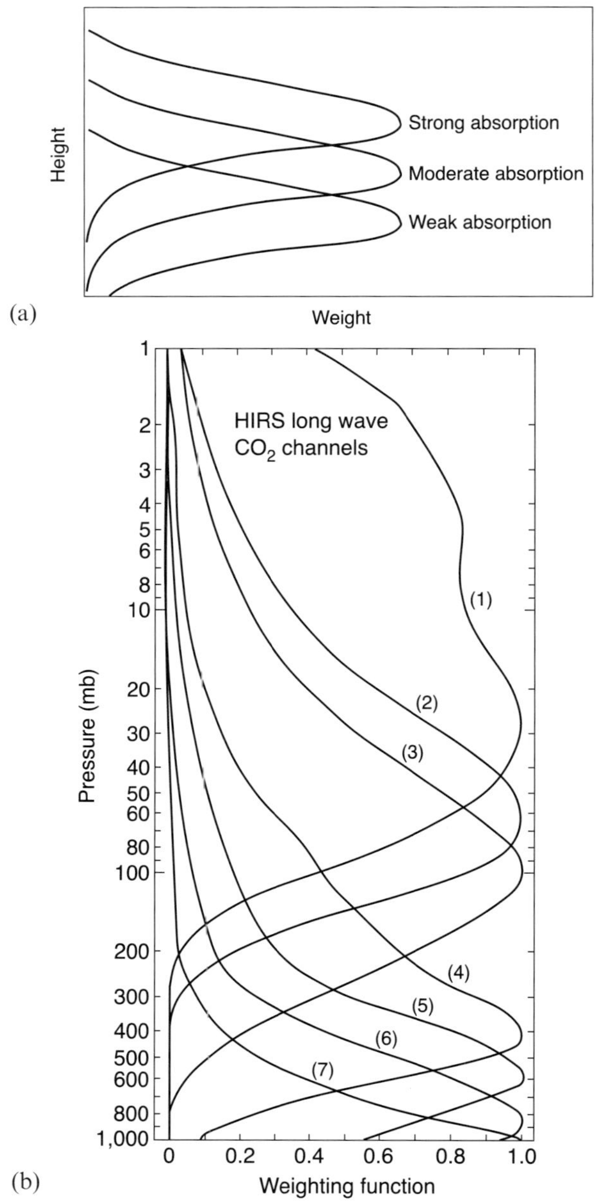

Figure 7.8. (a) Schematic weighting functions for strong, moderate and weak absorption; (b) weighting functions for HIRS sounder.

the surface, is equal to 1. The first term on the right-hand side of this equation represents the contribution from the atmosphere and the second term represents the contribution from the surface. We introduce $y = -\ln p$, where p is the atmospheric presure, so that:

$$I_\lambda = \int_0^\infty B_\lambda(T)(\mathrm{d}\tau/\mathrm{d}y)\,\mathrm{d}y + \mathrm{SFC} \tag{7.7}$$

$$\tau = \exp(-k_\lambda \int_z^\infty \rho\,\mathrm{d}z) = \exp(-k_\lambda cpg^{-1}) \tag{7.8}$$

$$\mathrm{d}\tau/\mathrm{d}y = -k_\lambda cg^{-1} \exp(-k_\lambda cpg^{-1})(-p) = W(p) \tag{7.9}$$

where $W(p)$ represents the relative contribution of each layer to the satellite-observed radiance. It is a maximum at $\mathrm{d}W/\mathrm{d}p = 0$:

$$\frac{\mathrm{d}W}{\mathrm{d}p} = -k_\lambda cg^{-1}(1 - k_\lambda cpg^{-1})\exp(-\delta) \tag{7.10}$$

so that for $(\mathrm{d}W/\mathrm{d}p) = 0$, we find that $\delta = k_\lambda cpg^{-1} = 1$ (i.e. the optical depth = 1).

The HIRS has been the mainstay of atmospheric temperature and water-vapour sounding from the National Oceanic and Atmospheric Administration (NOAA) polar-orbiting spacecraft since 1979. The instrument has undergone various developments over the years. It forms part of the TOVS (the TIROS Operational Vertical Sounder).

7.5 PHYSICAL BASIS OF RETRIEVALS FROM MICROWAVE-EMISSION MEASUREMENTS IN ATMOSPHERIC WINDOWS

Measurements at microwave-window wavelengths are used to determine precipitation, cloud water/ice, snow/ice and sea-surface wind. By observing at wavelengths where atmospheric oxygen absorbs and emits microwave radiation, it is possible to determine the temperature profile, as with the CO_2 absorption band in the infrared. By observing at wavelengths where water-vapour absorbs, it is possible to measure the water-vapour profile.

Microwaves have one major advantage over infrared for atmospheric soundings. Clouds (except thick precipitating systems) are transparent to microwaves, thus permitting observation through the clouds to the surface. On the other hand, infrared sounders have better horizontal spatial resolution. There is another problem with microwave sounders; at microwave wavelengths the Earth's surface is not a perfect emitter. The emissivity is variable, ranging from 0.5 for oceans to 0.95 for land.

7.6 OPERATIONAL WEATHER SATELLLITES

The countries of the world have co-operated in establishing a global system of weather satellites. These consist of geostationary satellites which are at an altitude of 36,000 km above the equator and rotate with the Earth. They constantly monitor the entire disc of the Earth below them. Polar-orbiting satellites orbit the Earth at an altitude of about 850 km and, with scanning instruments, can observe each area of the globe at least twice a day. Because of their global coverage and more sophisticated instruments, polar-orbiting satellites have provided most of the satellite data for climate monitoring.

7.7 GLOBAL WARMING

The Earth's mean temperature would be about 30°C lower were it not for the presence of water vapour and CO_2 in the atmosphere. By absorbing the radiation emitted by the Earth and then re-emitting energy to space at a lower temperature, these gases cause the atmosphere to act like a greenhouse to retain the energy absorbed from the Sun.

Figure 7.9 shows how the atmosphere's CO_2 concentration has varied over the geological past. The concentration of CO_2 in the atmosphere has increased by about one-third since the beginning of the Industrial Revolution, largely as a result of the burning of fossil fuels. Because of its greenhouse effect, increasing CO_2 levels may be causing global warming.

To measure global-temperature trends, climatologists rely on weather-station records. But weather stations are concentrated in populated areas, thus leaving large gaps in geographical coverage, especially over the oceans and introducing heat-island effects into the observations. Global-temperature trends computed from such data may have errors because of this sampling problem.

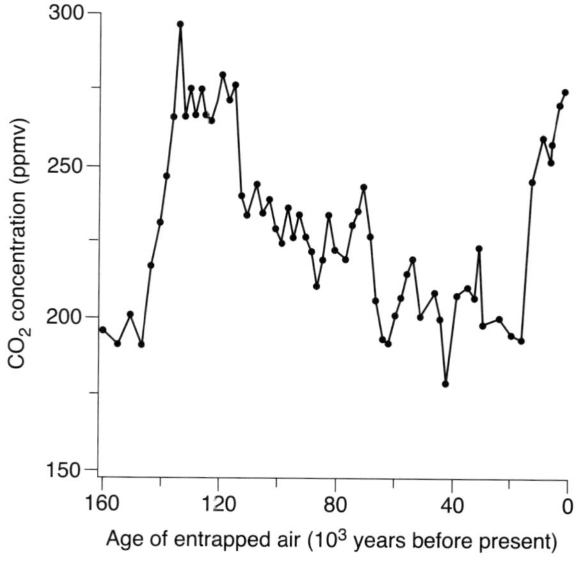

Figure 7.9. 16,000-year CO_2 record from Vostok ice core; air-bubble analysis shows the highest previous CO_2 amount as 300 ppm.

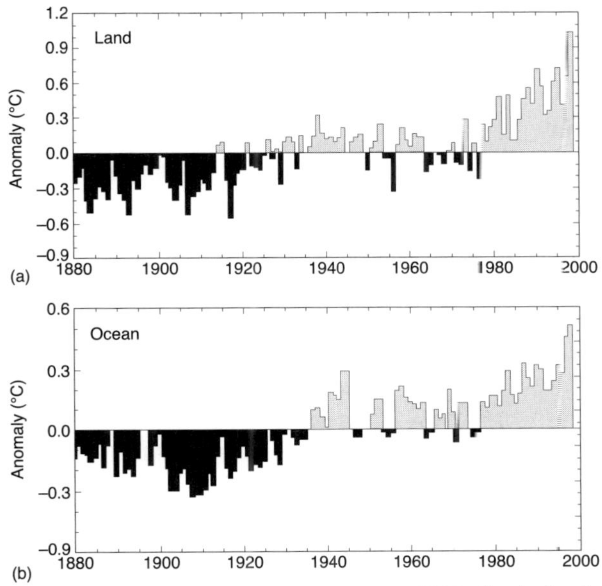

Figure 7.10. Surface-temperature trends from 1880 to 1998 (both land and ocean have warmed over the last 100 years by about 0.6°C).

Climatologists have determined trends in global upper-air temperatures from radiosonde balloon observations. Global coverage is much less than for surface weather stations – causing a significant sampling problem. Radiosondes provide data from the surface to about 30 km.

Polar-orbiting satellite observations provide the global coverage necessary for monitoring the entire globe. In a 24-hour period almost the entire Earth would be sampled. Satellite observations can fill in gaps in ground-based observing networks or be used on their own.

Calculations with computer climate models indicate that increasing CO_2 (and other man-made greenhouse gases) can be expected to cause the Earth to warm at a rate of about a few tenths of a degree C per decade. A major impediment to observing such small changes is the natural variability in temperature. This variability is sometimes called climatic noise since it masks the signal – the long-term trend – that we want to detect, see Chapters 1 and 2.

Global mean surface temperatures since 1880 are shown in Figure 7.10. These graphs are based on weather-station observations over land and on-ship observations for the oceans. Prior to about 1880, there were not enough weather stations to compile a reliable global mean surface temperature.

The Microwave Sounding Unit (MSU), flown since 1979 on NOAA's polar-orbiting satellites, has proved to be a precise thermometer of atmospheric temperatures. Because of the nature of its weighting function, the derived temperatures refer to a broad layer of the atmosphere rather than to a single level. The tropospheric-temperature time series (Figure 7.11) shows no significant trend, if we neglect the temporary warming caused by the strong El Niño of 1998.

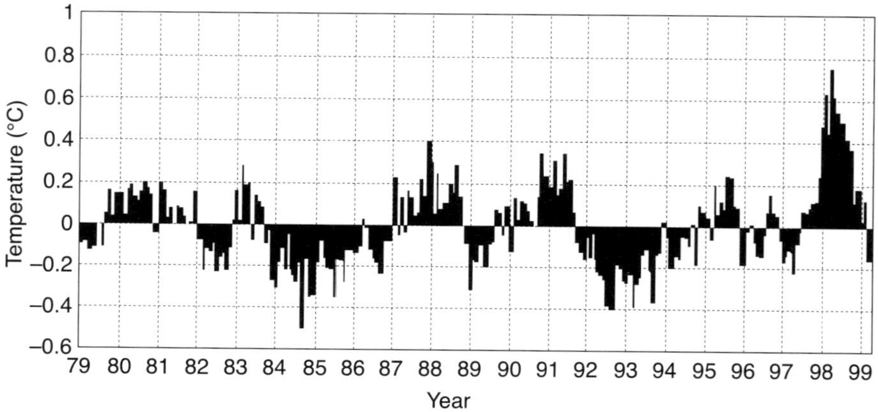

Figure 7.11. Global atmospheric temperature trends from MSU; global lower-tropospheric temperature anomalies (January 1979–March 1999).

The occurrence of El Niño events such as this adds to the climatic noise that tends to hide any small global-warming signal. Only when the greenhouse-warming signal climbs out of the noise will it be reliably detected. At a rate of tenths of a degree/decade, this may take decades.

Stratospheric temperatures from the MSU show a decreasing trend, as expected, from ozone depletion – which leads to less heating from the absorption of solar radiation – and from increased CO_2 – which, in the stratosphere, leads to a cooling as a result of greater emission of infrared radiation to space. The stratospheric-temperature time series has large positive anomalies, which are due to the absorption of solar radiation by particles introduced into the stratosphere by the eruptions of El Chichón in 1982 and Mt Pinatubo in 1991. Credibility of the MSU observations has been established by comparison with radiosonde observations, which give tropospheric and stratospheric trends similar to those of the MSU. MSU temperatures have been shown to agree with the analyses of temperature of the global Numerical Weather Prediction model (NWP) of the US National Weather Service, which assimilates all available observations to produce analysed fields.

The observations seem to indicate that the Earth's surface has warmed and the lower atmosphere has not changed significantly since 1979. If these observations are correct, and, especially if these trends continue, they imply a steepening of the vertical temperature gradient, with various implications for weather phenomena, such as convective systems.

7.8 OCEAN-TEMPERATURE TRENDS

Conventional ocean-temperature observations are concentrated along the major shipping lanes, leaving vast areas of the oceans unobserved. Satellite observations can fill in these holes.

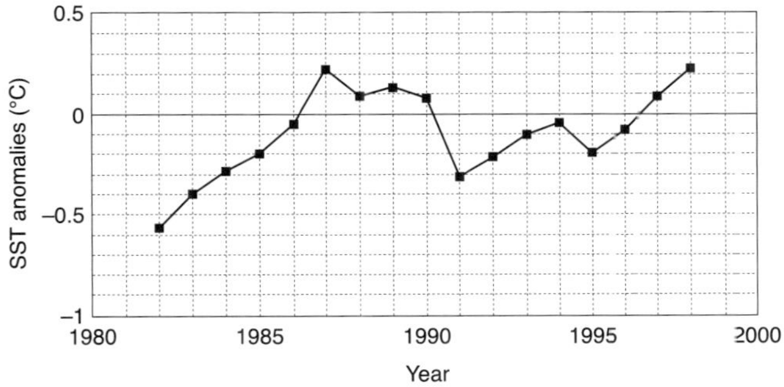

Figure 7.12. Annual average SST anomalies based on AVHRR measurements. High SSTs in 1987 and 1998 are related to El Niño events.

The El Niño phenomenon has a widespread effect on the world's weather. El Niño manifests itself by a warming of the eastern tropical Pacific and a cooling of the western tropical Pacific. Sea-surface temperatures (SST) in the eastern Pacific during the very strong El Niño event of 1998 were as much as 5°C above normal. During the opposite event – La Niña – the temperature anomalies for these regions are reversed. These large swings in ocean-surface temperatures have concomitant effects on precipitation and temperature in many parts of the world.

Annual average SST anomalies based on Advanced Very High Resolution Rdiometer (AVHRR) measurements are shown in Figure 7.12. High SSTs in 1987 and 1998 are related to El Niño events.

AVHRR satellite observations indicate that the tropical oceans have warmed at a rate of about 0.1°C per year over the last 15 years. Much of this may be the result of a strong El Niño at the end of the period, which would cause a positive bias in any linear fit to the time series.

One of the problems with satellite infrared measurements of SST is their sensitivity to dust in the atmosphere. Normally, there is not too much dust or aerosol in the atmosphere. However, in certain geographical regions (e.g., over the eastern Atlantic off the coast of Africa, where Saharan dust is found, and downwind of the industrial areas of eastern North America and Asia) large amounts of aerosol may be present. While the 'split window' technique accounts for atmospheric water vapour, it does not account for aerosols which, because of their absorption of Earth-emitted radiation, tend to lower the brightness temperature, leading to negative biases in SST.

More serious are the occasional injections of huge quantities of aerosols into the stratosphere by volcanic eruptions. As opposed to tropospheric aerosols discussed above, which are regional and fall out, or are washed out, of the atmosphere by rain rather quickly, volcanic aerosols spread throughout the entire globe and remain in the stratosphere for 1 to 2 years. Comparison of SSTs derived from AVHRR by shipborne observations immediately after the eruption of Mt Pinatubo showed large negative errors – more than 1°C – caused by the aerosols.

Because of the biases of satellite-derived SSTs, the NOAA National Centre for Environmental Prediction (NCEP) (Reynolds and Smith 1994) developed a method of combining satellite and ship observations in order to eliminate the negative bias.

High ocean temperature causes bleaching of coral reefs. NOAA/NESDIS (National Environmental Satellite Data and Information Service) (Strong *et al.* 1997) has developed a hot-spot index based on AVHRR SST observations. The index is the difference between the actual SST in a region and the average temperature of the warmest month in that region. If the index is greater than +1°C, a potential for coral bleaching exists. The 1998 El Niño caused the hot-spot index to signal danger to the major coral reefs of the Australasian region and reports indicate that severe bleaching affected many of the reefs in that area.

7.9 ATMOSPHERIC-OZONE TRENDS

A combination of circumstances has led to the formation of what is commonly described as the Antarctic ozone hole during the last 25 years. This is not actually a complete hole in the ozone layer, but the amounts of ozone do get depleted by up to 80 per cent of their normal values. Chlorofluorocarbons (CFCs) introduced into the atmosphere from aerosol spray cans, refrigerators, air conditioners, etc. lead to a set of photochemical reactions (chemical reactions involving sunlight) that tend to destroy the ozone (O_3) layer in the stratosphere. These O_3-destroying reactions occur more rapidly in the presence of stratospheric clouds, which occur primarily in the cold Antarctic stratosphere during the southern-hemisphere winter. With the return of the Sun at the end of the long Antarctic winter, conditions are most favourable for ozone destruction.

The Antarctic ozone hole has generally increased in size and become deeper over the last few decades, primarily because of the increasing amounts of CFCs in the atmosphere. However, as a result of the Montreal Protocol, which set strict limits on CFC use, the concentrations of O_3-destroying chemicals are stabilizing and should decrease over the next few decades. However, decreasing stratospheric temperatures may cause more polar stratospheric clouds to form, thus temporarily offsetting stabilized CFCs, and perhaps leading to deeper Antarctic ozone holes for some years to come.

The growth in size of the Antarctic O_3 hole, as measured by the NOAA SBUV instruments, is clearly shown in Figure 7.13. The size is given by the area covered by total O_3 values less than 220 Dobson Units (DU). DU are a measure of the total amount of O_3 in a vertical column of atmosphere. If all of the ozone in a vertical column of the atmosphere were to be reduced to standard temperature (273 K) and pressure (1,013 mb) conditions, it would have a certain thickness. This thickness is measured in DU, where 1 DU equals 10^{-3} cm.

SBUV instruments have been flown on several satellites. A study of the difference between each satellite's readings and those of ground-based observations (using Dobson instruments) shows that there is no drift of the satellite record with respect

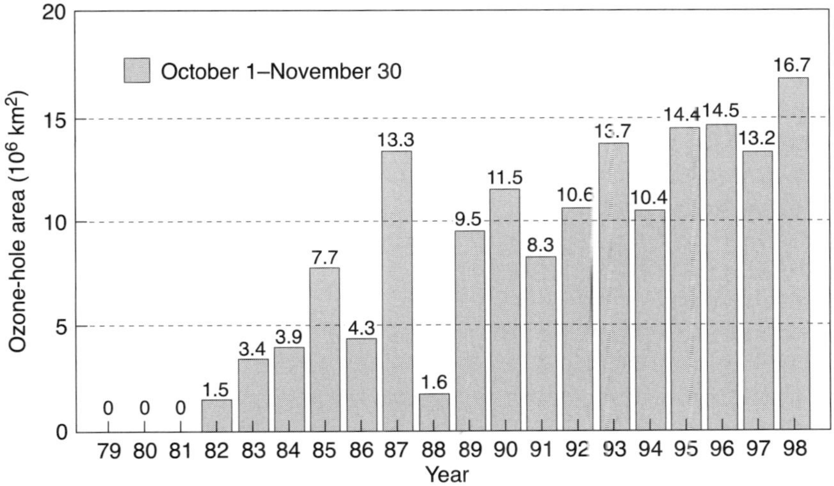

Figure 7.13. The growth in size of the Antarctic O_3 hole, as measured by the NOAA SBUV instruments between 1 October and 30 November, is clearly seen in this diagram.

to Dobson observations. Thus, satellite measurements can be used to study ozone-concentration trends.

Regular satellite measurements of total ozone concentration have been obtained from SBUV instruments flown on *Nimbus-7* SBUV (1979–1988), *NOAA-11* SBUV/2 (1989–August 1994), *NOAA-9* SBUV/2 (September 1994–June 1997) and *NOAA-14* (July 1997–present).

Observed ozone variations for the 5°N–45°N latitude belt from 1979–1997 are shown in Figure 7.14. After removing seasonal variations, solar-cycle effects and quasi-biennial oscillations, the ozone trends in the lower panel remain. Total ozone has decreased by about 10 DU (more in spring and less in autumn) over the 20-year period, or at a rate of about 2 per cent per decade (see Figure 7.14).

Ozone trends from different satellite instruments and the ground-based record agree reasonably well. In addition to the SBUV instruments, there is also the NASA TOMS (Total Ozone Mapping Spectrometer). Ozone concentrations can also be obtained from the infrared measurements in the 9.6-μm ozone-absorption channel of TOVS on the NOAA series of polar-orbiting meteorological satellites.

7.10 INTERANNUAL CLIMATE VARIABILITY – EL NIÑO EVENTS

The largest interannual climatic fluctuations are associated with El Niño events, which are characterized by warming of the eastern tropical Pacific Ocean. The total outgoing long-wave radiation (OLR) is estimated from AVHRR infrared-window channel measurements (there is a very high correlation between the two). The primary modulator of outgoing long-wave radiation is cloudiness. Recall that clouds radiate as black bodies at temperatures much lower than the clear Earth/

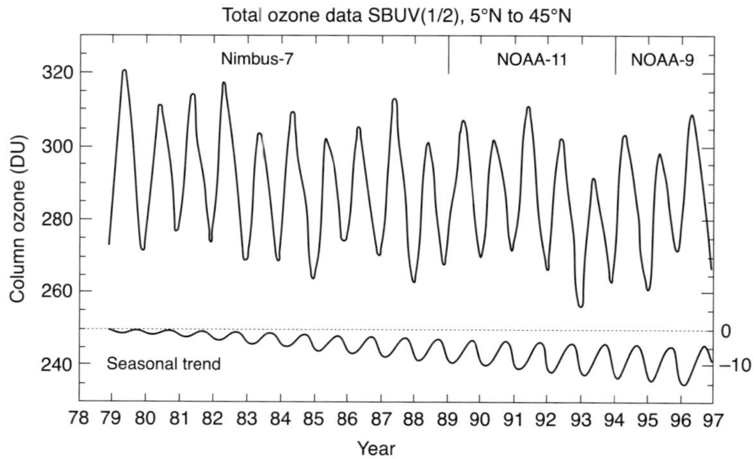

Figure 7.14. SBUV ozone variations (top) and long-term trend (bottom).

atmosphere system. Climatologists have used the outgoing long-wave radiation as an index of clouds and precipitation in the Tropics.

As the eastern Pacific warms during an El Niño event, the extensive cloudiness over the Indonesian area moves to the eastern Pacific. The opposite situation prevails during a La Niña event, when the eastern tropical Pacific Ocean cools.

There are significant changes to the rainfall pattern over the whole globe associated with El Niño/La Niña events and the details can be seen in the data from SSM/I (Special Sensor Microwave/Imager) (e.g. the 1997–1998 El Niño was associated with heavy rainfall in South America and near-drought conditions in Indonesia).

Less green vegetation and high temperature are characteristic of extreme drought conditions. Such conditions are indicated by a vegetation health index based on the AVHRR vegetation index and infrared-window brightness temperature, which is a measure of surface temperature. The 1997–1998 El Niño was associated with excellent vegetation health in South America and poor vegetation health in Indonesia.

7.11 CLOUDINESS AND AEROSOL TRENDS

Figure 7.15 shows a time series of global monthly cloudiness obtained from raw AVHRR data. The sawtooth pattern is an artefact and not a real trend; it arises because the dataset is based on observations from a number of successive satellites (see the extensive discussion in Chapter 6). When the spurious trends in the cloudiness time series are removed by taking out the average slope of the data with time, we obtain the time series shown in Figure 7.16. This dataset is much more stable, and no

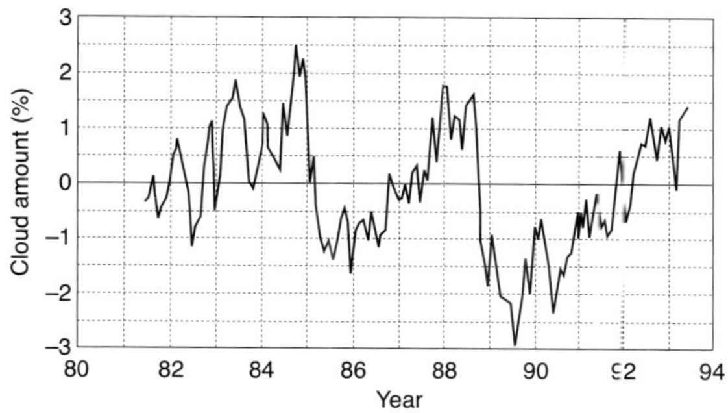

Figure 7.15. Global cloud amount anomalies from AVHRR raw data.

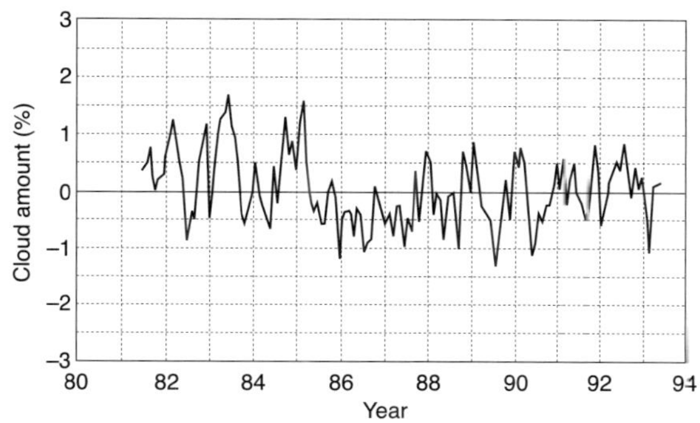

Figure 7.16. Global cloud amount anomalies from AVHRR: detrended.

long-term trend is apparent. The fluctuations in global cloudiness are very small (of the order of 1 per cent).

Aerosol concentrations are measured over the ocean from AVHRR data. But over the land the contrast between the reflectance of the land and that of the land plus aerosols is too low to be detected by satellite observations. In addition, the highly variable reflectance of land surfaces presents a problem.

The volcanic eruptions of El Chichón in 1982 and Mt Pinatubo in 1991 introduced large quantities of aerosols into the stratosphere. Dust from very explosive volcanic eruptions is the major signal in the atmospheric-aerosol time series. Note that the atmosphere does not return to normal until about 2 years after the eruptions. No long-term trends in aerosols are evident, but these measurements are not accurate enough to detect small trends.

The large amounts of dust from the volcanic eruptions cause an increase in the sunlight reflected back to space by the Earth. Other things being equal, an increase in reflected sunlight means less solar radiation to heat the Earth and, therefore, a decrease in the Earth's temperature.

The global surface temperature increases under El Niño conditions as shown in Figure 7.17(a). The El Niño index is based on SST anomalies in the Pacific. To

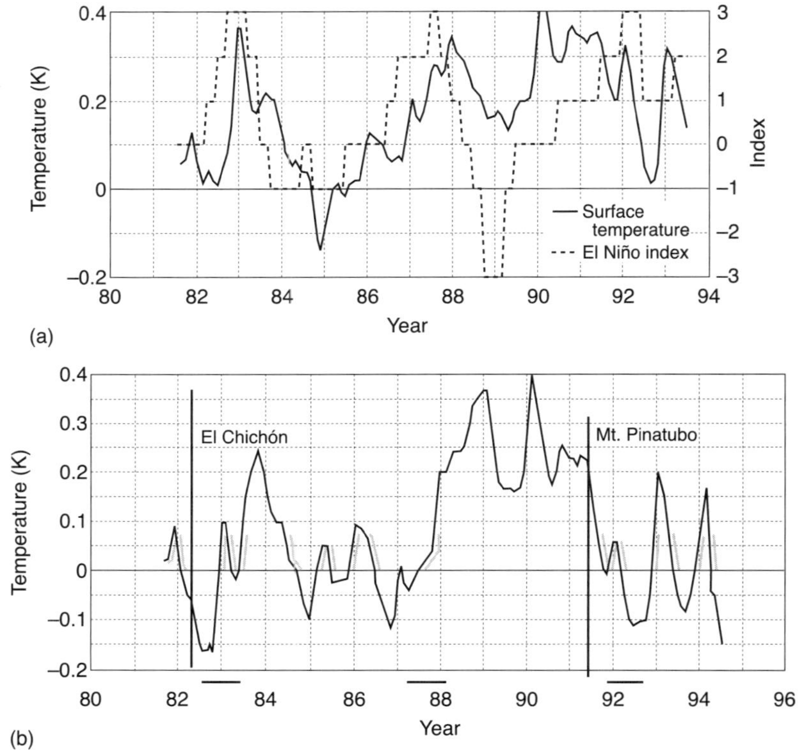

Figure 7.17. (a) Anomalies of global surface temperature and El Niño index and (b) residual temperature anomaly (actual – predicted from El Niño index).

extract the effect of the volcanic aerosols on surface temperature, we would like to eliminate this El Niño effect. By performing a linear regression between the El Niño index and surface temperature, and then subtracting the temperatures predicted from this relationship from the observed temperature, we obtain, in principle, a time series unaffected by El Niño, which is shown in Figure 7.17(b).

Decreases of several tenths of a degree C occurred in the years following the volcanic eruptions of El Chichón and Mt Pinatubo. These volcano-induced temperature variations are part of the climatic noise when monitoring global warming.

7.12 LAND SURFACE CHANGE

The Normalized Difference Vegetation Index (NDVI) is defined as:

$$(\alpha_2 - \alpha_1)/(\alpha_1 + \alpha_2)$$

where α_1 and α_2 are the observed reflectances in AVHRR channel 1 (visible) and channel 2 (near infrared), respectively (see Chapter 6). It is a measure of vegetation amount and health and is widely used in the study of vegetation.

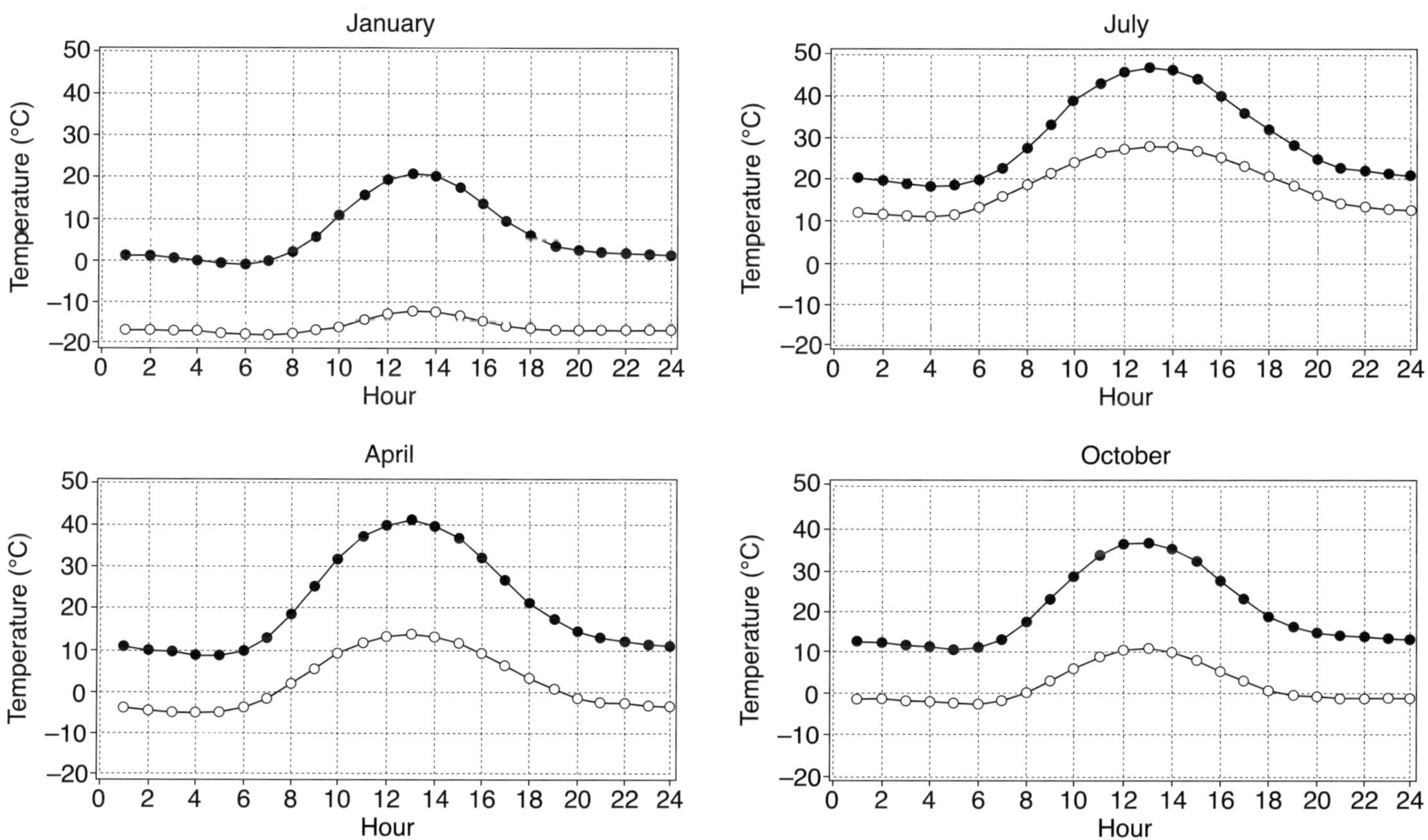

Figure 7.18. Diurnal cycle of land-surface-skin temperature (Ignatov and Gutman 1999) (based on ISCCP 7-year monthly dataset): 15°N < lat < 45°N (dots) and 45°N < lat < 75°N (circles).

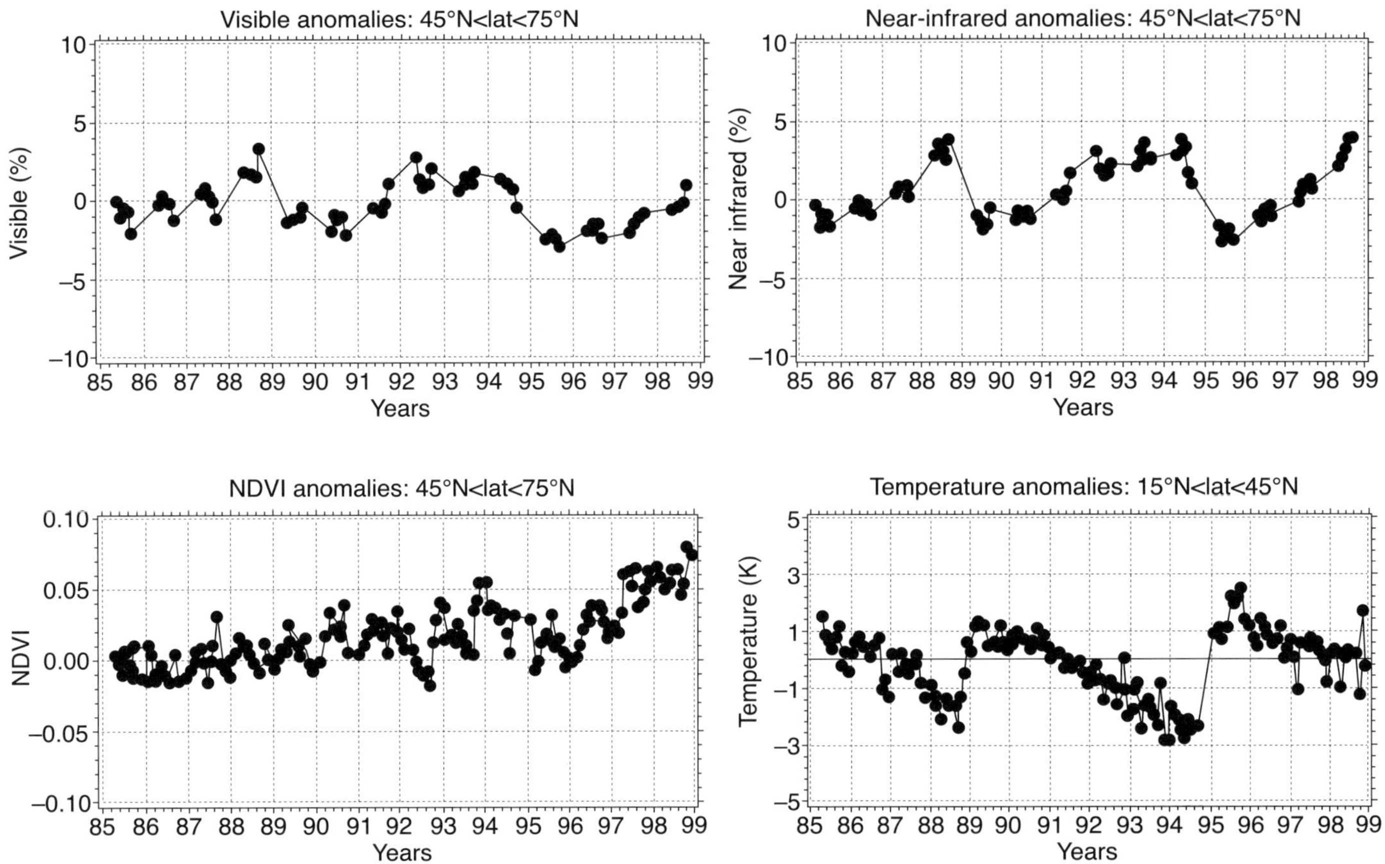

Figure 7.19. Anomalies of reflectances (visible and near infrared), NDVI and brightness temperature.

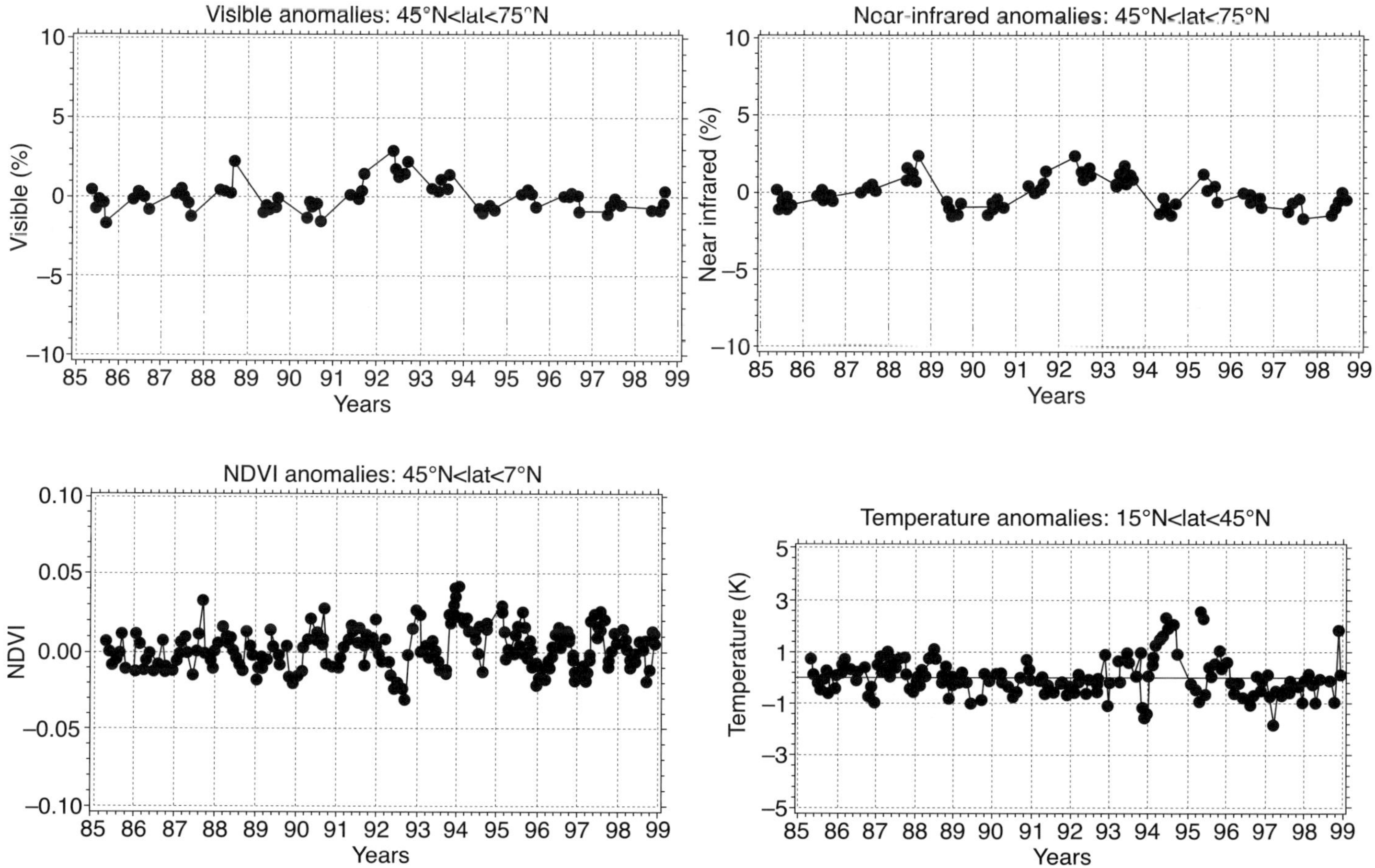

Figure 7.20. Anomalies corrected for satellite-orbit drift and calibration inconsistencies.

The fraction of the surface covered by green vegetation can be obtained from the NDVI. By averaging data from many years it is possible to construct a climatology of this greenness fraction. The annual cycle in vegetation is clearly shown by the differences between January and July at high latitudes.

The albedo (reflectance) of the Earth's land surface is derived from AVHRR visible and near-infrared channel observations of reflected sunlight under clear conditions. The albedo of the deserts is very high and the albedo of the tropical rain forests is very low. Since the surface albedo determines how much solar energy the Earth's surface actually absorbs, it is an important heat-budget quantity, both responding to climatic change (e.g. lower albedos because of less snow with greenhouse warming, and causing climate change (e.g. increasing albedos as a result of mankind's deforestation activities). However, so far no attempts have been made to determine trends in albedo.

Land-skin temperature can be derived from AVHRR using the same methodology as for SST. However, allowance will need to be made for the fact that the emissivity of the land surface is not necessarily equal to 1.

The diurnal cycles of surface-skin temperature for four different months are shown in Figure 7.18. They are derived from data of the International Satellite Cloud Climatology Project (ISCCP). Part of the ISCCP dataset includes estimates of the surface-skin temperature every 3 hours GMT. These are derived from the thermal infrared-channel observations of the global array of geostationary satellites. These satellites have only one infrared channel (the SST split-window method therefore cannot be used) and its brightness temperature has been corrected for atmospheric water-vapour attenuation to derive the surface-skin temperature. The diurnal temperature variation in the Tropics is largest in April and October, when the Sun reaches its highest point there.

Based on their analysis of AVHRR data, Myneni *et al.* (1997) have reported that the Earth has been greening up as a result of lengthening of the growing season. They attribute the lengthening growing season to global warming.

Figure 7.19 shows time series of AVHRR visible reflectances, near-infrared reflectances, NDVI and surface skin temperature prepared for this summer school by Gutman (NOAA) based on Gutman (1999a, b). The visible and near-infrared reflectances, and the skin temperatures have the same sawtooth pattern that we have just seen in the global cloudiness from AVHRR.

When the visible and near-infrared reflectances are combined into an NDVI time series, there appears to be a positive trend – a greening – similar to that reported by Myneni *et al.* (1997). But changes in instrument sensitivity and the drift in local observing time during the satellite's lifetime cast some doubt on this finding; it is important to take into account post-launch calibration.

When this was taken into account Gutman obtained the results shown in Figure 7.20, which was prepared especially for the summer school. The corrected time series shows no evidence for the greening trend for NDVI which had been found by Myneni *et al.* (1997). Although the time periods are not exactly the same (1982–1991 versus 1985–1999), Gutman's analysis illustrates the problems associated with deriving trends from these data.

7.13 SNOW COVER

The NOAA/NESDIS snow dataset is the longest available satellite dataset; it now covers some three decades. Snow cover over the northern hemisphere is mapped daily from observations of geostationary satellites and polar-orbiting satellites and from ground reports. The snow maps were originally produced by human analysts looking at satellite visible images, but they are now generated using an interactive (analyst-computer) multi-sensor method. Over the last 30 years, the fraction of the northern hemisphere covered with snow appears to have shrunk from 24 per cent to 22 per cent. If this is true it may be an indication of global warming.

8

Validation of climate-model outputs

J. P. Palutikof

8.1 CLIMATE CHANGE AND CLIMATE VARIABILITY

The weather is a short-term phenomenon, varying from day to day, while we think of climate as being the long-term average of weather. But the question arises in relation to climate as to how long is long term. It is a characteristic of climate change that it takes place on all timescales, from extremely long geological timescales through to much shorter timescales of a year or even less. There is a continuum when discussing climate change and the separation is quite blurred. At one extreme there is climate change and at the other extreme there is climate variability. It is important to distinguish between what is climate change and what is climate variability.

Let us say that climate change is a trend in a climate variable, such as temperature or precipitation, or a step jump which is persistent over a substantial length of time such as several decades, and for which it is possible to identify a forcing mechanism. Therefore we have two important criteria: the first is that climate change is something which is persistent over a timescale of decades and the second is that we can identify a forcing mechanism. Climate variability, on the other hand, is a step jump or a trend of short duration, after which the climate reverts to its original condition, and for which we cannot identify a forcing mechanism. Climate variability is part of the internal stochastic variability of the atmospheric system. An important distinction is the timescale: climate change is long term, climate variability is generally more short term.

Before discussing the consequences of human activities on climate, it is necessary to be able to distinguish anthropogenically induced effects from natural effects. Natural climate change has a whole range of timescales. Perhaps the one which we know most about, and which is most widely studied, is climate change related to the orbital parameters of the Earth. The timescale of climate change as a result of these effects is long. On the other hand, anthropogenic climate change because of the enhanced greenhouse effect is operating at short timescales of the order of tens or hundreds of years (see Section 1.3 and Goodess *et al.* 1992 for general background).

Let us say a little about natural climate change linked to the Earth's orbital geometry, otherwise known as the Milankovich cycles. As the Earth orbits the Sun, there are variations in those orbital parameters which change the amount of radiation received at the top of the Earth's atmosphere. It is thought that these variations in the amount of received radiation are responsible for ice ages. The three parameters of the Earth's orbit of concern are the tilt of the Earth's axis, the eccentricity of the Earth's orbit and precession of the equinoxes. We consider each of these in turn. The tilt, or obliquity, which is the tilt of the Earth's axis relative to the plane of the orbit, varies and this affects the amount of radiation that the northern and southern hemispheres receive in the winter and summer. The absolute amount of radiation is unaffected, but the relative amounts received in the southern and northern hemispheres in their respective summers and winters are affected by changes in the tilt of the Earth's axis. The eccentricity of the Earth's orbit also affects the amount of radiation which is received at the top of the atmosphere. If the orbit is almost circular then the amount of radiation that is received as the Earth moves around the Sun is close to constant throughout the year. But if the orbit is in an eccentric phase then there will be strong variation in the amount of radiation which is received at different points along the orbit. The last of these three factors is the precession of the equinoxes. The positions of the equinoxes (20 March and 22 September) and of the solstices (21 June and 21 December) shift slowly around the Earth's elliptical orbit and complete one full cycle about every 22,000 years (see Figure 8.1). Some 11,000 years ago the winter solstice occurred near the apogee of the orbit (i.e. when the Earth is at its furthest point from the Sun). Today the winter solstice occurs near the perigee (i.e. when the Earth is at its closest point to the Sun). This steady change in the orbital positions of the equinoxes (and solstices) alters the amount of radiation received throughout the year, because the Earth's orbit is eccentric. The difference between the amounts of radiation received in summer and winter is affected by this precession of the equinoxes (and solstices).

The upper part of Figure 8.2 shows the changes associated with each of the three parameters separately. The graphs are plotted from 200,000 years before present to 200,000 years after present. The longest characteristic timescales are associated with the eccentricity of the Earth's orbit, with a cycle length of approximately 100,000 years. The obliquity (or tilt) has a much shorter characteristic timescale, while the precession has the shortest timescale of approximately 22,000 years. The temporal variation in the parameters of the orbit of the Earth around the Sun is well understood and so it is possible to predict the harmonic variation in insolation as a result of this variation in the Earth's orbit. The lower part of Figure 8.2 shows the characteristic timescales of the variation in insolation at 65°N because of the Earth's orbital variation.

It is possible to correlate variation in solar insolation with the occurrence of ice ages, with some success. However, the forcing which is generated by this variation in radiation is insufficient on in its own to create ice ages. There must be some amplifying factor. Much research is being undertaken to pinpoint what the amplifying factor might be. It appears to be widely accepted that it lies within the carbon cycle, but where exactly within the carbon cycle is not entirely clear. There are close similarities

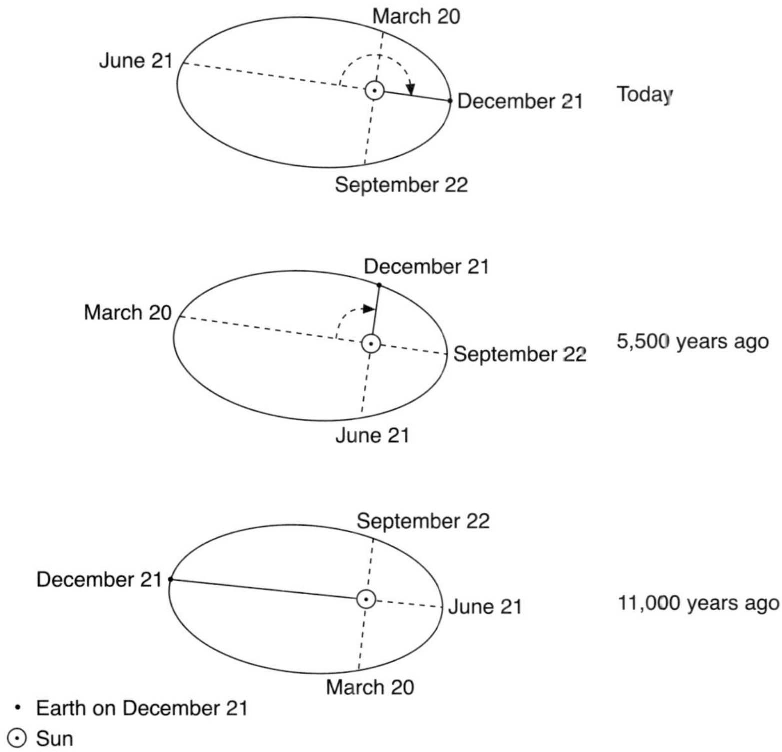

Figure 8.1. The precession of the equinoxes (from Goodness *et al.* 1992)

observed between the time series of global (or northern hemisphere) ice volumes and of atmospheric concentration of CO_2. However, the two time series are not exactly in phase and, depending upon which palaeoclimatic record is examined, it is not clear whether the CO_2 concentration is leading the global ice volume in time, or whether the global ice volume is leading the CO_2 concentration. This is an unresolved issue in an area of research in climate-change science which is extremely active at the present time.

To summarize, then, climate change as a result of natural forcing factors can occur over very long timescales. An example is the response to orbital variation, leading to substantial changes in global temperature and ice volumes. Climate change because of human activity (or anthropogenic climate change) is the focus of this chapter and is superimposed on longer term natural fluctuations.

8.2 BACKGROUND TO ANTHROPROGENIC CLIMATE CHANGE

So far we have considered climate change which is forced by factors entirely external to the Earth. But when considering anthropogenic climate change, interest is focused

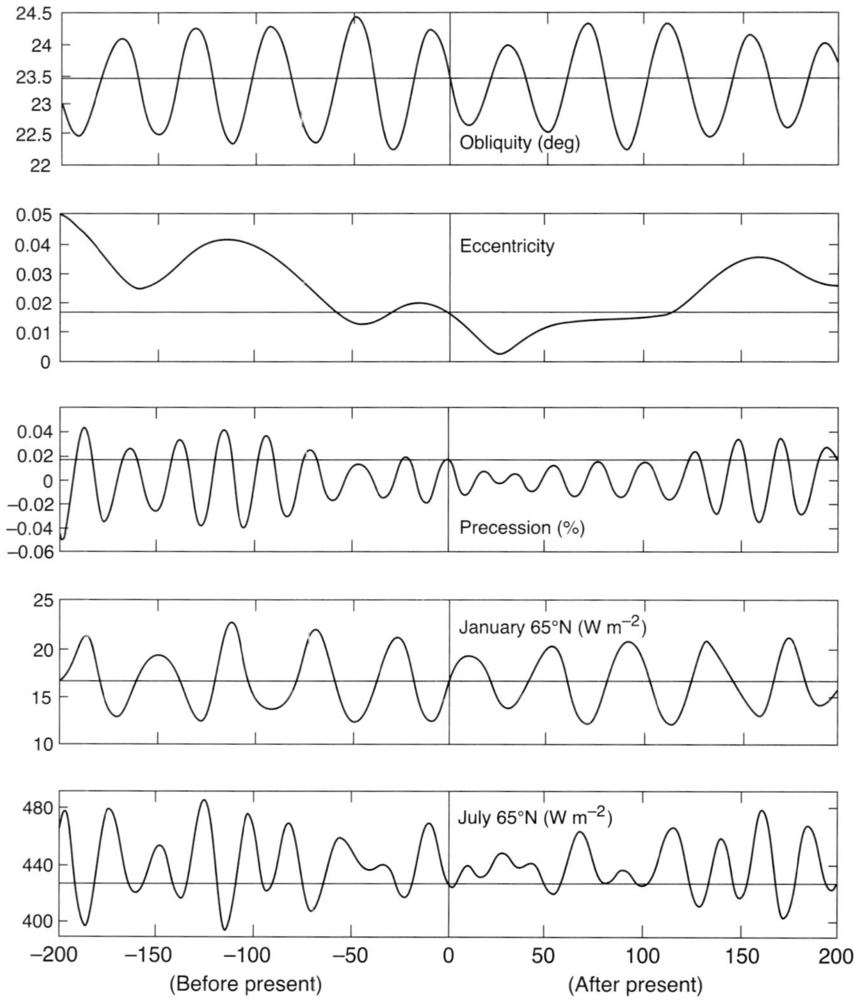

Figure 8.2. Variations in the Earth's orbit and the associated top of the atmosphere insolation (from Goodness *et al.* 1992).

on what is happening within the atmosphere. And, within the atmosphere, we are not concerned with stratospheric variation or the ozone hole, but with what is happening in the troposphere.

Figure 8.3 shows the behaviour of radiation which arrives at the top of the atmosphere, and what happens to it as it passes through the atmosphere. Incoming solar radiation is short wave, whereas radiation emitted from the surface of the Earth is long wave; this is because the wavelength is inversely proportional to the temperature of the emitting body. The incoming radiation passes through the atmosphere, but part of it is reflected by clouds, part is backscattered by air and part is absorbed by water vapour and by clouds. On reaching the surface,

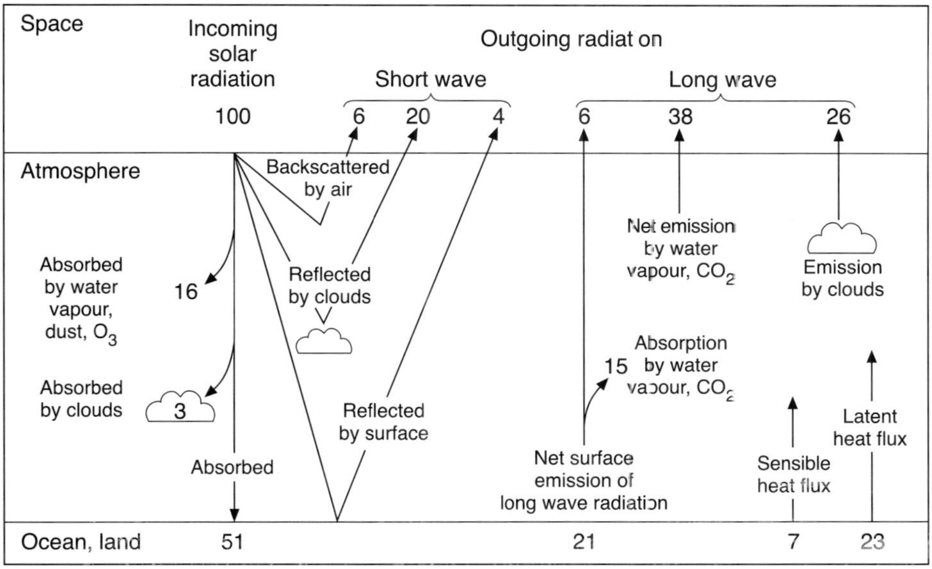

Figure 8.3. Mean annual radiation and heat balance of the atmosphere relative to 100 units of incoming solar radiation (from Barry and Chorley 1998).

a further part is immediately reflected back in the form of short-wave radiation. For every 100 units received at the top of the Earth's atmosphere, only 51 (typically) reach the surface. Then we have to consider what happens to the 51 units which arrive at the bottom of the troposphere. Some of this radiation is returned to the atmosphere in the form of sensible heat flux (i.e. by conduction and convection), some is returned in the form of latent heat flux (i.e. evaporation followed by condensation) and some radiation is directly emitted from the surface as long-wave radiation. Of that long-wave radiation some is absorbed by gases in the atmosphere, especially water vapour and CO_2, and some is absorbed by clouds.

We use the term radiation balance to describe the balance between the upward and downward propagation of radiation. For a general review see Barry and Chorley (1998). Human activities have disturbed the typical radiation balance described above. Several of the gases in the atmosphere behave in a way that is analogous to the glass of a greenhouse; without this natural greenhouse effect the Earth would be a very much colder place and human life, as we know it, would be impossible. The main greenhouse gases are CO_2, methane, some of the oxides of nitrogen and especially water vapour. Human activities, principally in the form of industrial processes but also because of changes in agricultural practice and especially deforestation, are adding CO_2 and other greenhouse gases to the atmosphere. The greenhouse gases are largely transparent to short-wave radiation coming from the Sun, and so they have virtually no effect on incoming radiation, but they are opaque to some long-wave radiation emitted by the Earth. Thus, by adding greenhouse gases to the atmosphere we reduce the amount of long-wave radiation which is radiated from the surface of the Earth back into space. The effect of this is to warm the surface and the lower troposphere. Thus human activities are causing global warming, or what is called the enhanced greenhouse effect, which arises from the emission of additional greenhouse gases into the atmosphere. What we are concerned with here is actually the enhanced greenhouse effect.

Some characteristics of the main components of anthropogenic forcing are shown in Table 8.1. This comes from the Intergovernmental Panel on Climate Change (IPCC) Working Group 1 report of 1995 (Houghton *et al.* 1996). The IPCC is largely responsible for bringing together in one place the research which is done on climate change in various institutions and various nations around the world. The left-hand column of this table lists the principal greenhouse gases and shows their 1994 concentrations. For CO_2 the present concentration is 358 parts per million by volume and the pre-industrial concentration was about 270 parts per million by volume. So we have already seen an increase of around 80 to 90 parts per million by volume. The generally accepted date for CO_2 concentration in the atmosphere to double is towards the end of the new century (i.e. somewhere in the region of 2090). This figure depends, of course, very much on how economies and human societies evolve in the future. The atmospheric lifetime for CO_2 is long, of the order of 50 to 200 years, and the current rate of increase is about 0.04% per annum or 1.5 parts per million by volume. Another very important greenhouse gas is methane. This has a shorter residence lifetime in the atmosphere than CO_2. However, its current rate of increase is higher than that of CO_2. Sulphates (see the last row of the table) are important because sulphate aerosols have an anti-greenhouse effect (i.e. they tend to lead to cooling because they are particulate matter and so reflect short-wave radiation). Increasingly, therefore, climate modellers are trying to consider the effects of both greenhouse gases and sulphate aerosols in their experiments. An important characteristic of sulphate aerosols is that they have a very short residence time in the atmosphere, of the order of days to weeks. This is in contrast to the long residence times of greenhouse gases. A consequence of the short residence times is that sulphates tend to have concentrations which are very variable spatially. Sulphates come mainly from industrial processes and so high concentrations of sulphate aerosols are found in industrial regions; these are mainly in the northern hemisphere and over land areas. On the other hand, increases in greenhouse gases, because of their long atmospheric residence times, are evenly spread around the globe and there is much less tendency to excessive concentration over the generating regions. So more localized sulphates have a very different effect on the radiation balance from globally well-distributed greenhouse gases. This is an important fact to note when we come to look at climate-model information.

The principal anthropogenic sources of CO_2 in the atmosphere are fossil-fuel combustion, cement production and other industrial processes. Measuring in terms of the carbon content of CO_2, the carbon budget of the atmosphere is made up from 5.5 Gt of carbon per year coming from fossil-fuel combustion and from cement production principally, and 1.6 Gt of carbon per year contributed by changes in tropical land use (although the error bars on these estimates are wide). Altogether this means that (7.1 ± 1.1) Gt of carbon per year are being added to the atmosphere.

Table 8.1. Present-day components of anthropogenic forcing from IPCC '95 (Houghton *et al.* 1996).

Gas/substance	Direct radiative forcing ($W\,m^2$) (best estimate/range)	1994 concentration	Atmospheric lifetime (years)	Current rate increase (per year)	Notes
CO_2	+1.56	358 ppmv	50–200	0.4% (1.5 ppmv)	0.6 ppmv in 1991/1992
CH_4	+0.47	1,720 ppbv	12	0.6%	
N_2O	+0.14	312 ppbv	120	0.25%	
Halocarbons	+0.25 (total forcing from CFCs/HCFCs)				Net forcing = 0.1 $W\,m^2$ because of O_3 feedback effects
Typical examples:					
CFCs: CFC-11		268 pptv	50	0	
CFC substitute: CFC-22		110 pptv	12	5%	
PFCs: CF_4	~0.01 (total PFCs)	72 pptv	50,000	2%	
O_3 tropospheric	+0.4 (0.2 to 0.6)		Days–weeks	Increase rate has slowed/stopped	
stratospheric	−0.1 (−0.05 to −0.2)				
Aerosols:	0.4 (0.2 to −0.8)		Days–weeks		Balance of direct aerosol forcing = −0.5 $W\,m^2$ (−0.25 to −1.0)
Sulphates (direct effect)	+0.1 (0.03 to 0.3)				
Soot	−0.2 (−0.07 to −0.6)				
Biomass burning	−0.8 (0 to −1.5)				
Clouds (indirect effect)					

The destination of this carbon is not totally clear. There is a lot of research literature on the carbon cycle, but the best estimate of IPCC '95 (see table 2 of the technical summary in Houghton *et al.* 1996) is that 3.3 Gt go into the atmosphere and stay there with a residence time of 50 to 200 years, 2 Gt are taken up by the oceans and (0.5 ± 0.5) Gt (i.e. anything between 0 and 1 Gt) go into northern-hemisphere forest regrowth. Adding all these contributions fails to match the estimated carbon input to the atmosphere and there is a missing sink of 1.3 Gt of carbon per year. However, the error bars on that missing sink are extremely wide (± 1.5 Gt) so that in fact there may be no missing sink at all. A considerable amount of effort has been expended in trying to trace where this inferred missing sink of carbon might be and various theories exist. Ocean sedimentation (e.g. rapid sedimentation in areas such as off the Nile Delta in the Mediterranean) is one possible theory at the present time.

8.2.1 Present-day climate change: myth or reality?

As part of the effort to detect climate change in present-day observations, the temperature record shown in Figure 8.4 has been compiled. It currently runs from 1856 to 1999, and is being continually up-dated and improved (Jones *et al.* 1999). Because of the addition of greenhouse gases to the atmosphere, which in turn is due to human activities such as industrialization and deforestation, we would expect an increase in temperature in the lower atmosphere.

The record is compiled from both land and marine observations of temperature. The land component is based on around 2,000 records, all of which have been assessed for homogeneity, including urbanization effects. The marine data are from ships and buoys, with corrections for changes in management practice over time. The following points about the global temperature record can be made:

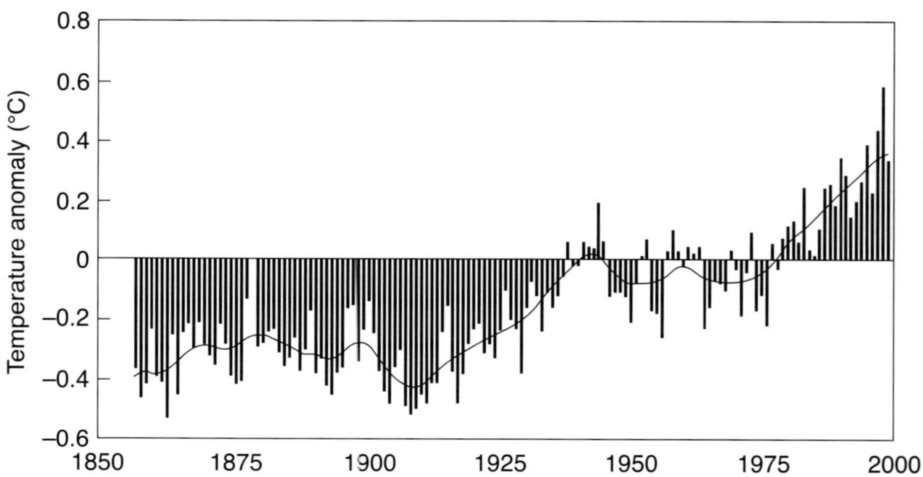

Figure 8.4. The annual combined land-surface and sea-surface temperature record for 1860–1999, relative to the 1961–1990 period (Jones *et al.* 1999).

- the two periods of greatest warming were 1925–1944 and 1978–1997, when global temperatures rose by 0.37°C and 0.32°C respectively;
- night minimum temperatures have increased faster than daytime maximum temperatures, so that between 1950 and 1993 the diurnal-temperature range has decreased by 0.08°C per decade.

Recently, some doubt has been cast on the veracity of the surface record which, by comparison with satellite measurements of the temperature of the lower troposphere, has shown a slight cooling since 1979. This is an issue which remains to be resolved (National Research Council 2000). The satellite archive is still inadequate for the detection of trends and also it measures lower tropospheric rather than surface temperatures.

The 1990s were the warmest decade in the global temperature record. The warmest 2 years of the decade were 1997 and 1998, with the latter the warmest at 0.57°C above the 1961–1990 mean. The seven warmest years globally all occurred in the 1990s. They are, in descending order, 1998, 1997, 1995, 1990, 1999, 1991 and 1994.

How does this record sit in the longer term context of climate? The last millennium is the period for which we know most about the pre-industrial past. Information comes from proxy data sources, such as tree-ring densities and widths, ice cores and coral records, etc. These records (Jones *et al.* 1999) suggest that the twentieth century was the warmest of the millennium, and that the amount of warming which has occurred within the century is unprecedented. It is likely that 1998 will prove to have been the warmest year of the millennium (1601 was probably the coolest year).

8.2.2 Future trends

Thus we come to look at what the climate might be as a result of global warming. Three possible estimates are given in Table 8.2. These are with respect to three scenarios of future emissions of greenhouse gases. IS92e, IS92a and IS92c are different scenarios produced by the IPCC (see Table 1.1). IS92a is widely known as the best-guess emissions scenario. It assumes that we shall continue more or less as we do at the present time with respect to fossil-fuel consumption and that rates of population growth and economic development will be on similar scales or slightly lower than at the present time. It also assumes that there will be a medium uptake of green technologies, renewable energy and energy conservation. The high scenario IS92e assumes that the population will increase at a rate similar to the present day and that economic growth rates will be high. It is also assumed that countries like India and China, which are poised for economic take-off, will indeed take off and that there will be a very high use of fossil fuels and a low take-up of green technologies. At the other end is what may be termed the optimistic scenario (IS92c), which assumes low population growth, low economic growth and high take-up of green technologies. Table 8.2 shows climate predictions of these three different scenarios for three future decades. These estimates are based on a one-dimensional energy-balance model. Although this is far less sophisticated than the three-dimensional

Table 8.2. Climate predictions from three climate-change scenarios (Hulme 1999).

	High (IS92e)	Medium (IS92a)	Low (IS92c)
Decade of the 2020s			
temperature (°C)	1.04	0.96	0.84
sea level (cm)	21	20	19
CO_2 (ppmv)	452	434	413
Decade of the 2050s			
temperature (°C)	1.96	1.68	1.27
sea level (cm)	41	38	34
CO_2 (ppmv)	591	528	452
Decade of the 2080s			
temperature (°C)	2.96	2.37	1.50
sea level (cm)	65	58	47
CO_2 (ppmv)	805	637	468

climate models which will be used later to look at regional climate change, this simplicity means that a large number of model simulations can be performed very quickly and the sensitivity of the system to different forcings can realistically be explored. Table 8.2 presents the estimated changes in two quantities: mean global temperature and mean sea level. In the medium scenario (IS92a) there is a global temperature increase of about 1°C by the 2020s, increasing to 1.68°C by the 2050s and to 2.3°C by the 2080s. The corresponding rises in mean sea level predicted in this scenario for these three decades are 20, 38 and 58 cm, respectively. These changes in temperature and sea level are averages and they may seem quite small. An increase in mean sea level of 58 cm by the 2080s may appear insignificant. However, as has been indicated already in Section 1.6.1, this can be very serious. There is the fact that the shape of the surface of the Earth is not completely rigidly fixed for all time and some parts of the land surface may rise or fall over a period of time. There may be relative displacements as a result of slippage at faults, landslides, volcanic eruptions, movements of the major tectonic plates, hydrostatic effects because of variation in the load of ice on the land surface or the extraction of groundwater or oil from beneath the surface, etc. There is also the question of extreme events such as storm surges; an increase of 58 cm in mean sea level will increase the vulnerability to storm surge in low-lying coastal regions. The relationship between changes in the mean and changes at the extreme can be non-linear, with a relatively small change in the mean leading to a much greater change in the occurrence of extremes (Easterling *et al.* 2000). Apart from changes in the vulnerability of low-lying areas to storm surges because of change in the mean sea level, the situation may be made worse if the enhanced greenhouse effect leads to an increase in storminess. Results from climate models with respect to changes in storminess are conflicting, although there is some indication that an increase may be expected over north-western Europe (Ulbrich and Christophe 1999). It would be prudent to make the assumption for risk management

Table 8.3. Contributions (cm) to sea-level change over the last 100 years from modelling and observations (Warrick *et al.* 1996).

Component contributions	Low	Middle	High
Thermal expansion	2	4	7
Glaciers/small ice caps	2	3.5	5
Greenland ice sheet	−4	0	4
Antarctic ice sheet	−14	0	14
Surface/groundwater storage	−5	0.5	7
Total	−19	8	37
Observed	10	18	25

that storminess will increase and, if that is the case, then of course this 58-cm change in sea level could have more serious implications than we might otherwise anticipate.

Table 8.3 lists the principal contributing mechanisms to sea-level change over the last one hundred years (these have already been mentioned briefly in Section 1.6.1). During this time sea level has risen by 10–25 cm (see the summary for policymakers in Houghton et al. 1996). The major contributing factor is thermal expansion which is simply due to higher temperatures. The other principal causes are contributions from glaciers and small ice caps and from the two major global ice sheets on land masses, the Greenland Ice Sheet and the Antarctic Ice Sheet. Finally, there is a small contribution coming from other sources of surface groundwater, such as permafrost-type storages. The contribution of the ice sheets is poorly understood; for example, for the Antarctic ice sheet the estimates of the contribution range from −14 cm to +14 cm. One of the reasons for the very wide error margin is, of course, how little we know about Antarctica and this situation is unlikely to change in the near future, given how very large it is and how very hostile it is to scientific research. If global warming leads to an enhancement of the hydrological cycle, such that precipitation over Antarctica increases, and if that precipitation falls as snow, then accumulation will dominate, leading to a negative contribution to sea-level rise. On the other hand, if temperatures rise sufficiently to cause melting to dominate, this will lead to a positive contribution to sea-level rise. However, apart from difficulties with the models we must also remember that it is not always completely straightforward to make observations of sea level because the place where we are standing is not absolutely stationary in space.

Table 8.4 gives estimates of what is likely to happen in the future up to 2100. It is based on results from fairly simple one-dimensional radiation models (see also Figure 1.7). Table 8.4 compares results from Wigley and Raper (1992) with those from IPCC '95, but both sets of estimates are based on the emissions scenario IS92a. It can be seen that, generally speaking, the estimates from 1995 are lower than the estimates from 1992. However, in both cases thermal expansion is the largest contributor to the change in sea level expected by 2100, followed by glaciers and ice caps and then the Greenland ice sheet. The Antarctic ice sheet is seen as accumulating

Table 8.4. Contributions (cm) to and best estimates of sea-level rise by 2100 (Warrick *et al.* 1996).

	Wigley and Raper (1992) IS92a	Other IPCC '95 state-of-the-art models
Thermal expansion	28	15
Glaciers and ice caps	16	12
Greenland Ice Sheet	6	7
Antarctic Ice Sheet	−1	−7
Best estimate	38	27

rather than melting so that it should make a negative contribution to sea-level rise.

8.3 THREE-DIMENSIONAL CLIMATE MODELS

We now turn to consideration of three-dimensional climate models: how good they are, what their weaknesses are and what they can tell us about future climate change both at the global and regional scale. Figure 8.5 depicts a very typical state-of-the-art three-dimensional coupled ocean–atmosphere model. This is the current version of the Hadley Centre model; it has nineteen layers in the vertical in the atmosphere and twenty in the vertical in the ocean, and a grid resolution of 2.5° latitude by 3.75° longitude, which is very typical for grid resolution of a global model This is a state-of-the-art model (Gordon *et al.* 2000). Horizontal resolution is an important factor in determining the performance of a model, because many of the atmospheric processes which the models are attempting to simulate take place over areas which are smaller than the grid resolution of the model. Taking the example of Mediterranean countries, the models are unable to simulate small-scale convective processes which are an important source of rainfall. Clouds and cloud processes are poorly modelled by three-dimensional global climate models, because processes within clouds operate at resolutions which are much finer than the horizontal resolution of the models. The models also have difficulty simulating cloud in the vertical dimension. The various types of cloud, stratus clouds, cumulus clouds, alto-cumulus clouds and cirrus clouds, concentrate at different heights above the surface of the Earth, and although the Hadley Centre model has nineteen atmospheric layers in the vertical, it is unable to represent properly the full structure of cloud climatology.

Another problem to which a great deal of effort is being devoted at the present time is land-surface processes. Most models are coupled ocean–atmosphere models and the interaction between the atmosphere and the ocean surface is well handled. But the modelling of the interaction at the interface between the atmosphere and the land has until recently been much more crude. Processes such as evapotranspiration, and hence latent fluxes, have often been very crudely modelled.

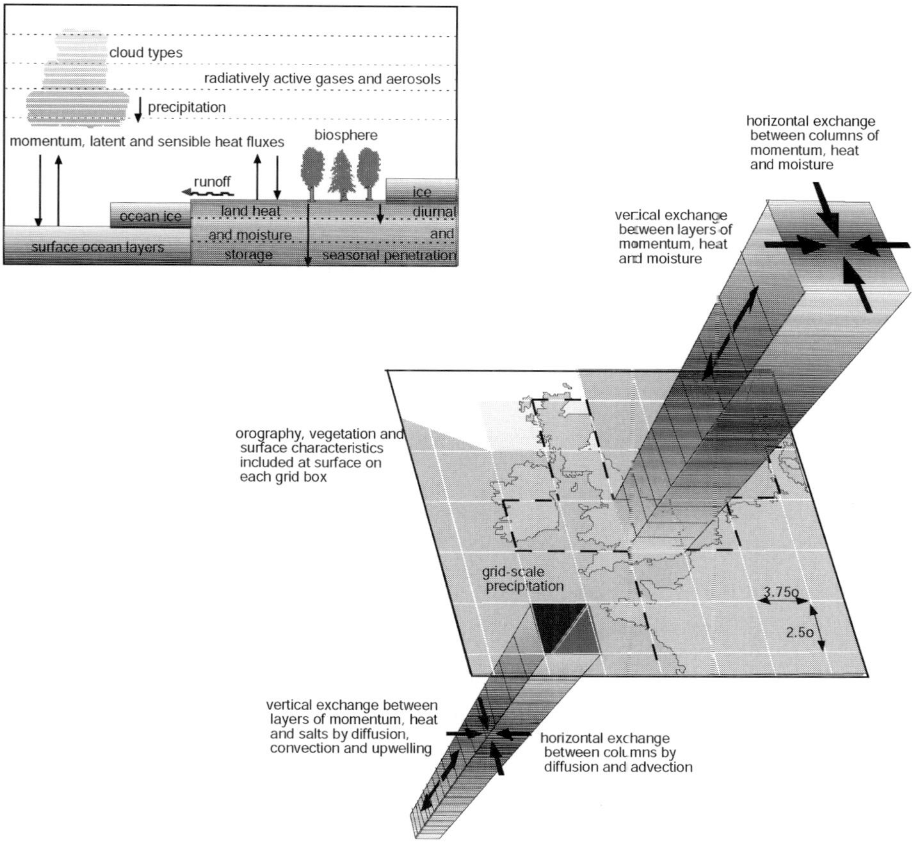

Figure 8.5. The configuration of a typical coupled ocean-atmosphere General Circulation Model (courtesy David Viner, Climatic Research Unit, University of East Anglia).

Feedback processes in climate models are extremely important. Water-vapour feedback in particular is a major cause of the temperature increases observed in climate models and hence it is very important that feedback should be accurately simulated. However, feedback, particularly that related to land-surface processes, is not well handled in climate models and this is an area of intensive research effort, along with cloud climatologies, at the present time.

Figure 8.6 shows just one feedback mechanism (from Charney 1975). This was a classic paper, still widely cited, showing a chain of events starting with a regional reduction in rainfall caused by climate change, which then feeds back on itself leading in turn to lower and lower rainfall. This is an example of positive feedback. This type of feedback is likely to apply to a semi-arid region such as the Sahel. Climate models are not able to properly handle feedback involving land-surface processes because these processes are only crudely modelled. Figure 8.7 shows some other feedbacks involving land-surface processes and which the current generation of climate models would find difficult to simulate. The purpose

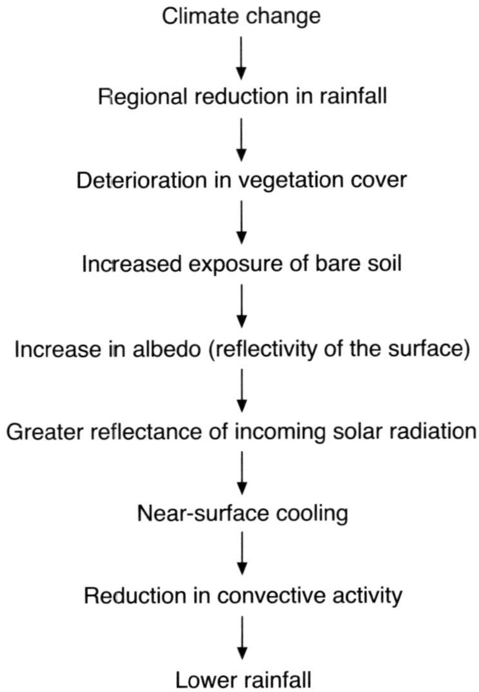

Figure 8.6. An example of positive feedback in semi-arid regions (Charney 1975).

of large climate experiments such as GEWEX (Global Energy and Water Cycle Experiment) is to use surface observations and remotely sensed data to understand these feedbacks properly, so that their simulation in climate models can be improved (Lawford 1999).

How do these models work? Typically, because of computing restrictions, 10 years ago they would have been run in what is called the equilibrium response mode. This means that a model would be run for two different periods, each with a fixed concentration of greenhouse gases. First, it would be run for the present day, or for the pre-industrial era, for a period of, say, 10 years in order to determine the average climate for this initial CO_2 concentration. Then the model would be run a second time with double the original atmospheric concentration of CO_2 in order to determine a new 10-year average. This is known as the equilibrium response mode because the model is always run until the climate is stable for the prescribed amount of atmospheric CO_2. However, this is quite an artificial way to consider the problem because natural variability is damped and because CO_2 concentrations are actually increasing all the time.

Today, with enhanced computing capacity, transient response experiments are the commonest form of model simulation. These typically begin in the mid-eighteenth century with historical atmospheric concentrations of the greenhouse

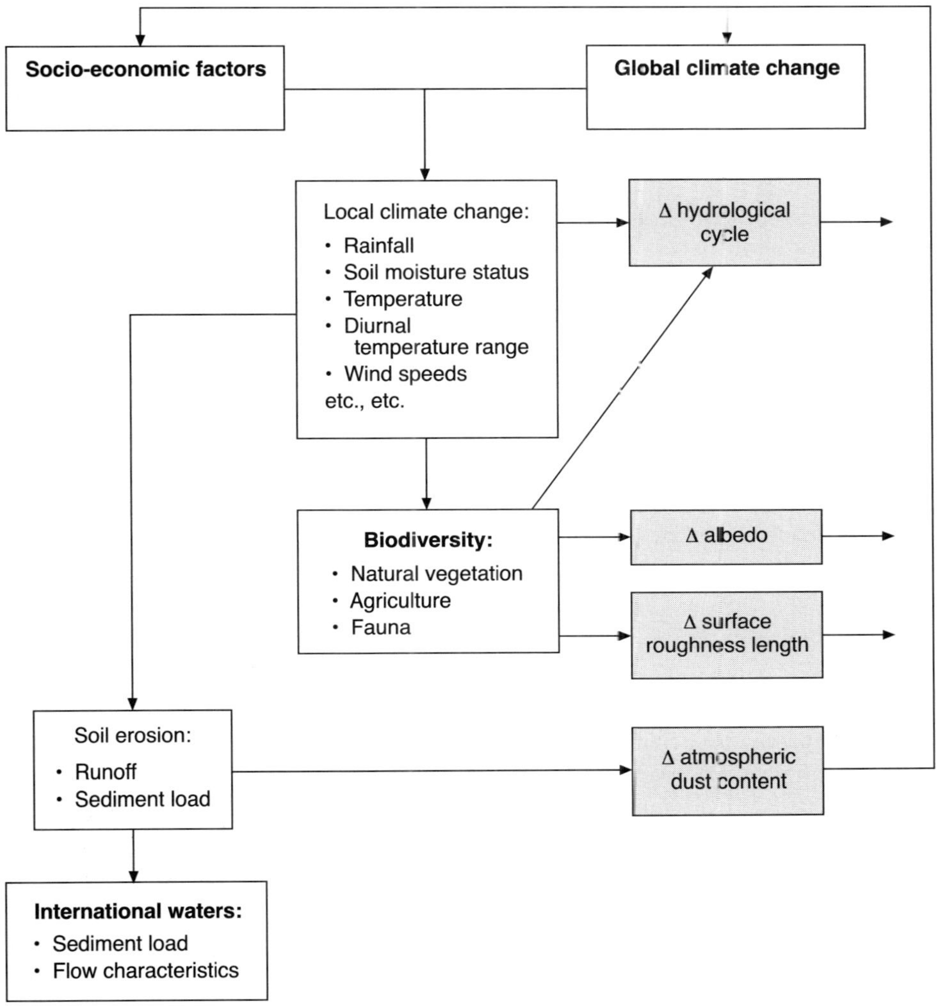

Figure 8.7. Examples of feedbacks related to land-surface processes. Feedbacks are shown in shaded boxes.

gases and run through to the present day. Then from the present day into the future they will assume some scenario of CO_2 concentration such as IS92a or, very commonly, a compound increase of 1 per cent per annum in concentration of greenhouse gases (the two being very similar). Typically such a model would be run one hundred years into the future. Figure 8.8 shows the results of a typical model run based on the Hadley Centre global climate model (HadCM2). This is from 1860 through to 2100 and shows the evolution of northern Mediterranean precipitation in

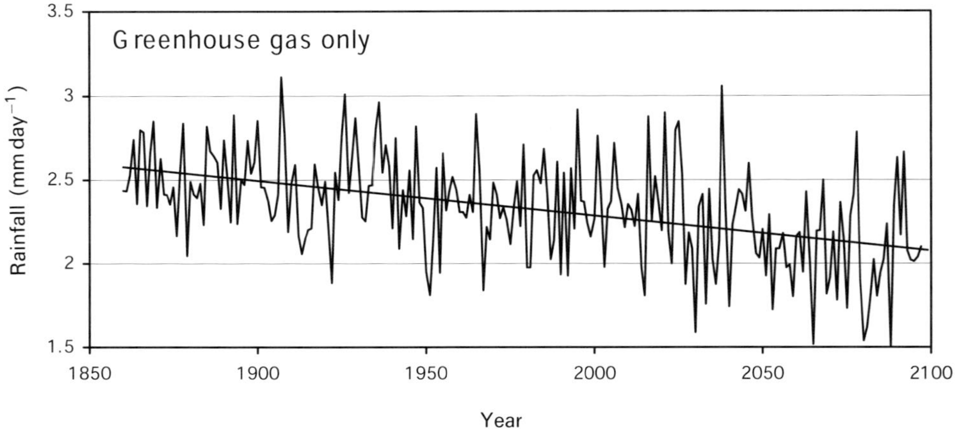

Figure 8.8. HadCM2 simulated rainfall (mm/day) for the Northern Mediterranean hydrological year (September–August).

that period, forced by historical observations of greenhouse-gas concentrations in the atmosphere up to the present day followed by an increase of 1 per cent per annum for the future.

Figure 8.9 shows mean global temperatures for two experiments using HadCM2 and compares these with observations. If we just take increasing concentrations of greenhouse gases into account, then the figure shows that simulated global temperatures are too high compared with observations. When sulphate aerosols are also included, the resulting temperatures are in much better agreement with observations. Sulphate aerosols have an anti-greenhouse effect and their inclusion in climate models is a relatively recent development. When we look at the spatial patterns of warming in climate models, a number of features can be observed (Hulme et al. 1999):

- Warming is greatest over land areas. This is not unexpected, because the thermal capacity of the land surface is much less than the thermal capacity of the ocean.
- There is an increased warming at high latitudes in the northern hemisphere, which is mainly due to ice albedo feedback (i.e. once the ice cover over the north pole begins to melt, then the albedo changes very rapidly from the very high reflectivity for ice to the rather low reflectivity for open water). When that happens, much more radiation is absorbed and therefore there is a very rapid increase in temperature.

Thus we have a concentration of warming over the continents and in the high northern latitudes, but really rather small amounts of warming over the southern ocean.

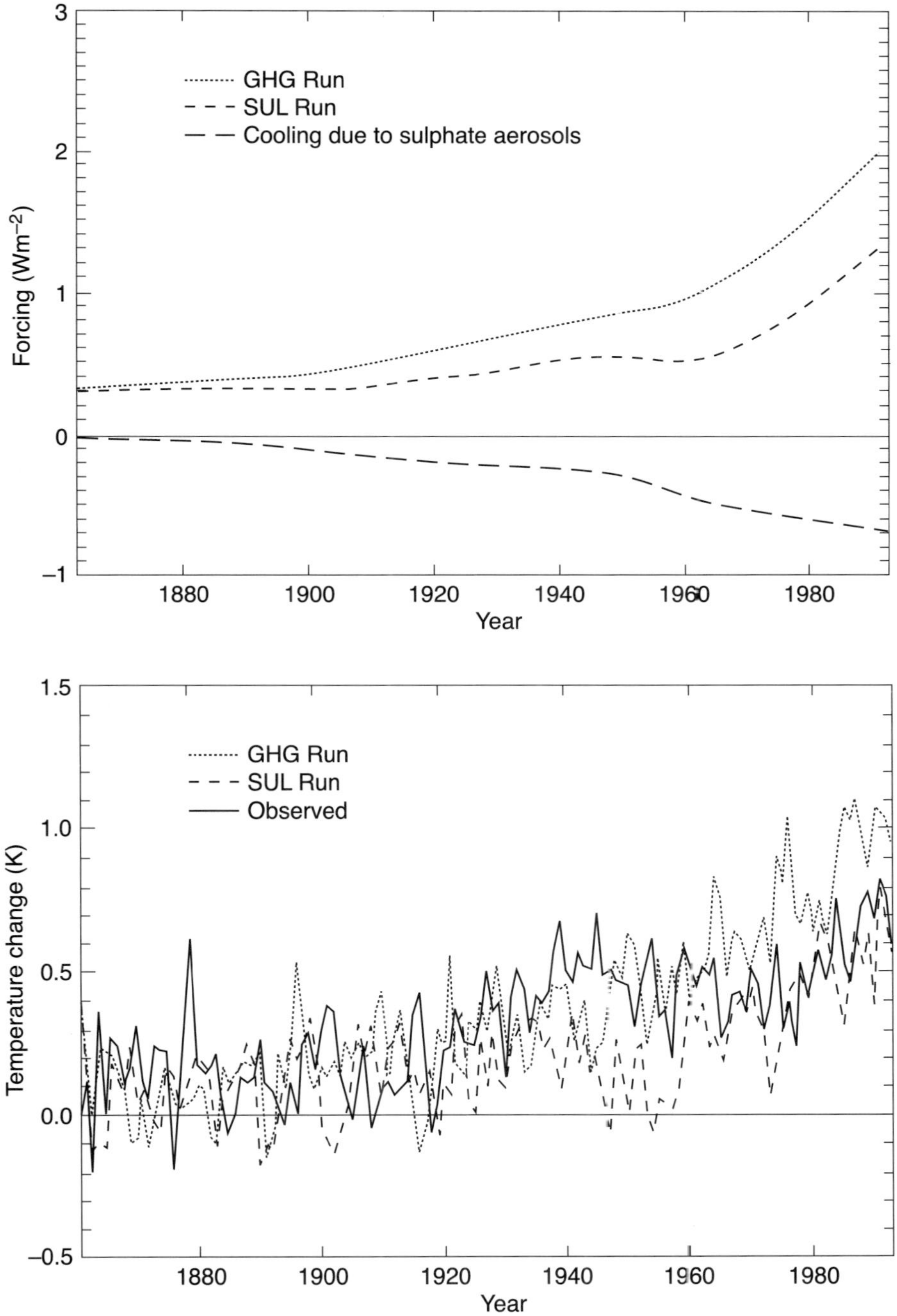

Figure 8.9. Changes in global annual mean surface air temperature (lower diagram) in two experiments with the UK Hadley Centre climate model HadCM2. GHG = greenhouse gases only; SUL = greenhouse gases + sulphate aerosols. Upper diagram shows the respective forcings (from Johns *et al.* 1997).

By comparison with temperature, precipitation is much more poorly simulated by climate models. This is partly because precipitation is highly variable in space, which in turn is related to the fact that precipitation-generating processes such as convection operate at subgrid scales. That being the case, there is less confidence in model prediction of rainfall than in temperature predictions. In general, models predict that rainfall will increase globally as a result of the enhanced greenhouse effect, in response to a more active hydrological cycle.

8.4 CONCLUSIONS

The IPCC '95 report from Working Group 1 concluded that 'the balance of evidence suggests a discernible human influence on the climate system'. In this chapter, an attempt has been made to set the context for these climate changes. Generally speaking, the changes discussed here relate to the best estimate of the future state of mean climate and sea level. We have not looked in any depth at issues of uncertainty. And yet, large error bars are associated with all the estimates of future climate presented here. These uncertainties concern, for example:

1. The likely future patterns of emissions of greenhouse gases – enormous uncertainties surround our understanding of the controlling factors, which are to do with economic and technological development, and political willingness (or otherwise) to limit emissions.
2. Biogeochemical cycling – the amount and rate of partitioning of greenhouse gases between the various sinks and the atmosphere.

These two together determine the atmospheric concentration of greenhouse gases. The sensitivity of the climate to the forcing from these increased concentrations is a further source of uncertainty. Sensitivity is generally defined as the global equilibrium temperature increase resulting from a doubling of CO_2 concentrations, and is estimated to be in the range 1.5 to 4.5°C, with a best guess of 2.5°C. Finally, uncertainties in the prediction of future climates arise from inadequacies in the climate models, for example:

1. Inability to properly model subgrid scale processes (e.g. related to clouds).
2. Weaknesses in modelling feedback, especially that related to land-surface processes.

Related to uncertainty is the possibility of surprises. These are outcomes of anthropogenic climate change which are considered to have a very low probability of occurrence, but which could have very severe impacts. By their very nature, we cannot describe the full range of possible surprises. They include the collapse of the west Antarctic ice sheet, which would cause a sudden and catastrophic increase in sea level, and the collapse of the thermohaline circulation in the North Atlantic. This latter event would imply the disappearance of the Gulf Stream, bringing much colder conditions to NW Europe.

All these issues are central to the work of the IPCC. The latest publications of the IPCC, the Third Assessment Reports, are due in 2001. For a comprehensive overview of the subject, the technical summaries of these reports are required reading.

9

Cooling off on global warming – the continuing debate

R. A. Vaughan

9.1 INTRODUCTION

The existence of global warming is not disputed, nor is the increase in production of greenhouse gases, notably CO_2 (carbon dioxide), over the past century. What may, perhaps, be in dispute is whether anthropogenic factors play a significant part in these phenomena. In the 1980s global warming was high on the political agenda and scary scenarios were driving politicians to make unwise political decisions on, for example, limiting the use of fossil fuels.

> Global warming is a political issue. Information provided by scientists is used to inform policy. Decisions about whether to take action and what should be done are taken in the public arena. Such decisions are political; they are subject to international diplomacy and the democratic process. Public-choice theory recognizes that participants in this process cannot help but bring with them their own private aims and incentives. This is an unavoidable, but not unassailable, problem except when it is forgotten; when it is assumed that political action is altruistic. In this way, policies have been adopted in the name of averting damage to the planet from global warming which will not have the desired effect, even if society complies fully, because the aims of those influencing and deciding policy were not those stated.

Thus begins the Foreword to *The Global Warming Debate* (Emsley 1996). This is one of a series of books published to 'put the other side of the case and to refute current misconceptions concerning the existence of global warming'. The first was entitled *Global Warming – Apocalypse or Hot Air?* (Bate and Morris 1994). The authors were not scientists but members of the Institute of Economic Affairs (IEA) Environment Unit, which was set up in 1993 to apply market analysis to environmental problems and to bring the results of that work to the attention of the general public. It should be pointed out that this book reflects its authors' own views as the IEA claims to have no corporate view. The authors set out to expose what they call the shoddiness of the apocalyptic predictions, and to try to 'breach the barriers to the dissemination of good news and the media's natural preference for reporting only the disasters'.

The current political interest in global warming probably stems from the Rio Summit which took place in 1992 and has been fuelled by the reports of the Intergovernmental Panel for Climate Change (IPCC) (see Chapters 1 and 2). The first reports (Houghton et al. 1990, 1992), which were based on the predictions of rather crude models, predicted a warming of between 1.5 and 4.5°C by 2050. By 1995, using slightly more sophisticated models (i.e. including such things as the presence of sulphate aerosols), this had fallen to a predicted warming of between 1.0 and 3.5°C by 2100 (Houghton *et al.* 1996). Such temperature rises would be expected to produce rises in sea level of between 15 cm and 1 m.

Anthropogenic emissions of greenhouse gases are suggested to be the main cause of global warming, and politically and economically unpopular measures to reduce or at least to stem them have been proposed (the so-called protocols) at a number of summit meetings (see Chapter 2). This cause must be seen in context with other natural causes of warming, particularly solar activity, and it is interesting to note the change in the tone of the IPCC over the last 10 years as their models have been improved – in the 1990–1992 reports they were talking about a doubling of the CO_2 concentrations by 2100 resulting in a warming of 2°C. In 1996 the conclusion stated 'The balance of evidence suggests a discernible human influence on global climate through emissions of carbon dioxide and other greenhouse gases'. But now the 1999 Draft Report of the IPCC contains scenarios in which even a doubling of economic activity does not increase emissions and goes on to say that the future is so inherently unpredictable that 'views will differ on which of the scenarios could be more likely'. Paul Stalpman of the US Environmental Protection Agency appears even to dismiss the need to enquire whether the Earth is warming. He told the 1999 Earth Technologies Forum 'the real debate about climate change should not be about whether it exists or not, but what to do about it'.

Global warming now seems to be slipping down the agenda for both politicians and the general public, as the gloomy apocalyptic predictions of temperature rises and sea-level rises have failed to materialize and the enormous economic and social costs of international intervention for implementing the social and technological changes suggested to 'stem the tide' are seen to be unpalatable or even dangerous. Much publicity has been given to the apocalyptic agenda, but little attention has been paid to the other side of the story. The danger is that the overreaction of the early 1990s to the apocalyptic reports may have backfired and will now produce a sense of complacency. What is certain, though, is the fact that the alarms have achieved one thing at least – a great deal of government research funding has been diverted towards the environmental sciences, particularly to buy supercomputers on which to run ever more elaborate models. Perhaps it should be asked whether the millions spent this way might not have been better spent solving real environmental problems by supporting research in biology, ecology and agriculture (Emsley 1996).

This chapter does not set out to make a closely argued case against the doom scenarios but merely to arouse the awareness of the reader to the now sizeable, sceptical science community whose point of view is rarely reported in the media and whose ideas are dismissed out of hand by the Establishment. It will draw widely on points made in the publications by Bate and Morris (1994), Emsley (1996) and Bate (1998) and on the discussions now taking place on the World Wide Web (WWW).

9.2 CONSENSUS OVER GLOBAL WARMING

From reading the newspapers, watching television or listening to politicians there would appear to be no doubt that there are going to be rising temperatures and rising sea levels and that these are being caused by increased pollution of the atmosphere by humankind. Anyone who tries to argue that these events will not occur, at least not on the scales predicted, or that maybe the changes would be welcome, or that humans are not mainly to blame is dismissed as either ignorant or complacent. It would appear that there is consensus even among scientists. However, if you look more closely there is evidence of considerable disagreement – indeed violent disagreement from some eminent scientists – particularly in some heated web-based discussions. As early as 1992, in the run-up to the Rio Summit, Greenpeace carried out a survey of over 400 of the world's leading climatologists and found that only 15 admitted to believing in global warming although relying on it for their employment. It gave very little publicity to this finding! The Leipzig Declaration of 1995, signed by 1,500 scientists from around the world, disputed the IPCC's assertions about man-made global warming.

The world's climate always has been in a state of constant flux. On timescales of days, years, decades, centuries, millennia and tens of millennia, global and regional temperatures, wind and rainfall patterns are changing. Ice ages have come and gone, sea levels have risen and fallen, all without the help of humankind. They have all been the consequence of the combined effects of changes in the Earth's orbit, changes in the composition of the Earth's atmosphere and changes in the output of the Sun. The reasonable stability of the Earth's climate in the face of such perturbations suggest that there is a strong negative feedback mechanism (e.g. increased insolation leads to increased surface heating which in turn leads to increased evaporation and therefore cloud formation which in turn reduces the solar heating of the surface). It is perhaps arrogant for humans to believe that their paltry efforts can have any noticeable long-term effect on the Earth.

One thousand years ago, vineyards were found in the south of England and average temperatures were about 2°C higher than they are today and temperatures were warm enough to allow the colonization of Greenland. By 1500, the northern hemisphere was in the grip of a little ice age and the temperatures were about 1°C lower than at present. Temperature records since the beginning of the twentieth century have shown a net increase of about 0.5°C in the first 40 years followed by a slow fall until about 1980 (superimposed on these trends are short-term fluctuations of 0.2–0.3°C). In 1975 the editorial in *Science* magazine, quoting C. C. Wallen of the WMO (World Meteorological Organization), said 'The cooling since 1940 has been large enough and consistent enough that it will not be soon reversed, and we are unlikely to quickly regain the "very extraordinary period of warmth" that preceded it.' Yet little more than 15 years later, the IPCC was set up to investigate global warming!

Let us look first at this concept of 'consensus'. There are few certainties in climatology, but there are several features which few dispute. One is the existence of the 'greenhouse effect'. The Earth's atmosphere certainly does keep the surface about 30°C higher than it otherwise would be and makes life as we know it possible. Second, there does seem to be an upward trend in 'global' temperatures, however that is defined. Third, humankind's activities have led to an increase over the last two centuries of between a quarter and a third in the concentration of atmospheric CO_2. What is in dispute is the interconnection between these factors and the magnitude of the effects. Do increased levels of CO_2 necessarily increase global temperatures, or is the converse likely to be true? Is the warming which is detected due solely to humankind's activities or is it natural? Will warming raise the sea level or lower it? Is warming necessarily a bad thing?

The climate is a coupled, non-linear dynamical system. It is such a complex system that it is not surprising that different inputs to different models produce very different answers. Early zero-dimensional models assumed the Earth to be a point and simply analysed net flows of radiation. These suggested that a doubling of CO_2 would lead to a 5°C increase in temperature, but there was no way to account for regional impacts which are crucial. One-dimensional models consider the vertical distribution of the components of the atmosphere and produce not dissimilar results. Three-dimensional models, considering land–atmosphere interactions over, say, a 500-km grid, should give better results, while including the effects of the ocean and using finer grids should improve things even more. But this is now such a complex problem that it is not surprising that even to get approximate answers has required the advent of super-computers.

The models considered by the IPCC in its first report (Houghton *et al.* 1990) were rather crude and computing power was limited by today's standards. Simulations took no account of such key factors as oceans, polar ice or clouds and predicted temperature rises of 5.2°C by 2100. This was toned down in the 1992 report, but the gloom scenario was still being spread by the uncritical supporters of the global-warming hypothesis. After a few more years of criticism, the effect of sulphate aerosols was included and the latest calculations suggest a much-reduced temperature rise over the next century of as little as 0.5°C, which is about the size of recent natural fluctuations. There have been suggestions that opponents of the warming hypothesis were deliberately excluded from the inner workings of the IPCC and that the political demand for consensus stifled public debate. For detailed criticism of the politics of the IPCC see, for example, Singer (1996) and Boehmer-Christiansen (1996). Two thought-provoking quotations will finish this section, one from Bate and Morris (1994) on the workings of the IPCC: 'When confronted with this type of data ... the response by ardent global-warming advocates is typified by Dr Stephen Schneider – "looking at every bump and wiggle of the record is a waste of time ... so I don't set very much store in looking at the direct evidence"' and the other from Dr Chris Folland of the UK Meteorological Office and a contributer to the IPCC policymaker's summary, 'who went even further by saying "the data don't matter ...besides, we [the UN] are not basing our recommendations (for immediate reductions in CO_2 emissions) upon the data; we're basing them upon the climate models"'!

9.3 MEASUREMENT PROBLEMS

Predictions of global warming have been based mainly on computer simulation, and their magnitude has decreased steadily as the models have become more refined. Even so, even these reduced predictions have not been borne out by scientific evidence. For example, models predict that a 50 per cent increase in greenhouse gases should lead to a warming of 2°C. In fact, over the last century there has been an increase of 50%, but measured temperatures have risen by only 0.5°C. But, just as we may criticise the deficiencies in the models, we must also consider the validity of the scientific measurements. It is not so much the accuracy of individual measurements, though, as the way in which they are obtained, extrapolated and interpreted that may cause confusion.

Let us take as an example the measurement of 'global temperature'. First of all, what does it mean? Temperatures vary widely both in space and time. There are vertical variations in temperature as well as differences across the surface of the world. There are short-term differences a a result of diurnal variations as well as longer-term differences as a result of seasonal effects. How do you take an 'average' of these? Do all measurements have the same significance, and, if not, on whose judgement are they weighted? Do all these measurements measure the same thing? Does 'global' temperature have any scientific significance anyway?

The traditional method of measuring temperatures is by using thermometers situated at different stations dotted around the world. Problems associated with this method relate to station inhomogeneity. Historically these stations tended to be densest around the higher populated regions. There were no stations in the whole of Africa prior to 1860, only one from then till 1900 and by 1920 there were still only nineteen. Large portions of the land surface were not covered (and still are not covered), and most tended to be in or near urban areas where the heat-island effect is not only significant but has increased with time. Soon *et al.* (1999) present a graph of temperature trends per decade between 1940 and 1996 against the population of the county (in California) over which they were measured. There is a linear increase from about 0.1°C for counties of about 20,000 inhabitants to over 0.5°C for counties having ten million inhabitants. The results from the best-situated rural station in the state gave a corresponding trend of –0.005°C per decade. Rural station coverage has fallen from 20 per cent of the Earth's area in the 1970s to 7 per cent in 1998. These varying factors must present a significant uncertainty in global average calculations. Attempts have been made to factor out some of these variables (e.g. Weber 1996), but the final result will inevitably depend on human frailty.

Another problem arises with the way in which temperatures are measured and their accuracy. Although there are standard heights and conditions which should apply, we cannot be sure how well these criteria are satisfied nor how well equipment is maintained. Then again how are these temperatures obtained? Is a straight average of maximum and minimum taken, or is the average of values at certain times of the day used? Little 'metadata' is available for many of these measurements to enable meaningful comparisons to be made. Different parts of the world have different time series of recorded temperatures, so any inferred historical trends will have inherent regional biases which again must be accounted for.

Sea-surface temperatures have traditionally been measured from opportunistic measurements made from ships. The distribution of such measurements will therefore follow closely the populated shipping lanes and virtually ignore inhospitable regions and seasons. The traditional method of measuring the temperature is to throw a bucket over the side and dip a thermometer in it, but nowadays it is more usual to use the temperature of the intake of cooling water. During the Second World War, because of the hazards of stopping ships to take measurements, it was noticed that there was a sudden jump in the temperature records coinciding with the change in measuring technique. It is tempting to use a blanket adjustment parameter in order to relate the two types of measurement – but how universally valid is it? Sea-surface temperatures can also be measured from space, but because of the 'skin effect' the value measured can differ considerably from the bulk value obtained from ships.

For longer-term trend determinations, indirect methods must be used. On a geological scale, ice-core records have been used. Gases trapped in air bubbles have been analysed for CO_2 content, the changes in concentration of this in the atmosphere being then related to changes in temperature. Not only is this relationship suspect (see Section 9.4) but so is the chemistry; dubious assumptions have to be made about the solubility of CO_2 and its rate of diffusion in ice. On a shorter time scale, indirect methods – such as tree-ring diameters and records of, for example, the date on which certain flowers bloomed in China 2000 years ago, changes in migratory patterns of birds and diary entries of weather observations – can also contribute to the picture of climate trends.

More recently, satellite measurements have been able to provide a real measure of 'global' temperatures. The problem here, though, is that again a different parameter is being measured, namely the temperature of the lower troposphere. However, the fact that satellites can give frequent, uniform global sweeps, oceans included, with no problems because of urban heat-island effects, and accurate to one-hundredth of a degree, suggests that this is the most meaningful and accurate method of all. It is therefore significant that temperatures measured in this way over the last 30 years show small fluctuations about a constant mean value (i.e. the 'warming' is zero!). These measurements are confirmed by those made from radiosonde balloons, in contrast to the small but significant rise inferred from 'terrestrial' measurements.

9.4 SEA LEVEL CHANGES

One predicted manifestation of global warming which the media like to promote is a catastrophic rise in sea level as a result of thermal expansion and the melting of the polar ice caps. Even at the time of writing this section, a headline appeared in *The Times* 'Oceans to rise 20 ft if ice sheets melt' (Nuttall 1999). This article reports Dr Geoff Jenkins of the Hadley Centre as saying that there is enough extra energy in the Earth's system to raise sea levels for many centuries to come, even if big cuts in greenhouse gases were made. Final sea levels would depend on the extent to which the Arctic and Antarctic ice melts. It goes on to report that 'new findings' from the Free University of Brussels indicate that a rise of 'only a few more degrees' in global temperature could trigger a total melt of the Greenland ice sheets. The findings indicate that this melting could add up to 70 cm a century of extra sea-level rise to the 20 cm already forecast from 2100. And this is on top of the 50 cm forecast over the next 100 years as global temperatures climb from 0.5°C above the average now to 3°C above average by 2080. The total loss of the Greenland ice sheet could add 6 m to ocean levels in the next 1,000 years.

Let us look at the fallacy of that argument. The oceans, in fact, act as a huge, heat-energy buffer, and the global climate is primarily governed by the enormous amount of heat stored in the oceans. The total internal energy of the whole ocean is about 1700 times larger than the total internal energy of the whole atmosphere. The large latent heat of melting of the whole cryosphere is the hypothetical energy equivalent to cooling the entire oceans by about 2°C (Segalstad 1996), which is more than 1,000 times the energy required to heat the entire atmosphere by 1°C. It would therefore be impossible to melt the Earth's ice caps and thereby increase sea level just by increasing the heat energy in the atmosphere by a few per cent. Indeed, in July 1997 an article in *Nature* reported a study which showed that moderate warming might thicken the ice shelves rather than melt them by bringing cooler water into contact with the ice shelf. Such a counter-intuitive result highlights the complexity of the ocean–atmosphere–cryosphere system. This report was not mentioned by the press!

The fact that there is no scientific evidence that this catastrophic sea-level rise is happening nor that massive melting of polar ice is yet taking place is another reason why such a scenario should be questioned. It is not clear, because of the regional variability of any global temperature rise, that polar temperatures will necessarily go up to the same extent as elsewhere. One of the predicted consequences of global warming is an increase in extremes of weather conditions in different places at different times of the year. Certainly, ice shelves have been seen to break up only recently, but satellite measurements seem to suggest that there is no significant loss of polar snow and ice, and that the Greenland ice sheet is relatively stable. There appears to be virtually no warming in the polar regions even though this is where computer models predict warming should be the greatest. In 1997 it was reported that the South Pole was experiencing its coldest weather since records began 40 years ago. Some small changes in sea level have been observed, but most of them are due to geological or tectonic changes (e.g. the eastern side of Britain is sinking and the western side is rising, producing the *effective* raising of sea levels along the east coast and the associated erosion and flooding consequences). Information from the WWW (and there is little point in giving the URL here as such items are ephemeral) gives the latest sea-level-change trends in the Pacific Ocean over the last 3 to 5 years, based on tide gauge measurements. These vary from −41.3 mm/yr in the Solomon Islands, to +29.1 mm/yr in Tonga. There is absolutely no consistency between the measurements, but there are certainly more negative than positive values, and the negative values are larger. Taken as a whole, the data are meaningless. A broad average of the eleven values quoted would give a value of −11.2 mm/yr. The average for the Australian coast, which is geologically stable and fronts onto three oceans, according to the National Tidal Facility in Adelaide, is a sea-level rise of only +0.15 mm/yr, compared with the IPCC claim of +6.0 mm/yr.

Satellite measurements of global sea-level variations (Soon et al. 1999) show a global rate of rise of less than 1 mm per year, with short-term fluctuations of up to 10 mm. The trends in rise and fall in various regions have a wide range of about 100 mm per year with most of the globe showing a downward trend. Historical records show no acceleration in sea-level rise in the twentieth century. Another WWW article by John Daly reports his visit to the Island of the Dead in South-Eastern Tasmania and his discovery of an old Ordnance Survey benchmark. In 1841 Captain Sir James Clark Ross, commander of British Antarctic exploration ships, who gave his name to the Ross Sea, decided to fix a mark for the purpose of 'showing the mean level of the ocean at a given epoch'. This particular location was deemed ideal since the harbour opened out directly to the vast Southern Ocean where the tides circulate the world freely, unobstructed by continents, and where there was minimal local tidal effect to distort the elegance of what was to be the 'zero point of the sea'. In August 1999, Daly found that the benchmark was standing at 30 cm *above* the present-day mean sea level. Where, then, is the scientific evidence for the predicted sea-level rise?

9.5 ROLE OF GREENHOUSE GASES

The Times article referred to above (Nuttall 1999) starts off by saying that 'scientists were claiming that a 70 per cent cut in greenhouse gas emissions was needed to save the planet from dangerous climate change'. Anything less would see sharp rises in drought, the spread of diseases such as malaria and the dying-off of huge swathes of the tropical forests in South America and Africa, as well as the dire consequences of sea-level rise. But will cutting emissions really have any discernible effect on our climate? Current models are still so crude, and contain so many 'adjustable parameters' that the truth is that we can prove anything from them. They fail miserably to predict past events, so are certainly not to be trusted to predict future events. Calculations of the climatic impacts of increases in CO_2 concentrations are not robust. Major components of the climate system are not satisfactorily represented in climate models because of a lack of good understanding of climate dynamics both on theoretical and observational grounds. It is far better to look at the scientific facts and see if any correlation between greenhouse-gas concentration and global temperature exists.

Atmospheric CO_2 concentrations rose from about 295 ppm in 1900 to about 307 ppm in 1940. It then increased with an increasing trend to its present value of about 360 ppm. The argument put forward is that if this concentration continues to increase almost exponentially, the concomitant temperature increases will be cataclysmic. But, on closer examination, there may be flaws in this line of logic. During this century, there was a small increase in temperature (however that was measured) up until 1940, followed by a decrease over the next 30 years, since when temperatures have increased again, but at a slower rate. The early increase coincided with the period when CO_2 levels were increasing only slowly. The decrease coincided with the period of maximum increase in CO_2. Approximately 80 per cent of the CO_2 rise during the twentieth century occurred *after* the initial major rise in temperature.

Detailed study of ice-core data covering hundreds of thousands of years, and also analysis of weather and climate records covering this century, both show that changes in temperature generally precede changes in CO_2 levels rather than the other way round. Typically, short-term changes in CO_2 levels lag behind changes in temperature by about 4 months (Corbyn 1996). One reason suggested for this is that the ocean acts as a sink to mop up excess CO_2 and Henry's law states that gases are less soluble in water at higher temperatures. This means that, as the water temperature increases, it cannot keep as much CO_2 in solution and so it is degassed to the atmosphere.

The other question is whether the increased concentration of greenhouse gases in the atmosphere will in turn cause a rise in temperature. Barrett (1996) suggests that most of the energy that can be absorbed by the atmosphere is already being absorbed under present conditions (the so-called wavelength saturation effect), and that any effect from the predicted 25 per cent increases in CO_2 would not be distinguishable from background variability. Atmospheric concentrations of CO_2 are reported to have varied widely over geological times, with peaks, according to some estimates, some twenty-fold higher than at present and troughs at approximately eighteenth century levels (Soon *et al.* 1999). The carbon budget is extremely complex and is certainly only partially understood. The IPCC only considered direct solar forcing rather than the total, which, it has been argued, has the effect of underestimating the effect by a factor of about 4, thus considerably reducing the CO_2 climate sensitivity.

We need also to ask whether beneficial effects of the increase will outweigh any possible damaging effects. Rising levels of CO_2 in the atmosphere promote plant growth and at the same time reduce their demand for water. These effects should lead to a greening of the Earth, and there are signs that this has already begun. Plants that barely subsist on the borders of barren deserts should gradually acquire the ability to grow and reproduce where it is currently too dry for them. It is also found that they would grow with less light. This will obviously affect the water-vapour feedback parameter in any climate model (e.g. increase in cloud cover could well have a negative feedback effect), but exactly how is only speculative at present.

9.6 SOLAR EFFECTS

Studies have shown that most of the long-term trends since 1850 in sea-surface temperature can be explained by variations in solar irradiance (Posmentier *et al* 1998). A number of the articles in Emsley (1996) are devoted to studies which have taken place to demonstrate this conclusion. For example, changes in the Scandinavian climate since the last ice age have been estimated from such information sources as tree rings, ^{14}C dating, glacial sediments and carbon dating of glacier moraines. Both the long-term and short-term variations show good correlation with solar-activity variations. A close inverse correlation has also been found between the length of the sunspot cycle and mean northern hemisphere air temperatures over the period 1795–1990. Indeed, most reports show that there is considerably better correlation between climate and sunspot activity than there is between climate and CO_2 levels.

9.7 POLITICAL ISSUES

There are a large number of interest groups involved in the climate-change debate, the result of which is a political process driven by diverse and perverse incentives. The general public can base their opinions only on the information which is fed to them by the media, and good news does not sell newspapers. Businessmen, transport managers, insurance companies, etc. have a vested interest in the outcome of the debate, and green pressure groups would have no need to exist if there were no crisis. The people who have most to gain are the politicians and the scientists. Any politician facing credibility problems at home may seek to become a statesman in the international arena where they are less accountable for their actions (Bate 1998). Those politicians fortunate enough to represent countries which will meet their emissions targets can score additional green points and can use the situation as a rod with which to beat their enemies.

Scientists are normally considered to be above suspicion, but it has been suggested that the action of some in this debate has been anything but altruistic. Reference was made to criticism of the IPCC in Section 9.1, but objectivity comes from open debate, and credibility from peer review. Those many scientists working in disciplines related to global warming have prospered in the last 10 years; indeed, in the UK more is spent on global climate research than on cancer research (Bate 1998). Journals rarely carry articles where the findings are largely negative – the results are just not exciting. An analysis of about 2,000 papers published between 1989 and 1995 picked at random from forty biological journals has shown that fewer than 9 per cent contained 'non-significant' results – the more prestigious journals fared the worst. Since credibility is linked to publishing in these prestigious journals, the temptation is to publish 'selectively', suggesting that the conclusion to be drawn is that results are biased in favour of global warming. In a zero-sum economy, it is very tempting not to kill the goose that lays the golden egg! Science will remain a servant of politics, and should take great care in what it offers and how it responds to opportunities. Short-termism may not only be the fate of politicians' (Boehmer-Christiansen 1996).

9.8 CONCLUSIONS

As promised in the Introduction, this chapter has been but a pot-pourri of ideas and observations, unsubstantiated by hard evidence. But a close study of the literature will produce such evidence, in great quantity. The references cited in this chapter provide a good start – they contain a number of closely-argued articles with reference to published scientific results. There is much relevant material on the WWW, but some of it is ephemeral and we always need to be circumspect as to its quality. But if this chapter has stimulated even a slight sense of uncertainty or doubt, and perhaps some further debate, then it has not been without value.

10

International co-operation; a prerequisite for global environmental observations

A. S. Belward

10.1 INTRODUCTION

We discussed Agenda 21 in Section 2.4.4; this document was intended to set out a master plan or international programme for achieving sustainable development. Section 40.9 of Agenda 21 states 'Relevant international organizations should develop practical recommendations for co-ordinated, harmonized collection and assessment of data at the national and international levels.' Earlier chapters in this book have emphasized the need for global environmental datasets and shown that Earth observation has the capacity to provide some of the required information. But, 'the range of global observations needed to understand and to monitor Earth processes, and to assess human impacts, exceeds the scientific, technical and financial capability of any one country' (IGOS 1999). Indeed, the situation is even more complex than this because 'neither a surface-based and/or in-situ observing system nor a space-based system have the capacity to singularly observe the Earth system' (Unninayer and Schiffer 1997).

The need for international efforts to study climate and climate change has long been recognized. It is impossible to trace this process back to a single starting point and this chapter makes no attempt to do so. Instead this chapter provides an overview of the development of international efforts in this arena, and provides one vision of the future, the emerging Integrated Global Observing Strategy (IGOS).

10.2 INTERNATIONAL CO-OPERATION IN SPACE

As we have seen, Earth observation from space provides unique spatial and temporal perspectives and unique information. However, the missions are usually the initiatives of individual nations or space agencies. But space is still an expensive business and the missions or programmes of individual players are tailored accordingly. As such, no nation can unilaterally satisfy all of its observation requirements. Recog-nizing the need for co-operation, the Committee on Earth Observation Satellites (CEOS) was formed in 1984 under the aegis of the Economic Summit of Industrialized Nations' Working Group on Growth Technology and Employment. CEOS membership includes all the world's civil agencies responsible for Earth-observation satellite programmes along with international user organizations. The overall goals of CEOS are:

- to promote co-operation so as to maximize the benefits of spaceborne Earth observations;
- to aid members and users by acting as a focal point for international co-ordination of space related Earth-observation activities;
- to exchange policy and technical information.

CEOS established working groups to address issues needed to meet these overall objectives. There are two working groups – the Working Group on Information Systems and Services (WGISS) (http://wgiss.ceos.org) and the Working Group on Calibration and Validation (WGCV) (http://wgvc.ceos.org).

The working groups have made important contributions to the process of international co-operation ensuring long-term confidence in the accuracy and quality of Earth-observation data and products and improving access and uniformity of formats concerning these products. Users benefit from better-quality data, better documentation and sustained commitments to provide calibration updates; they get improved access to products and tools, and services to locate these products. Belward (1999) describes the work of the WGCV in more detail.

Debate at the working-group level results in recommendations to the CEOS Plenary. These can be highly specific concerning best practice for calibration of specific instrument-types, or more programmatic, such as data sharing for joint experiments. The CEOS Members and Associates in Plenary then judge these recommendations and, if accepted, act upon them.

10.3 RESEARCH AND CLIMATE STUDIES

There are three key international organizations responsible for research issues concerning climate – the World Climate Research Programme (WCRP), the International Geosphere–Biosphere Programme (IGBP) and the International Human Dimensions Programme on global environmental change (IHDP). They have complementary, but different research agendas, see Chapter 2.

10.3.1 World Climate Research Programme (WCRP)

Research activities in WCRP (http://wmo.ch/) are characterized as:

- Studies aiming to understand and model the 'fast climate processes', mainly in the atmosphere and at the Earth's surface, that regulate the flow and

transformation of energy, radiation and heat as well as the near-surface hydrological processes. The Global Energy and Water Cycle Experiment (GEWEX) provides the scientific focus for studies of atmospheric thermodynamic variations that determine the global hydrological cycle and energy budget;
- Studies aiming to understand and model the slow response of the ocean and ice to changes in atmospheric forcing, in order to predict natural climate variations and the transient response of climate to progressively-changing environmental conditions.

10.3.2 International Geosphere–Biosphere Programme (IGBP)

The IGBP (http://igbp.kva.se/index.html) focus is acquiring basic scientific knowledge about the interactive processes of biology and chemistry of the Earth as they relate to global change. The objective is:

- to describe and understand the interactive physical, chemical and biological processes that regulate the total Earth system, the unique environment that it provides for life, the changes that are occurring in this system and the manner in which they are influenced by human actions.

IGBP research currently focuses on six key questions that are addressed by eight core projects:

1. How is the chemistry of the global atmosphere regulated and what is the role of biological processes in producing and consuming trace gases?
2. How will global changes affect terrestrial ecosystems?
3. How does vegetation interact with physical processes of the hydrological cycle?
4. How will changes in land use, sea level and climate alter coastal ecosystems, and what are the wider consequences?
5. How do ocean biogeochemical processes influence and respond to climate change?
6. What significant climate and environmental changes have occurred in the past and what were their causes?

There are also three cross-cutting activities:

- IGBP Data and Information System (IGBP-DIS);
- Global Analysis, Interpretation and Modelling (GAIM);
- Global Change System for Analysis, Research and Training (START).

10.3.3 International Human Dimensions Programme on global environmental change (IHDP)

The IHDP (http://uni-bonn.de/ihdp) focus is very clearly on the interactions of individuals and societies with the global environment. The programme addresses four key questions:

1. How do humans interact with the environment?
2. What are the consequences?

3. How can individuals and societies mitigate or adapt to environmental change?
4. How will policy responses to such changes influence present and future economic and social conditions?

10.4 GLOBAL-SCALE OPERATIONAL OBSERVATIONS

Another group of international organizations exists that have as a goal the provision of global observations in an operational sense rather than being research oriented. There are three such systems, the Global Ocean Observing System (GOOS) established in 1991, the Global Climate Observing System (GCOS) established in 1992 and the Global Terrestrial Observing System (GTOS) established in 1993 (collectively these are referred to as G3OS). The following UN agencies sponsor one or more of the G3OS:

- Food and Agriculture Organisation of the United Nations (FAO);
- Intergovernmental Oceanographic Commission of UNESCO (IOC);
- International Council for Science (ICSU);
- United Nations Educational, Scientific and Cultural Organization (UNESCO);
- United Nations Environment Programme (UNEP);
- World Meteorological Organization (WMO).

To improve synergy and information exchange, the G3OS sponsors group was established in 1997.

10.4.1 Global Ocean Observing System (GOOS)

GOOS (http://ioc.unesco.org/GOOS) has the goal of a permanent global system for observations, modelling and analysis of marine and ocean variables needed to support operational ocean services worldwide. GOOS will provide:

- accurate descriptions of the present state of the oceans, including living resources;
- continuous forecasts of the future conditions of the sea for as far ahead as possible;
- the basis for forecasts of climate change.

10.4.2 Global Climate Observing System (GCOS)

GCOS (http://wmo.ch/web/gcos/gcoshome.html) objectives are to ensure acquisition of data to meet requirements for:

- climate-system monitoring, climate-change detection and response monitoring especially in terrestrial ecosystems;
- application of climate information to national economic development;
- research toward improved understanding, modelling and prediction of the climate system.

10.4.3 Global Terrestrial Observing System (GTOS)

GTOS (http://fao.org/gtos/) mission is to provide policy makers, resource managers and researchers with the data they need to detect, quantify, locate and understand changes (especially reductions) in the capacity of terrestrial ecosystems to support sustainable development. There are five foci:

- changes in land quality;
- freshwater-resources availability;
- pollution and toxicity;
- loss of biodiversity;
- climate change.

10.4.4 Global observing systems, Earth observation and the Framework Convention on Climate Change (FCCC)

The reporting standards agreed upon for the FCCC are those recommended by the Intergovernmental Panel on Climate Change (IPCC). IPCC produces detailed reports that are used as a basis for discussion by the policy makers and implementers at the conferences of the parties via the Subsidiary Body for Scientific and Technological Advice (SBSTA). As pointed out in Chapter 12 the SBSTA introduced a recommendation making explicit reference to G3OS information-acquisition strategies. In parallel to getting political acceptance of these strategies there is the need to ensure that the technical means of obtaining them are in place.

It is obvious from the objectives of the research organizations and operational observing systems presented in Sections 10.2 and 10.3 that there is great potential for overlap, redundancy and possibly (but less obviously) gaps in the data collected and measurements made. Figure 10.1 illustrates the complexity and interrelationships of existing international initiatives. Figure 10.2 shows how they fit into the policy-making arena with reference to the Kyoto Protocol.

To improve the process of optimization and harmonization the G3OS have established common groups to assess particular facets of their work. These include the joint GOOS/GCOS Ocean Observing Panel for Climate (OOPC), the joint GCOS/GTOS Terrestrial Observing Panel for Climate (TOPC) and the Global Observing Systems Space Panel (GOSSP) common to all three. GOSSP will then, for example, act as a common interface to CEOS to ensure dialogue with the space agencies. This process should expand beyond G3OS to include end-users (i.e. the policy makers and implementers, UN agencies and research groups) and data providers (including the space agencies). IGOS was started in just this frame.

10.5 INTEGRATED GLOBAL OBSERVING STRATEGY (IGOS)

IGOS (http://igospartners.org/) unites the major satellite and ground-based systems for global environmental observations of the atmosphere, oceans and land, in a framework that delivers maximum benefit and effectiveness in their final use. The

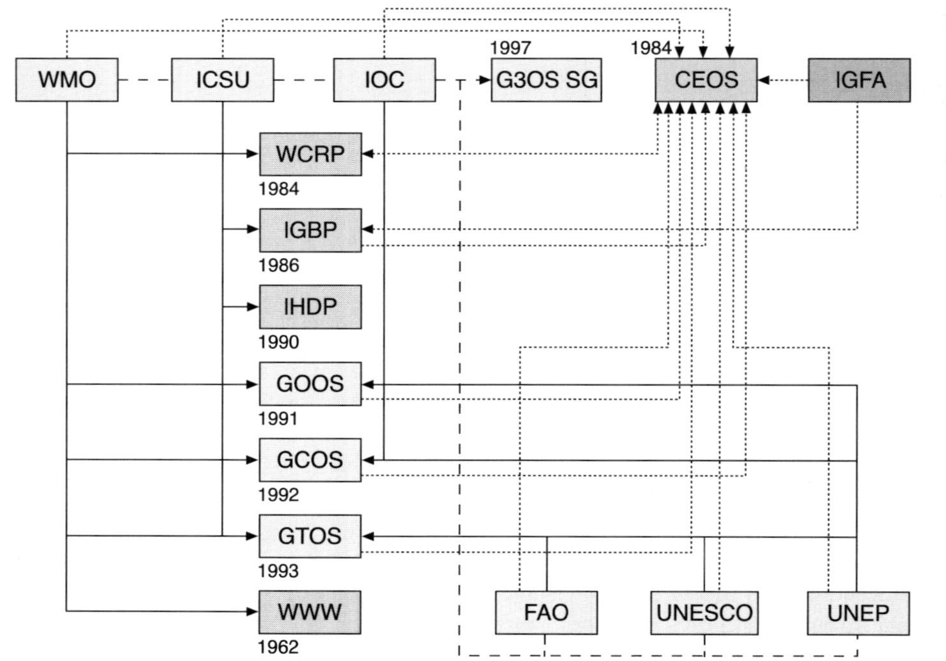

Figure 10.1. Links among international organizations concerned with the global environment.

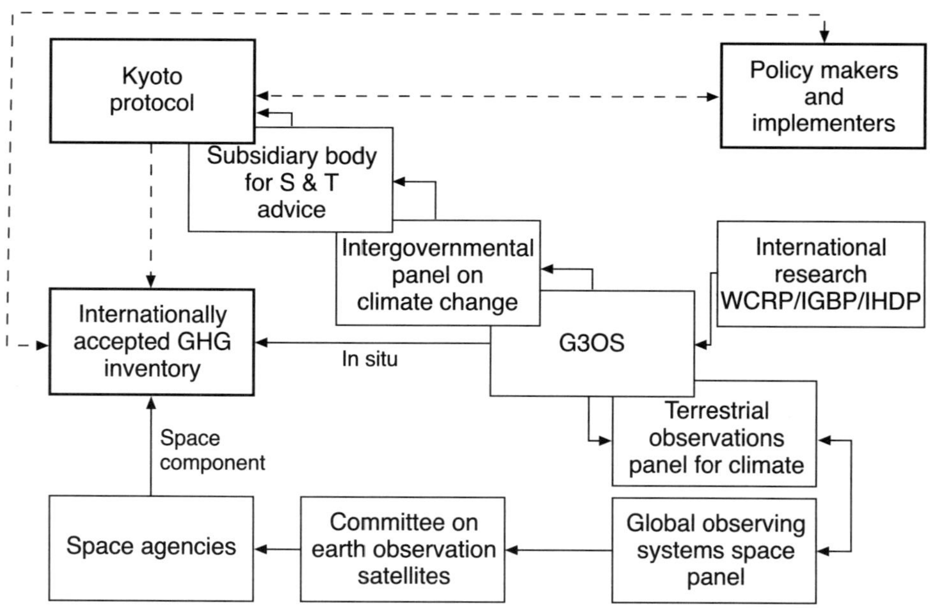

Figure 10.2. International organizations and the development of the Kyoto Protocol.

overall concept is set out by Townshend and Williams (1998). IGOS has the following characteristics:

- it is an initiative to develop a comprehensive strategy for enhanced levels of support to scientific, operational and research communities;
- it is a strategy to enable agencies to plan missions and activities better;
- it would enable space agencies to plan Earth-observation projects with the minimum of unnecessary overlap and to devise joint strategies for addressing serious gaps in their observation capabilities;
- ultimately, IGOS should be the joint product of all groups involved in the collection and analysis of space-based and *in-situ* data.

The need for IGOS arises out of recognition that no single nation can satisfy all of its observation requirements, that contemporary data/information products often require the integration of multiple observations for multiple sources and that performance measures should be adopted congruent with agency needs for continuous improvement. IGOS will:

- achieve outcomes that are beyond the capabilities of existing systems;
- be the joint product of all agencies involved;
- build on current achievements with additional efforts directed to areas where satisfactory international arrangements and structures do not currently exist;
- address the collection of both space-based and *in-situ* data.

The development and implementation of IGOS is supported by a partnership among space agencies grouped in CEOS, the G3OS sponsors group, the programme offices of all three global observing systems, two international research programmes, IGBP and WCRP, and major national agencies funding global-change research grouped in the International Group of Funding Agencies (IGFA). The IGOS partners' website (IGOS 1999) provides all the background documents. IGOS is in a development phase and the site is regularly updated.

11

Global dataset creation; experience of the IGBP DISCover Project

A. S. Belward

11.1 INTRODUCTION

Since 1992 (Townshend 1992) the Data and Information System (DIS) of the International Geosphere–Biosphere Programme (IGBP) has been working towards the completion of a validated global land-cover dataset, (DISCover). The 1-km resolution dataset consists of seventeen cover classes identified on the basis of the science requirements of the IGBP's core projects. DISCover was created from Advanced Very High Resolution Radiometer (AVHRR) data. The data were processed and assembled into a coherent set of monthly Normalized Difference Vegetation Index (NDVI) composites (April 1992 to April 1993) and classified using unsupervised techniques with post-classification refinement. The first global land-cover classification was completed in July 1997. The IGBP-DIS' Land Cover Working Group (LCWG) convened a validation working group to define and implement a validation method. This provides statistical statements concerning the accuracy of the global land-cover product and allows estimation of the error variance in areal totals of classes globally and within regions. The validation workshop was completed in September 1998 and preliminary analysis concluded in March 1999. This chapter describes the process by which DISCover version 1.0 was created.

11.2 DATASET DEVELOPMENT

The science requirements for a global land-cover dataset as specified by IGBP (see Chapter 10) made the use of Earth-observation data indispensable. Even so, the notion of regular global coverage from Earth-observation satellite systems has until very recently been largely erroneous. Of the systems in operation when the DISCover project began the best match, in terms of spatial, temporal and spectral characteristics, was AVHRR on the polar-orbiting meteorological satellites of the US National Oceanic and Atmospheric Administration (NOAA).

Even so, AVHRR data at full resolution (i.e. around 1 km at nadir) were not available on a global daily basis. They are only available when the satellite is in sight of a ground receiving station or if recorded on-board. In April 1992 the EROS Data Center (EDC) of the United States Geological Survey (USGS), the European Space Agency (ESA), the National Oceanic and Atmospheric Administration (NOAA), the National Aeronautic and Space Administration (NASA) and the Australian Commonwealth Scientific and Industrial Research Organization (CSIRO) began to co-ordinate data collection from twenty-three receiving stations from around the world (Eidenshink and Faundeen 1994).

More than 4.4 Tbyte of 1-km resolution AVHRR data have been archived. Data collection has been continuous, with the exception of the period September 1994 to February 1995 when there was no operational afternoon satellite.

There are over 126,230 satellite scenes available in the project's archive. Information and data are available from http://edcwww.cr.usgs.gov/landdaac/1KM/1kmhomepage.html. The establishment of the dataset is a major achievement and these data are now the basis of other global terrestrial products (e.g. see Stroppiana *et al.* 2000; Hanson *et al.* 2000).

11.3 DISCover CLASSIFICATION SCHEME

The DISCover categories use the climate-independence and canopy-component philosophy presented by Running *et al.* (1994b) but modified to be compatible with classification systems currently used for environmental modelling to provide, where possible, land-use implications and to represent landscape mosaics. The classification scheme also embraced modifications proposed as the validation strategy developed. DISCover contains seventeen land-cover classes. The scheme is based on definitions of three canopy components: above-ground biomass, leaf longevity and leaf type (Table 11.1).

11.4 DISCover CLASSIFICATION METHODOLOGY

LCWG based the DISCover mapping exercise on an adaptation of the method developed by the USGS *EROS* Data Center and the University of Nebraska-Lincoln (Loveland *et al.* 1991, Brown *et al.* 1993).

The approach involves mapping land cover for each continent in turn. The first mapping step involves unsupervised classification of NDVI composites to identify greenness classes using an unsupervised clustering algorithm run on monthly NDVI maximum-value composites from April 1992 to March 1993. Each cluster was assigned a preliminary cover-type label on the basis of the spatial patterns and spectral or multi-temporal statistics of each class, and comparison with ancillary data. Ancillary data include descriptive land-cover information, NDVI and

Table 11.1. IGBP DISCover dataset land-cover classification system.

Class	Class name	Description
1	Evergreen needleleaf forests	Lands dominated by needleleaf trees with a canopy cover >60% and height exceeding 2 m – almost all trees remain green all year – canopy is never without green foliage
2	Evergreen broadleaf forests	Lands dominated by broadleaf trees with a canopy cover >60% and height exceeding 2 m – almost all trees remain green all year – canopy is never without green foliage
3	Deciduous needleleaf forests	Lands dominated by trees with a canopy cover >60% and height exceeding 2 m – consist of seasonal needleleaf-tree communities with an annual cycle of leaf-on and leaf-off periods
4	Deciduous broadleaf forests	Lands dominated by trees with a canopy cover >60% and height exceeding 2 m – consist of seasonal broadleaf-tree communities with an annual cycle of leaf-on and leaf-off periods
5	Mixed forests	Lands dominated by trees with a canopy cover >60% and height exceeding 2 m – consist of tree communities with interspersed mixtures or mosaics of the previous four forest-cover types – no one forest type exceeds 60% of landscape
6	Closed shrublands	Lands with woody vegetation less than 2 m tall and with shrub canopy cover >60%. The shrub foliage can either be evergreen or deciduous
7	Open shrublands	Lands with woody vegetation less than 2 m tall and with shrub canopy cover between 10–60% – the shrub foliage can either be evergreen or deciduous
8	Woody savannahs	Lands with herbaceous and other understorey systems, and with forest canopy cover between 30–60% – the forest cover height exceeds 2 m
9	Savannahs	Lands with herbaceous and other understorey systems, and with forest canopy cover between 10–30% – the forest-cover height exceeds 2 m
10	Grasslands	Lands with herbaceous types of cover – tree and shrub cover is less than 10%
11	Permanent wetlands	Lands with a permanent mixture of water and herbaceous or woody vegetation that cover extensive areas – the vegetation can be present in either salt, brackish or freshwater
12	Cropland	Lands covered with temporary crops followed by harvest and a bare-soil period (e.g. single and multiple cropping systems) – Note that perennial woody crops will be classified as the appropriate forest or shrub land-cover type
13	Urban and built-up	Land covered by buildings and other man-made structures – note that this class will not be mapped from AVHRR imagery but will be developed from the populated-places layer that is part of the Digital Chart of the World
14	Cropland/natural vegetation mosaics	Lands with a mosaic of croplands, forest, shrublands and grasslands in which no one component comprises more than 60% of the landscape
15	Snow and ice	Lands under snow and/or ice cover throughout the year
16	Barren	Lands with exposed soil, sand, rocks or snow that never have more than 10% vegetated cover during any time of the year
17	Water bodies	Oceans, seas, lakes, reservoirs and rivers – can either be fresh or saltwater bodies

elevation statistics, and class relationships to other land-cover legends. Related single-category classes are then grouped under DISCover labels using a convergence of evidence approach, in which all available documentation, including class attributes, image maps, atlases and existing land-cover/vegetation maps are compared with the spatial distribution of each DISCover class.

Identification of DISCover classes water, snow and ice, and barren from greenness classes is inappropriate, and these were identified from AVHRR 0.72–1.10-μm data also retained by the IGBP-DIS processing chain, and a 12-month maximum-value NDVI composite. The urban class was added from the urban-layer information from the Digital Chart of the World (Danko 1992). Loveland et al. (1999) provide full details of the classification methodology.

11.5 DISCover VALIDATION STRATEGY

Accuracy assessments of digital land-cover classifications are typically based on contingency tables where accuracy is expressed in terms of errors of omission and commission, or in terms of agreement analysis using the kappa-test statistic (Stehman 1996; Congalton 1991; Rosenfeld and Fitzpatrick-Lins 1986). The contingency table is created by comparing, on a class-by-class basis, the land-cover classification with an independent data source – field observations, existing maps, higher resolution imagery – collected using a statistically valid sampling strategy (Robinson *et al.* 1983, Rosenfeld 1986).

These methods have generally only been applied to local-scale classifications, occasionally to regional-scale work and only in isolated instances on scales greater than them (Hay 1988; Estes and Mooneyhan 1994; Jeanjean and Achard 1997). LCWG convened a specific subgroup, the Validation Working Group (VWG), to examine how previous approaches to regional work could be adapted to the global scale of DISCover. Their mandate was 'to specify and co-ordinate the implementation of a practical and achievable methodology for the validation of global land cover products'. Validation was defined as 'the process that verifies the accuracy of land-cover information provided by the Land Cover Working Group in as independent a means as practical' (Belward 1996a).

The IGBP core projects asked that the final global land-cover map have 85 per cent correct classification for each class. Developing a validation strategy to check if this target was met by the DISCover classification involved two assumptions. First, that at 1-km resolution the thematic and spatial accuracy of the classified datasets cannot be separated, and second that higher resolution satellite images from sensors on *Landsat* or *SPOT* will provide accurate independent reference data describing the true land-cover class.

The use of high-resolution data as a surrogate for reference data has some precedents. Visual interpretation of higher resolution satellite imagery, either from hardcopy prints, or on screen using image processors has been used for operational crop-yield forecasting on a pan-European basis. It has proved to be cost effective, capable of delivering consistent results over large areas and easy to apply in an operational setting (De Boissezon *et al.* 1993). This approach was thus to form the basis of the VWG validation strategy.

Using high-resolution data as a reference has obvious limitations. The higher resolution data will not have the same temporal characteristics as AVHRR. The higher resolution characteristics will also vary within the classes depending on the time of year that the data were obtained. To ensure consistent visual interpretation decision rules must be provided to the interpreters. Ambiguity in class definition must be avoided. The questions of what percentage of a given class must occupy a given sample site for that sample to be designated as that class must also be clear. Finally, independence of interpretation must be maintained. In this case, using a team of three interpreters was deemed acceptable.

11.6 SAMPLING STRATEGY

The principal objective of the sampling strategy was to provide a set of statistical statements that: (1) characterize the accuracy of the global land-cover product and (2) allow estimation of the error variance in areal totals of land-cover types. Four approaches to sampling were identified.

Random sampling or random sampling stratified by continent (on the basis of percentage total land area only) are the simplest approaches and would allow sample points to be reused for validation of future land-cover classifications. However, without stratification and subsampling on the basis of class area, less frequent classes would be sampled less frequently, and thus the accuracy of their labelling would be known less well than for other, more frequently occurring classes (Rosenfeld *et al.* 1982). Accuracy statistics would only be valid on a global basis, confidence intervals would be unequal and would vary widely as the number of sample points per class would vary. There would be strong bias towards the larger classes as these will tend to be sampled more frequently; some classes may not be sampled at all.

Random sampling stratified by class would provide roughly equal confidence ranges a priori, and produce unbiased estimates of cover areas and proportions at the global scale. The optimum sampling strategy would be random sampling stratified by class and processing region (where processing region refers to all areas classified in the same production run, using the same ancillary data, same clusters, etc.). This would highlight any systematic variation in error structure between processing regions and the errors of omission and commission would also be applicable on a regional scale. Though optimum, the cost is high as the number of samples per class is then multiplied by the number of processing regions. Because of the high cost of this optimum solution the VWG recommended random sampling stratified by class.

The sampling strategy aims to estimate the probability p that a 1-km square, truly type I, gets classified according to AVHRR data as type I. In terms of conditional probabilities this is expressed as:

$$p = \mathrm{P}\ (\text{classify as I} \mid \text{truly type I}) \tag{11.1}$$

that is, p is the probability of classification as I, given truly an I.

To achieve this, n squares of type I will be sampled (known to be of type I on the basis of reference data from high-resolution imagery) and a check made of how many are allocated to type I by the classification scheme being validated. If r is the number correctly classified by the scheme, then r/n is a natural estimate of the accuracy probability above, and a 95 per cent confidence interval around r/n tells us how reliable this estimate is. Ideally the accuracy of each class would be calculated with the same confidence interval. For practical reasons the sample size per class needs to be kept down to a minimum. Standard charts for binomial-confidence intervals (Pearson and Hartley 1966) confirm that a sample of 25 would be adequate, but this gives wide confidence intervals ($\pm$14.3 per cent) and the accuracy statements are thus somewhat limited. Problem classes can be identified but the wide confidence interval means it is impossible to say if a class has truly been classified to 85 per cent correct. Bearing in mind the VWG's key word 'practical', this limited validation was judged to be of far greater value than no validation, and the sample of 25 accepted.

11.7 VALIDATION RESULTS

The core sites were identified using a stratified random sample of 25 points per class. A number of *Landsat* Thematic Mapper (TM) scenes (285) and *SPOT* High Resolution Visible sensor (HRV) scenes (150) were acquired by the EDC, the European Commission's Joint Research Centre, the IGBP-DIS office in Toulouse, *SPOT* Image and the Centre National d'Etudes Spatiales CNES (France). The high-resolution reference data were then prepared by the University of California–Santa Barbara and transported to EDC where they were analysed at a dedicated workshop involving participation of expert interpreters from around the world. Table 11.2 presents the results, and Scepan (1999) provides a more detailed analysis.

11.8 CONCLUSIONS

This is the first time a global scale land-cover product has been statistically validated, and the DISCover project has been a forerunner to many new global land-cover products (DeFries and Belward 2000).

No unique permanent infrastructure was put in place to accomplish the development and validation of DISCover. Instead, through the auspices of the IGBP-DIS' working groups a wide variety of participating organizations were involved on the basis of their already-established activities. What is remarkable about this effort

Table 11.2. Confusion matrix (DISCover V1.0).

Mapped class		True class																	total	% class accuracy
		1	2	3	4	5	6	7	8	9	10	11	12	13	14	15	16	17		
rgreen needleleaf forests	1	15				1						1	1		1		1		20	0.75
rgreen broadleaf forests	2		21									1			1		1		24	0.88
iduous needleleaf forests	3			5								2	2						9	0.55
iduous broadleaf forests	4				10	1							3		1			1	16	0.63
Mixed forests	5	1			1	15		3					1		1				22	0.68
Closed shrublands	6						15				1	1					1	2	20	0.75
Open shrublands	7							21			1						2		24	0.88
Woody savannahs	8								18	4	1		3						26	0.69
Savannahs	9								3	11			1		2				17	0.65
Grasslands	10							1			15		2		1		1		20	0.75
Permanent wetlands	11	1						5			1	5					1		13	0.38
Croplands	12										1		18	1	1				21	0.86
Urban and built-up	13					2							2	16	2		1	1	24	0.67
pland/Natural-vegetation mosaics	14	2	3			1							3	1	13				23	0.57
Snow and ice	15																			
Barren	16																27		27	1.00
Water bodies	17																			
TOTAL		19	24	5	11	20	15	30	21	15	20	10	36	18	23	0	35	4		

was that participation was by and large voluntary and self-funded. Only $500,000 was allocated specifically to fund DISCover related development and validation activities. Many organizations and individuals involved in this effort used existing operational funding or funding from projects they considered related to ensure their participation in the project. More than 100 individuals from thirty nations participated. These participants came from seven international organizations, twenty-six research institutes and academic organizations, twenty-one national agencies, centres or laboratories and seven industrial groups.

In common with other global datasets the DISCover product is already dated. Yet the need for global terrestrial information continues to grow. As outlined in various other chapters, the importance of the global perspective now has new significance as environmental-information input to the policy-making, policy-development and policy-implementation processes grows. Moreover, as new products emerge, new validation strategies are required. The infrastructure needed to support the development of DISCover was put in place through international co-operation and co-ordination. This process must continue (Belward 1999). A framework for this was discussed in Chapter 10.

12

A policy context for global environmental-information systems

A. S. Belward

12.1 INTRODUCTION

There are more than 400 multilateral environmental treaties currently listed in the World Conservation Union database (CIESIN 1998). The legal and economic implications of these are such that reliable and accurate information concerning the global environment will soon be at a premium. Earth observation is only one part of the information-acquisition chain, but one with important strategic characteristics in a global context. This chapter discusses the policy drivers and outlines the nature of a global environmental-information system.

12.2 REGIONAL INTERESTS

From a policy perspective the case for global environmental monitoring is not immediately obvious. Current environmental-policy implementation is largely served by information gathered at a national level. Regional assessments are made through aggregation.

The European Environment Agency (EEA), for example, has a mandate defined by the European Union's Council* Regulation (No. 1210/90) to ensure the supply of objective, reliable and comprehensive environmental information. The geographical scope is European, rather than global. EEA acts as an interface between data producers and information users. Data handling, analysis and assessment is performed at national monitoring centres and European Environment Information and Observation Network partners. Though mandated to collect and supply the

* The Council of Ministers is the principal decision-making body within the European Union.

data, EEA itself has no regulatory role that demands provision of specific information. Thus, EEA aggregates national information to provide a regional overview.

Combinations of regional areas of interest collectively call for a global picture. Examples from the European Commission concerning forestry interests illustrate this. Directorate General XI (DGXI) (Environment) requires forest-monitoring systems to support its work on the preparation of the International Forest Convention and for its work implementing the Convention on Biological Diversity (see Chapter 2). Directorate General VIII (DGVIII) (Development) runs forest-management programmes throughout Africa, the Caribbean and Pacific countries and needs regular information for its environmental-protection programmes associated with the International Convention to Combat Desertification. The External Relations DG is concerned with forest-management programmes in the Commonwealth of Independent States, South-East Asia and Latin America. Collectively these DGs call for global-scale monitoring of change in the world's forest resources to support policy planning and to assess policy impact.

Non-governmental and intergovernmental organizations (such as the World Resources Institute, the Environmental Law Institute, the United Nations Environment Programme and the Organization for Economic Co-operation and Development) have long maintained a global perspective. But global environmental-information systems tend to be somewhat fragmentary. There are many reasons why this should be so. The lack of a comprehensive global environmental picture reflects concerns for national security and national sovereignty, the lack of appropriate institutional infrastructures, and technological and scientific constraints. But of overriding importance is the high cost of establishing any such system, and the lack of political will.

The cost involved in establishing and operating an infrastructure for gathering global environmental information is far from negligible. The World Weather Watch (WWW; the most comprehensive, operational, global, environmental-monitoring programme at present) combines observing systems, telecommunication facilities and data-processing centres to make available the meteorological and related geophysical information needed to provide weather services on a day-to-day basis. Even then, the programme itself does not provide these weather services. That is the responsibility of the individual members. The facilities are owned and operated by the 185 member countries of the World Meteorological Organization. The scale and completeness of the system are such that the entire WWW took almost three decades to put into place. The last piece was the geostationary satellite placed in orbit over the Indian Ocean in the late 1990s. This completed the network of geostationary satellites providing complete global cover. WWW runs at a cost of around $5million/day.

Logistical and financial challenges such as these obviously call for strong political backing. In May 1999 the European Parliament published its resolution on Environment, Security and Foreign Policy (EC 1999). This document 'calls on the Commission to present to the Council and Parliament a common strategy ... which brings together the common, foreign and security-policy aspects of EU policy with its trade, aid, development and international environmental policies so as to tackle the following individual issues and the relationships between them:

(a) agricultural and food production and environmental degradation;
(b) water shortages and transfrontier water supply;
(c) deforestation and restoring carbon sinks;
(d) unemployment, underemployment and absolute poverty;
(e) deforestation, desertification and population growth;
(f) sustainable development and climate change;
(g) the links with all of the above and global warming and the humanitarian and environmental impact of increasingly extreme weather events.

This illustrates just how broad the policy dimensions of global environmental monitoring are. Access to global environmental information touches the independence and security of nation states and economic groupings (such as the European Union [EU]), reflects their global responsibility, and affects trade, industrial development and employment.

The resolution also 'Considers that environmental strategies should be able to include monitoring the world environment, assessing the data thus collected, co-ordinating scientific work and disseminating information, exploiting relevant data from national observation and monitoring systems to give a continuous and comprehensive picture of the state of the environment.' This yet again reinforces the message that the global environmental picture is missing, and that once more aggregation of national observations is seen as one way of obtaining this picture.

Unfortunately, there is a conflict of interest at work here. On the one hand, we expect nation states to contribute selflessly to a global environmental-information system, and, on the other hand, there are clear strategic and economic benefits to be had from such information. Relying on national contributions is certainly an option within the bounds of political alliances, such as the EU, but has not been shown to work on the global scale.

This then raises the question as to just what political drivers there are that call for global environmental information, and that are strong enough to justify the time, effort and cost of putting such a system into place?

12.3 GLOBAL INTERESTS

Perhaps the key policy driver is the United Nations Framework Convention on Climate Change (FCCC) adopted in 1992 (UNFCC 1992); see also Chapter 2. This has the objective of:

> stabilization of greenhouse gas concentrations in the atmosphere at a level that would prevent dangerous anthropogenic interference with the climate system. Such a level should be achieved within a time-frame sufficient to allow ecosystems to adapt naturally to climate change, to ensure that food production is not

threatened and to enable economic development to proceed in a sustainable manner.

The Kyoto Protocol to the FCCC (UNFCCC 1997) (see also Section 2.5) must be seen as a milestone in terms of environmental concern. This internationally-prepared document, for the first time, contains legally-binding commitments either to reduce or limit the emissions of six major greenhouse gases (carbon dioxide [CO_2], methane [CH_4], nitrous oxide [N_2O], hydrofluorocarbons [HFCs], perfluorocarbons [PFCs] and sulphur hexafluoride [SF_6]). The Protocol contains agreed targets for the so-called Annex I countries (essentially the already industrialized nations) that collectively amount to a 5 per cent reduction on 1990 levels.

The Protocol makes provision for the use of biological sources and sinks to meet commitments and understandably insists on national systems for verification, reporting and accountability.

This results in specific-information requirements for signatories to the Protocol. They will be required to:

- monitor measures promoting the protection and enhancement of greenhouse-gas sinks and reservoirs (Article 2);
- measure changes in carbon stocks resulting from human-induced land-use change and forestry activities (Article 3) and monitor such activities under the clean-development mechanism (Article 12);
- monitor transfer of emission-reduction units resulting from projects which enhance anthropogenic removals by sinks (Articles 6 and 17).

While these reporting mechanisms are essentially a national requirement, the clean-development mechanism and the joint implementation scheme do call for information beyond any one nation's boundary. It is worth noting, for example, that the EU collectively negotiated its commitment to reduce yearly emissions by 8 per cent compared with 1990 levels within the commitment period 2008–2012. Thus the member states must collectively report. This automatically raises the need for trans-boundary information, but alone does not call for global monitoring. This requirement though is embodied in the terms of Article 10, which states that signatories must:

- provide systematic observation and the development of data archives to reduce uncertainties related to the climate system (Article 10).

This is a task that cannot be achieved without global monitoring.

The final elements of the Protocol relevant to this discussion are Articles 5 and 17. These stress that any nation's reporting system:

- be internationally recognized and accepted (Articles 5 and 17).

Article 5 of the Kyoto Protocol calls for national reporting systems for sinks and sources that are accepted by the Intergovernmental Panel on Climate Change (IPCC 1996). The IPCC working with the Subsidiary Body for Scientific and Technological Advice (SBSTA) provides advice to the conferences of the parties concerning methodologies for estimating anthropogenic emissions by sources and removals by sinks. SBSTA recommended a draft decision on research and systematic observation that was adopted by the Fourth Conference of the Parties (COP) at Buenos Aires in 1998. The text of the draft decision includes the following article

> #7. Urges Parties to actively support national terrestrial networks including observational programmes to collect, exchange and preserve terrestrial data according to the Global Climate Observing System (GCOS) and the Global Terrestrial Observing System (GTOS) climate priorities and particularly hydrosphere, cryosphere and ecosystem observations.

These in turn recognize the importance of combined in-situ and spaceborne Earth-observing systems.

12.4 ROLE FOR EARTH-OBSERVATION SYSTEMS

Spaceborne Earth-observation systems provide wall-to-wall global observations. 'Carbon relevant' global-scale products derived from Earth-observation systems already exist; global land-cover maps, tropical deforestation-risk maps and wildfire maps, precise ocean surface temperature measurements, surface-wind speed and even sea-surface elevation and ocean-circulation measurements have all been produced on the global scale.

Datasets such as these will play a vital role in resolving the uncertainties related to the climate system, as called for by the Kyoto Protocol. The information provided by remote-sensing systems goes beyond the scientific, and will play a part in assessing compliance with other elements of the Kyoto Protocol too. Of particular importance will be the capacity to determine land-use change and forestry activities in countries other than the Annex I States. Interpretation of the clean-development mechanism (Article 12) by certain parties to the convention means that detailed information on changes in land use – especially afforestation, deforestation and reforestation – must be assessed worldwide, not just for Annex 1 states (WBGU 1998).

12.5 GLOBAL ENVIRONMENTAL-INFORMATION SYSTEMS

Central to any attempt at global environmental monitoring is the idea of an information system. It is no use building a sophisticated policy-driven monitoring system if the policy makers and implementers cannot access the information, or even worse, do not use the information.

There are a number of initiatives being put into place to bridge the gaps from data to information and from information to knowledge. A recent overview is provided by the UNEP (United Nations Environment Programme) Earthwatch strategic framework for environmental observing, assessment and reporting (UNEP 1999). This emphasizes the point that:

> the whole chain of environmental information flow must be included. From observing, monitoring and other data collection, through assessment, modelling and other value-added data processing, to delivery of reports and other information products to priority groups of users, including decision-makers and the general public.

Specific objectives of any such system as identified by UNEP are:

- to keep policy makers informed of the global environmental situation, particularly where it threatens human health and well-being and environmental sustainability;
- to provide adequate scientific information on the global environment – assembled, integrated and organized so that the current status and trends can be summarized for each necessary global report;
- to provide the basis for integrated assessments of the global environment;
- to measure the effectiveness of management actions;
- to maximize participation in all parts of the environmental-information system.

The Tropical Forest Information System (TFIS), developed by the European Commission, satisfies these objectives, but, of course, for only one single environmental variable – the state of the world's tropical forests. This system allows users to combine up-to-date tropical-forest maps and deforestation statistics derived from satellite imagery with other sources of information, ranging from spatial datasets – such as global digital-elevation models – to anecdotal information – such as local newspaper reports documenting specific deforestation events. The system is being developed so that it runs models of deforestation risk. The success of TFIS is in no small part due to the fact that it is installed directly in the headquarters of those responsible for implementing environmental policy (in this case DG Environment). Yet it is maintained and updated by the scientists making the measurements and analysing the data (the Space Applications Institute of the Joint Research Centre). The next logical development of TFIS is to transfer it to a World Wide Web based system. This will allow access (where appropriate) by the general public, and will allow the Directorates General to place systems in the field, thus using TFIS for project management in an operational sense.

12.6 CONCLUSIONS

Global environmental information is increasingly seen as an essential commodity in the light of international environmental treaties, protocols and conventions. Access to such information will provide a strategic advantage in international negotiation and trading, industrial development and employment. Policy makers need this information to support their own negotiating positions and to counter or corroborate the policy positions of others. Industry will need this information to demonstrate compliance with the terms of such treaties, and thus maintain a competitive role in the global market place.

Global environmental monitoring and the creation of global environmental-information systems will never be cheap. The pressure on our environment though will continue to increase (at least until such point as we decide, or are forced, to adopt less wasteful and more ecologically aware lifestyles). Environmental information will provide the understanding on which to base environmentally sound policies, and to evaluate the effectiveness of such policies once implemented.

There is room for direct commercial involvement in global environmental monitoring but there is still a strong call on the public purse. The necessity for action, the capacity of science to respond and the political will to act seem to be growing almost by the day, and it can only be a matter of time before global environmental-information systems become a reality.

13

Envisat – the mission

R. A. Vaughan and S. T. Wilson

13.1 INTRODUCTION

Envisat, the flagship platform for the European Space Agency for the early part of this new century, was mentioned briefly in Chapter 3 in the context of Europe's contribution to satellite climatology. This mission, based as it is on the proven heritage of the two European Remote Sensing (*ERS*) satellites, will extend the time series of global environmental measurements, started in 1991, till at least 2006, thereby providing a useful 15 years of continuous monitoring. Changes in philosophy and funding mechanisms probably mean that *Envisat* will be Europe's last large, multi-purpose platform, the trend being towards less expensive, smaller, dedicated missions which have a much shorter lead time for development and implementation.

The European Space Agency (ESA) was formed in 1975 'to provide for and to promote, for exclusively peaceful purposes, co-operation among the European States in space research and technology and their space applications, with a view to their being used for scientific purposes and operational space applications systems'. Originally, there were ten European members plus Canada. Since then membership has increased to fifteen, with the accession of Portugal which joined in early 2000. In the early 1990s, some central European states – namely Hungary, Poland, Romania and the Czech Republic – concluded co-operation agreements with ESA, as did Greece in 1994.

ESA is a research and development (R&D) organization, pooling its member states' resources, scientific communities and industrial capabilities. The central and eastern European countries have yet to consolidate their capabilities. Criteria for admissibility of new applicants have recently been established by ESA and a specific status of 'European Co-operating State' will be created to accommodate the interests of new states not having the same objectives or capabilities as the existing Member states. ESA has recently set up a committee of three people to provide the Director General with independent advice on the links between potential enlargement and the direction in which ESA needs to evolve in order to respond to the expectations of the citizens of Europe.

In 1999 ESA had a budget of over €2,600 M, of which nearly a quarter was devoted to Earth observation. It is funded by direct levy from the Member States, originally on an ad-hoc basis but now on a rolling basis through the 'envelope programme' (see Chapter 3). *Envisat* itself is due to cost over 2.5 billion euros.

13.2 ERS PROGRAMME

ERS-1 was launched in July 1991 and provided nominal service until June 1996. After that it was on standby as a backup until finally decommissioned in March 2000. *ERS-2* was launched in April 1995 and it is hoped that it will continue to operate until after the launch of *Envisat* in 2001. Both satellites operated in tandem between April 1995 and June 1996 providing Synthetic Aperture Radar (SAR) image pairs for interferometric analysis. The payloads consisted of an Active Microwave Instrument (AMI), SAR and a wind scatterometer, a Radar Altimeter (RA), the Along-Track Scanning Radiometer (ATSR), a microwave sounder, the Precise Range and Range-Rate Equipment (PRARE), a laser Retroreflector, and, on *ERS-2*, the Global Ozone Monitoring Experiment (GOME). Although designed mainly for oceanographic applications, the instruments have been used in many other studies over land and ice fields. SAR and the wind scatterometer are routinely used for vegetation monitoring and even the altimeter has been used to produce topographic maps of deserts and ice fields (see Chapter 3).

13.3 STRATEGY

In 1988 an ESA overall 'strategy for Earth observation' was presented to the ESA member states. This proposed a series of complementary polar-orbiting and geostationary satellites to study the Earth's environment and resources, and to continue and improve meteorological observations. Based on this scenario, two ESA councils at ministerial level, held in Munich in November 1991 and Granada in November 1992, established a programme composed of two missions: the *Envisat-1* mission and the *METOP-1* mission preparatory programme.

While *METOP* is primarily an operational meteorological satellite (see Chapter 3), *Envisat* is a satellite dedicated to the study of the Earth and its environment. ESA has produced several publications summarizing both the *Envisat* mission and systems, and the opportunities for science and application of the ensuing data. The rest of this chapter will outline the principal instruments and the science which they are designed to address. Much of this material has been freely drawn from the various ESA publications listed in the References (at the end of the book), and from discussions with ESA personnel.

Table 13.1. Comparison of *Envisat* and *ERS-2*.

	Envisat	*ERS-2*
Launch mass	8,140 kg	2,510 kg
Payload mass	2,145 kg	710 kg
Solar power array	6,500 W (after 4 years)	1,940 W (after 3 years)
Instruments	10	5
Overall length	24.8 m	11.5 m
Payload length (excluding solar panel)	10 m	8 m

13.4 *Envisat* – THE BIG PLATFORM

Envisat will not only be expensive, but will also be one of the largest satellites ever launched. It will weigh over 8,000 kg (more than three times the mass of *ERS*), measure 26 m × 10 m × 5 m, and require more than 3 kW of power (see Table 13.1). Its mission objectives include an enhancement of the ERS missions and to provide continuity of the observations started in 1991, particularly the radar-based observations, to extend the range of parameters observed to meet the need to increase knowledge of the factors determining the environment, to make a significant contribution to environmental studies, notably in the areas of atmospheric chemistry and ocean studies (including marine biology), to allow more effective monitoring and management of the Earth's resources and to better understand solid Earth processes. Many examples of possible uses of *Envisat* data in a wide range of environmental contexts in studying the atmosphere, the oceans, the land and the cryosphere are illustrated using data from other earlier satellite systems in an ESA (1998e, f) publication.

The instruments carried will be of two types – those developed by ESA and the so-called 'announcement of opportunity' instruments, developed by national or multinational institutions (see Figure 13.1).

ESA-developed instruments comprise:

- Medium Resolution Imaging Spectrometer (MERIS) – a high-spectral-resolution imaging spectrometer operating in the solar reflective spectral range, to produce, *inter alia*, large-scale maps of ocean pigment concentration, clouds and water vapour and vegetation status and distribution.
- Michelson Interferometric Passive Atmospheric Sounder (MIPAS) – a high-resolution Fourier Transform infrared spectrometer designed to measure a wide range of parameters for stratospheric chemistry, global climatology, atmospheric dynamics and upper-tropospheric chemistry.
- Advanced Radar Altimeter (RA-2) – determination of the two-way delay of the radar echo from the Earth's surface to a very high precision, less than 1 ns. Operating over oceans, its measurements will be used to determine the ocean topography. Furthermore, RA-2 will be able to map and monitor sea ice, polar ice sheets and most land surfaces.

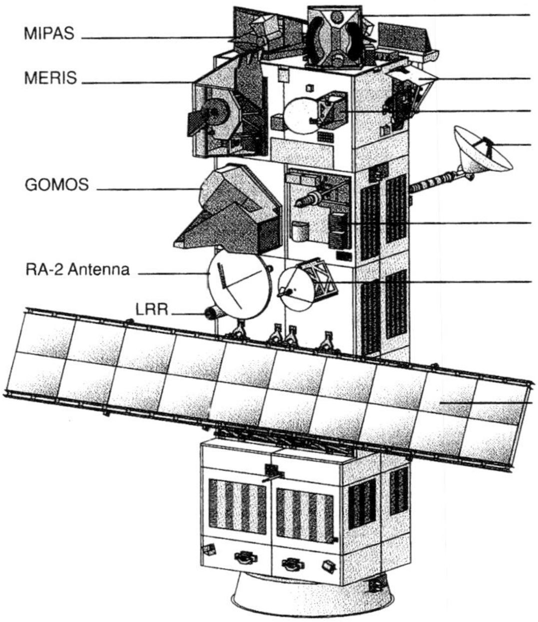

Figure 13.1. The *Envisat* scientific payload.

- Advanced Synthetic Aperture Radar (ASAR) – a high-resolution, wide-swathe imaging radar instrument that can be used for site-specific investigations as well as for land, sea, ice and ocean monitoring and surveillance. It will provide a continuation of the ERS-SAR missions.
- Global Ozone Monitoring by Occultation of Stars (GOMOS) – a UV-visible and a near-infrared spectrometer fed by a telescope looking at a target star for the measurement of profiles of ozone, NO_2 (nitogen dioxide), NO_3 (nitrate radical), OClO, temperature and water vapour, from the tropopause to 100 km with vertical resolution better than 1.7 km.
- Microwave Radiometer (MWR) – a dual-channel Dicke-type radiometer to determine column water-vapour and liquid-water measurement using passive microwave radiometry.

The opportunity instruments comprise:

- Scanning Image Absorption Spectrometer for Atmospheric Cartography (SCIAMACHY) – the primary mission objective is to perform global measurements of trace gases in the troposphere and stratosphere using nadir and limb viewing, and differential optical absorption spectroscopy.
- Advanced Along-Track Scanning Radiometer (AATSR) – the prime scientific objective is to establish continuity of the ATSR-1 and ATSR-2 datasets of precise sea-surface temperature thereby ensuring the production of a unique 15-year near-continuous dataset at the levels of accuracy required (0.3 K or

better), for climate research and for the community of operational as well as scientific users, which will have been developed through the *ERS-1* and *ERS-2* missions.

- Doppler Orbitography and Radio-positioning Integrated by Satellite (DORIS) – a microwave Doppler tracking system for precise orbit determination to an accuracy of the order of centimetres.

13.5 ORBIT AND MISSION

Envisat will be launched into a Sun-synchronous orbit with mean altitude of 800 km. It will have a 35-day repeat cycle, with 1–3-day subcycles for some instruments in order to give suitable information for meteorological and climate data-assimilation requirements. The global mission will be to provide continuous datasets for climate studies, meteorology, etc. This will involve all instruments (MERIS, ASAR, low bit-rate modes) and will make use of tape and solid-state recorders and the data will be dumped to primary ground stations either directly or via the *ARTEMIS* relay satellite. For regional missions – such as agricultural applications, marine-pollution monitoring, etc. – the full-rate MERIS (300-m resolution) and ASAR (all modes except wave mode) data will be required only when in line of sight and will be collected by local ground stations.

13.6 INSTRUMENT DETAILS

Envisat has a unique combination of sensors which will improve the range and accuracy of scientific measurements of the atmosphere, oceans, land and ice, particularly for environmental monitoring on a global scale.

The instruments of most relevance to remote sensing are ASAR, MERIS, AATSR and RA-2. Associated data for climate studies are, of course, provided by the other instruments – MIPAS, GOMOS, MWR, SCIAMACHY and DORIS. Further details of these are as follows (see also ESA 1998e, f).

13.6.1 ASAR

ASAR is a high-resolution, wide-swathe imaging radar that can be used for both site-specific research and for global land and ocean monitoring and surveillance. The main objectives are to monitor the Earth environment and collect information on, among other things, ocean-wave characteristics, sea-ice extent and motion, snow and ice extent, surface topography, land-surface properties, surface-soil moisture and wetland extent and deforestation and extent of forest areas.

Compared with *ERS-1/2* instruments, ASAR is a significantly advanced instrument employing a number of new technological developments. It uses an active phased array antenna system using distributed elements, allowing up to 400-km-wide scan coverage using ScanSAR techniques and the alternating polarization feature allowing targets to be imaged simultaneously in both vertical (V) anc horizontal (H) polarizations. It is a C-band SAR which can operate in a numbe of modes: image, wave, alternating VV/HH polarization (all 30-m resolution), wide swath (100 m) and global monitoring (ScanSAR) (1,000 m). Swathwidths vary from 5 km to 400 km depending on mode (see Plate 3).

13.6.2 AATSR

The main mission objective of AATSR is to extend the record of global sea-surface temperatures to an accuracy of better than 0.3 K for a period of 15 years. It will also produce cloud parameters and experimental vegetation products – such as biomass, moisture and health and growth stage. AATSR is an enhanced version of those carried on *ERS* satellites. It consists of a dual-view imaging radiometer with views at nadir and 55° forward of nadir. Two views of the same point through different path lengths of atmosphere are provided, enabling atmospheric corrections to be performed. Seven spectral bands will be used – 0.55 μm, 0.67 μm, 0.87 μm, 1.6 μm, 3.7 μm, 10.7 μm and 12 μm – (compared with only four on ATSR-1), at a spatial resolution of 1 km at nadir over a 500-km swathwidth.

13.6.3 RA-2

The radar altimeter will measure ocean topography, wind speeds and wave heights. It will also map and monitor sea ice, polar ice sheets and provide some information on land surfaces as well. It operates by determining the two-way delay of a nadir-pointing radar echo from the Earth's surface to a precision of better than 1 ns. It also measures the power and shape of the reflected pulse which contains the wind and wave information. Radar altimeters have been flying on satellites since 1973. RA-2 is an improved dual-frequency (K_u band and S band) version of those instruments flown on *ERS*, with a footprint of about 2 km (depending on frequency and mode of operation). It will also be able to map sea ice and polar ice sheets. The dual-frequency operation allows for ionospheric corrections to be made and a programmable range resolution will permit different surfaces to be tracked.

13.6.4 MERIS

Up to fifteen spectral bands can be selected by ground command, each of which has a programmable width and a programmable location in the 400–1,050-nm spectral range. MERIS is a push-broom instrument operating in the visible and near infrared. The 1,150-km-wide swathe is divided into five segments covered by five identical cameras having corresponding fields of view with a slight overlap between them and each receiving a 14° section of the instrument's total field of view. Full spatial resolution data (at 250 m at nadir) will be transmitted over coastal zones and land surfaces. Reduced spatial resolution data, achieved by on-board combination of 4 × 4 adjacent pixels across track and along track resulting in a resolution of approximately 1 km at nadir, will be generated continuously. These data will be processed in flight and further on the ground to provide atmospherically corrected data for ocean pigment concentration, clouds and water vapour and vegetation status and distribution. Fully programmable on-board processing will allow arbitrary selection of spectral bands (between 400 and 1050 nm) and the two spatial resolutions. CCD (charge-coupled device) arrays have a useful image area of 740 (spatial) × 520 (spectral) pixels with a spectral sampling interval of 1.25 nm. Fifteen spectral bands (central frequencies programmable) with bandwidths programmable from 1.25 to 25-nm will be available. Observations will be made with solar-elevation angles greater than 100°, and calibration will be performed at the orbital south pole, where the calibration diffusers are illuminated by the Sun by rotating a calibration mechanism.

Its missions include biooptical oceanography (assessment of ocean-surface optical properties and water constituents, leading to phytoplankton biomass and productivity estimates and pigment concentrations), atmospheric monitoring (investigation of cloud and aerosol parameters such as cloud-top height, water-vapour column content and aerosol load) and land-surface processes (global-scale monitoring of vegetation, assessment of its distribution and condition).

13.6.5 MIPAS

MIPAS is a high-resolution Fourier Transform Infrared (FTIR) spectrometer which is designed to measure concentration profiles of various atmospheric constituents on a global scale. It is a limb-sounding instrument, operating in the wavelength range 4.15–14.6 μm, providing observations of photochemically interrelated trace gases in the middle atmosphere, in the tropopause and in the upper troposphere. It is designed to measure simultaneously more than twenty relevant trace gases. MIPAS observations will also allow the derivation of atmospheric temperatures, the distribution of aerosol particles, tropospheric cirrus clouds and stratospheric ice clouds amongst others.

13.6.6 GOMOS

GOMOS has been designed to measure trace-gas concentrations and other atmospheric parameters in the altitude range between 20 and 100 km. The overall instrument specifications (coverage, spatial and spectral resolutions and accuracy) are far superior to previous systems such as TOMS (Total Ozone Monitoring Spectrometer), SBUV (Solar Backscatter Ultraviolet) or SAGE (Stratospheric Aerosol and Gas Experiment). It consists of a UV-visible spectrometer oriented towards the target star by means of a steerable mirror. It then tracks the star and observes its setting behind the atmosphere. About twenty five stars are bright enough to observe, and GOMOS will produce as much data as a global network of 360 ground stations. Parameters measured include vertical profiles of water vapour (which is a major contributor in ozone-destruction processes) and vertical temperature profiles (which are useful for deducing ozone-concentration profiles).

13.6.7 MWR

The main objective of the MWR is to measure atmospheric humidity as supplementary information for tropospheric-path correction for the RA signal. In addition these data are useful for determining surface emissivity and soil moisture over land and the surface-energy budget, and for investigations to support atmospheric studies and for ice characterization. The MWR is a two-frequency radiometer very similar to those on the *ERS* satellites.

13.6.8 SCIAMACHY

Global measurements of various trace gases in the troposphere and stratosphere can be retrieved from SCIAMACHY by observing the transmitted, back-scattered and reflected radiation from the atmosphere in the wavelength range 240–2,400 nm. The large wavelength range is also ideally suited to the measurement of aerosols and clouds. The nadir and limb-viewing strategy yields total-column values as well as profiles for trace gases and aerosols in the stratosphere. Measurements are performed by observing the atmosphere under different viewing angles, and also by using differential optical-absorption spectroscopy in Sun- and Moon-occultation measurements.

13.6.9 DORIS

DORIS is an orbit-determination instrument, providing data on the orbit of the satellite to an accuracy of a few centimetres. This is required for the accurate altimetry measurements made by RA-2. DORIS is based on the accurate measurement of the Doppler shift of radiofrequency signals transmitted from a worldwide network of ground beacons and received on board the spacecraft. The location of each ground beacon must be known to within a few centimetres. On-board measurements are performed every 7–10 s, enabling radial velocity values to be determined to an accuracy of 0.44 mm s^{-1}. This is combined with a dynamic model of the satellite trajectory to perform the precise orbit determination.

13.7 INSTRUMENT SYNERGIES

It can be seen that the range of instruments is remarkably diverse. Some instruments have been designed to work closely together (e.g. the RA-2 requires input from both MWR and DORIS for some of its basic geophysical products), whereas other instruments provide synergistic opportunities simply because several different types of observations are available for the same place and time. Table 13.2 shows the measurement capabilities of each of the instruments and the possibilities for combining instrument data within different applications.

There is clearly great scope for synergy and intercomparison of results with data from SCIAMACHY, GOMOS and MIPAS which are functionally quite different

Table 13.2. Measurement capabilities of the *Envisat* instruments.

	MWR	RA-2	ASAR	MERIS	AATSR	SCIAMACHY	MIPAS	GOMOS
Atmosphere								
Temperature/pressure				■		■	■	■
Trace gases						■	■	■
Cloud type/height				■	■	■	■	
Water vapour	■			■	■		■	■
Aerosols				■	■	■	■	■
Radiation budget	■			■	■	□	□	□
Ocean								
Marine geoid		■						
Global circulation		■			■			
Wave characteristics		■	■					
Ocean fronts			■	■	□			
Ocean colour and turbidity				■				
Sea-surface temperature					■			
Coastal dynamics			■	■				
Oil slicks			■					
Ship traffic			■					
Land								
Global vegetation monitoring			■	■	■			
Surface temperature	□				□			
Agriculture and forestry			■	□	□			
Geology and topography		□	■					
Hydrology parameters	□	□	■	■	■			
Flooding			■					
Fire					■			
Ice								
Temperature					■			
Ice-sheet dynamics	□	■	■	■				
Snow cover			■	■	■			
Sea-ice mapping	□	■	■	■				
Ship routing			■					

■ Principal contribution
□ Secondary or experimental contribution

instruments, but which measure similar atmospheric properties in partially overlapping profiles. Data from these instruments may also be integrated with surface measurements from MERIS and AATSR to examine atmosphere–biosphere interactions and to improve atmospheric corrections.

Envisat will be able to provide simultaneous measurements from AATSR/MERIS, and from MERIS/ASAR (see Figure 13.2). The ATSR swath of 500 km around nadir falls within the 1,150-km swathe of MERIS. Whenever ASAR is operating in daytime, it will be possible to obtain simultaneous MERIS coverage

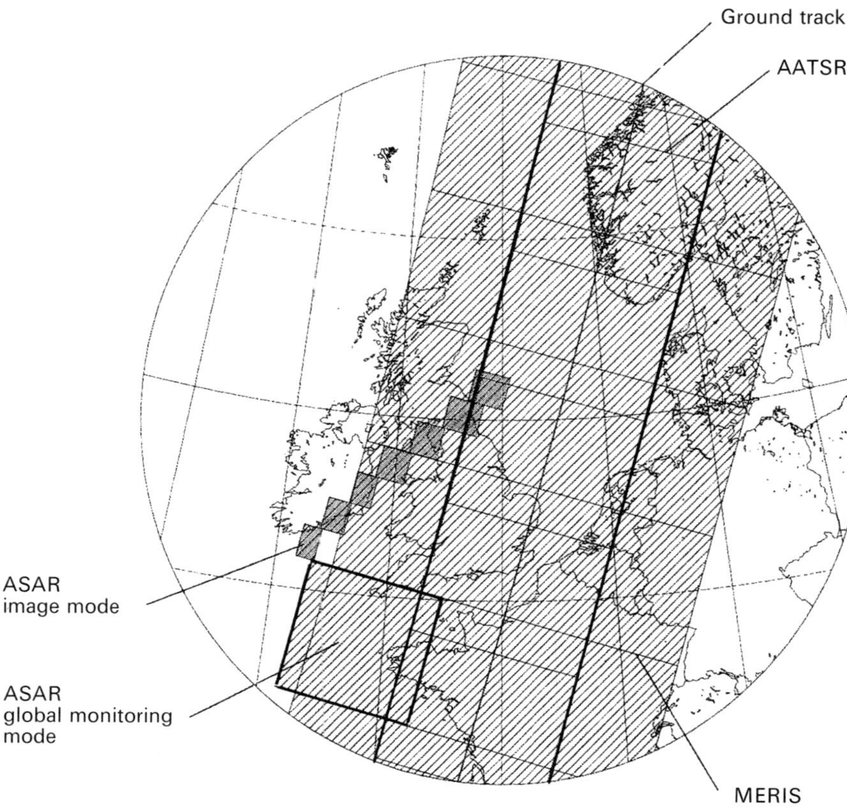

Figure 13.2. Comparisons of swath coverage for ASAR, MERIS, AATSR.

except at the highest incidence angles. It will be possible to acquire a global coverage of simultaneous ASAR low-resolution images and MERIS data every 8 days.

Over the oceans, these combinations of AATSR, MERIS and ASAR will provide a unique synergy for bio-geophysical characterization. Simultaneous measurements of comparable resolution at both microwave and optical wavelengths have not previously been available on a routine basis. These instruments permit small- and large-scale wave characteristics, ocean colour and temperature to be observed with frequent global coverage. It is possible therefore to study transient phenomena (ocean fronts, gyres, blooms, upwellings, surface films, etc.) and their dynamics in a way which was previously not readily possible. In addition, the combination of AATSR and RA-2 will provide coincident observations of ocean topography and sea-surface temperature along the satellite track.

On land there is great potential to exploit synergies between ASAR low-resolution products and MERIS data and also between MERIS and AATSR data. In addition, the combination of ASAR differential interferometry and RA-2 data will improve the potential for land-surface-height measurement.

The *Envisat* instruments will also complement instruments flying on other satellites at the same time. GOMOS, MIPAS and SCIAMACHY will complement *SAGE II's* TOMS and GOME. MERIS and AATSR complement SeaWiFS, MODIS and AVHRR. RA-2 complements *Topex-Poseidon's* successor *Jason* and the *Geosat* follow-on. Several of the instruments on these satellites will provide opportunities for improving the frequency of observation as well as for the intercalibration and comparison of measurements derived from independent instruments. In addition, some instruments will provide crucial auxiliary data to assist in the retrieval of more accurate geophysical measurements.

13.8 *Envisat* DATA POLICY

A flexible pricing policy applies to *Envisat* data in order to make it accessible to a range of users. The conditions attached to its distribution depend on the use of the data. Two categories of use are defined:

1. Research and applications development in support of the mission objectives, including research on long-term issues of Earth system science, and R&D in preparation for future operational use;
2. All other uses which do not fall into category 1 including operational and commercial use.

For category 1 ESA will retain overall policy and programmatic responsibility for the distribution of data and will be the direct provider of the service from its own facilities wherever possible. ESA will also fix the price, which will be set at or near the cost of reproduction of the data. Data from the 'atmospheric' instruments (GOMOS, SCIAMACHY and MIPAS) will always be provided under category 1. Category 1 users will include those principal investigators (PIs) under ESA announcement of opportunities, who will get the data free of charge so long as they provide confirmation of project funding. Other users – such as those from EU R/D programmes and large international programmes, and those involved in education and training – will be subject to peer review.

Strategic Data Sets (SDS) of ASAR and MERIS FR (full resolution) data are provided under certain conditions. These SDS collections must meet strategic needs, consistent with the mission objectives, but not be associated with distribution under categories 1 or 2. They must be precisely defined in terms of geographical coverage, acquisition strategy, time period, required completeness and timeliness. Examples of SDS are:

- ASAR European coverage in ERS-like mode;
- ASAR polarization combinations at various incidence angles on a set of sites distributed worldwide;
- ASAR wide-swathe world coverage of land and ice masses, the objective being to achieve at least one complete coverage dataset;
- MERIS FR coverage of Europe, the acquisition being systematic when flying over Europe;

- MERIS FR on a set of selected sites distributed worldwide;
- ASAR wave mode over oceans;
- ASAR wide-swathe over European coastal zones;
- ASAR global-monitoring mode over any land and ice surface.

13.9 CONCLUSIONS

Although originally flagged up in 1992 as *Envisat-1*, there would appear to be no firm plans for any more follow-up missions of this form. The present-day thinking of 'smaller, cheaper, faster' missions certainly has much to commend it – rapid response to changing situations, dedicated instruments not subject to compromises in orbit/repeat from other instruments, instruments targetted to specific scientific problems, etc. But, set against these advantages, is one major disadvantage – the loss of synergy between the instruments. *Envisat* will provide an opportunity to carry out major scientific investigations not achievable from dedicated platforms. Make the most of it – this may well be the last opportunity.

14

Supercomputing for the masses – the SETI@Home project

C. P. Cracknell

14.1 INTRODUCTION

The Search for Extraterrestrial Intelligence (SETI) has been a part of human activity for some time. A recent development is the SETI@Home project This project is not directly relevant to remote sensing (Earth observation) and climate change. However, it is interesting to try and see if the concepts involved might be capable of being adapted to some aspects of the processing of large remote-sensing datasets in the search for evidence related to climate change. We shall describe the SETI@Home project first and then consider the possible relevance of some of its concepts to remote sensing.

14.2 OBJECTIVE OF THE SETI@Home PROJECT

The SETI@Home project is concerned with analysing radio signals received from space to try to detect intentional transmissions that originate from some extraterrestrial civilization. The data are collected using the (US) National Astronomy and Ionospheric Center's 305-m radio telescope in Arecibo, Puerto Rico. The SETI@Home project utilizes a dedicated L-band receiver at the Arecibo telescope, which also carries a receiver dedicated to the earlier SERENDIP IV SETI programme. Using an independent feed and receiver permits SETI@Home observations to be carried out simultaneously with ongoing ionospheric and astrophysics research programmes at the telescope; in this way SETI@Home can obtain observations all the year round without interfering with the main scientific tasks of the telescope. This commensal technique, which is also called 'piggyback SETI' was developed by the SERENDIP SETI programme at the University of California, Berkeley (Bowyer *et al.* 1983; Werthimer *et al.* 1997). The fact that the receiver is carried as a passenger on the Arecibo radiotelescope while ordinary radio-astronomy work is carried out means that the sky is viewed in a pseudo-random fashion. However, any given patch of sky is typically revisited every 3-6 months and this is important for discriminating against man-made Radio-Frequency Interference (RFI). Data for SETI@Home can be obtained when the telescope is stationary or the Gregorian feed is tracking a source. Data cannot be obtained when a transmitter is on or when the telescope is slewing round the sky.

The problem in any SETI experiment involves trying to find weak signals in noisy data; this inevitably requires a large amount of processing power. There have been previous SETI projects which have looked for signals which might have originated from extraterrestrial intelligences. However, a more CPU-intensive approach could allow the detection of signals up to 10 times weaker than previously detectable, but the computer costs would be prohibitive. Therefore, instead of using an expensive supercomputer in a central research laboratory, the solution adopted by SETI@Home is to attempt to harness the spare CPU cycles of desktop computers around the world (Sullivan *et al.* 1997). Most home computers and many institutional desktop computers are idle for a considerable fraction of the time. The problems involved in harnessing this spare computer power include (a) providing some incentive for people to offer the spare capacity of their machines to the project, (b) organizing the distribution of the data (and processing software) to all the home computers involved, (c) receiving all the results back from the various home computers and (d) validating and checking the system. The incentive offered to individual computer users, by donating their spare computing capacity to the project, is the slight possibility that their computer may be instrumental in the detection of another technical civilization in the Milky Way. The SETI@Home project (a) has provided the opportunity for ordinary computer users to be able to participate in a major scientific experiment and (b) is the largest distributed computation ever undertaken. There have been other projects which have harnessed the spare computing power in partly idle desktop computers (e.g. a successful attempt to crack the exportable version of Netscape Navigator's security). However, SETI@Home is the biggest so far.

14.3 THE SETI@Home PROJECT

The major components of the SETI@Home project include:

- a sky-survey experiment running piggyback on the Arecibo radio telescope;
- a collection of many (as of 11 August 2000 more than 2.2 million) computers around the world co-operating to process the data;
- a server infrastructure to pre-process the raw data, prepare it for the individual processing units and collect the results of the processing.

The overall architecture of SETI@Home is shown in Figure 14.1. This figure indicates the relation between the radio-telescope facility at Arecibo and the company Big Science, Inc.

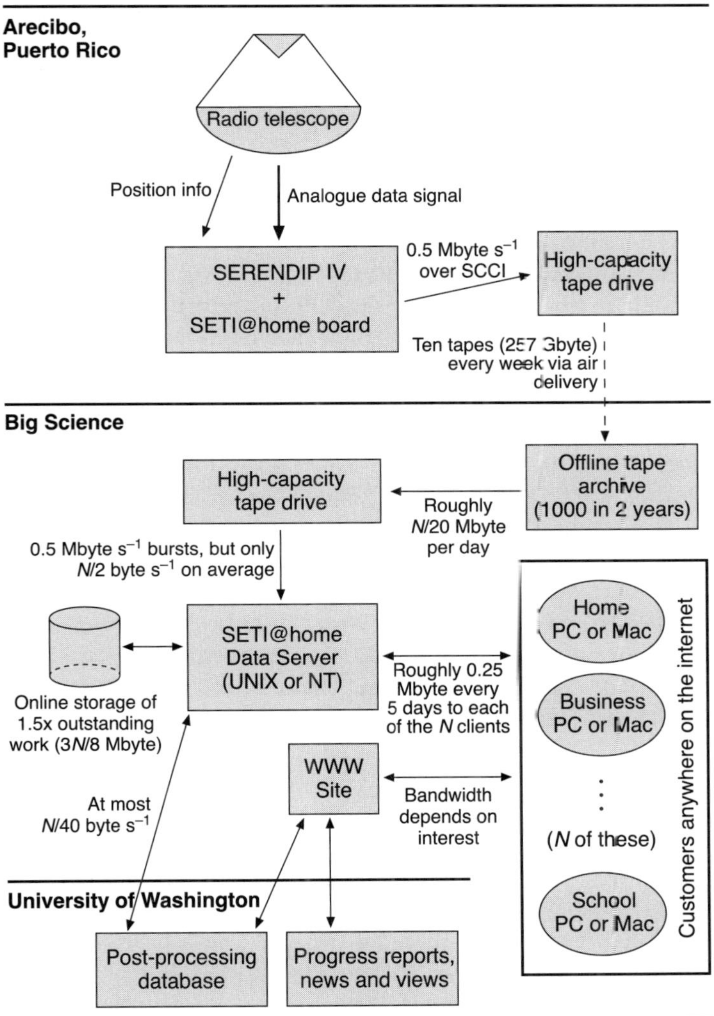

Figure 14.1. A schematic diagram of the architecture and data flow of SETI@Home. Note that N refers to the number of computers currently participating in the analysis.

The frequency chosen for the sky survey covers a 2.5-MHz bandwidth centred at 1,420 MHz. This frequency was chosen because it corresponds to the wavelength of an important line in the hydrogen spectrum (21 cm) which, it has been argued, is the wavelength at which intentional interstellar transmissions are most likely to be found. A bandwidth of 2 MHz is adequate to allow for Doppler frequency shifts as a result of the three possible velocity rest frames of heliocentric, galactocentric and cosmic background radiation (2.5 MHz includes a 0.5-MHz margin). The digitized raw data, along with housekeeping data, are acquired at the rate of 0.5 Mbyte s^{-1}

and recorded continuously onto tape. One archive tape contains the data from 1 day (35 Gbyte).

The initial processing of the data from SETI@Home archive data tapes involve dividing the data into small 'work units'. The original 2.5-MHz data are broken down into 256 subbands of approximately 10 kHz (9.765 kHz) by means of a 2,048-point Fast Fourier Transform (FFT) and 256 eight-point inverse transforms. Each work unit then consists of 107 s of data from one of these subbands; this amounts to 250 kbyte and the whole work unit sent to each client is 370 kbyte, which includes ancillary information. Adjacent work units overlap. Work units are then sent over the Internet to the user's computer for the bulk of the data analysis. The automatic use of the idle time of a user's computer is achieved by providing a client program which can be run either as a screen saver or as a daemon to process the work unit. Running the program as a daemon is most likely to be faster (in real time) but it is less interesting than a screen saver since there is nothing for the user or anyone else to watch. Programs are available for most operating systems and CPUs. The user is not able to alter the client program. The client connects to the Internet and downloads a work unit which takes about 2.3 min on a 28.8-kb s^{-1} modem. On completion, the client program can either automatically connect to the Internet or prompt the user to instigate the connection. The client then sends the results to the SETI@Home server and downloads a new work unit.

The bulk of the computing performed on a work unit of data is to carry out an initial filtering of the data to identify likely candidate signals for further investigation. Signals with the following characteristics are sought:

- 12-s Gaussian peaks;
- narrow-band signals;
- pulsed signals;
- chirped signals.

After the user's computer has automatically processed a complete work unit, which is likely to take a few days (of real time) on a typical home or office computer, a short message reporting candidate signals is generated for the client and is also returned to Big Science, Inc., and to the University of Washington for post-processing. The post-processing involves building up a database of all candidate signals reported back from the users' computers. This database is used to monitor progress with the survey, to monitor data quality and the RFI environment at Arecibo, and to decide if any changes in survey or operations procedures may be desirable. Most of the signals found by the users' computers turn out to be terrestrial-based RFI. A number of algorithms are used to reject the various types of RFI (Cobb *et al.* 1997). After the RFI is rejected, the remaining data are searched for multiple detections. It is extremely important that any high signal/noise candidates observed once are only considered highly tentative until they are positively confirmed by other (random) passes (usually 3–6 months later) of the beam over the same patch of sky. The post-processing will monitor for these subsequent passes and compare the consistency of the characteristics of the candidate signals for each pass. Higher weight is given to drifting or pulsed signals, those that repeat in the barycentric frame, that match the antenna beam pattern, or detections coincident with newly detected planets, nearby stars (from the Gliese catalogue) or globular clusters (see also Cobb *et al.* 1997). Candidate signals are compared with SERENDIP IV data and plans are made to follow up the most interesting candidates with dedicated observations using the radio telescope at Arecibo.

14.4 MOTIVATION OF USERS

Why and how do the individual users offer their computers to take part? The motivation of individual users will, of course, be variable but the reasons are likely to include the following:

- general interest in SETI;
- a share in the credit if anything is discovered;
- publicity in the press.

A record is maintained of the number of work units processed by each user. There is a World Wide Web (WWW) site to capture the interest of the users and foster their sense of involvement in the project. Information available includes regularly updated statistics, news and other information about progress of the SETI@Home project, as well as technical information and links to more detailed sources of information.

Some SETI@Home statistics are shown in Table 14.1. The first date, 30 July 1999, corresponds to about 3 months after the start of the project and the second date, 11 August 2000, is slightly over a year later than that. By the first of these dates, 38,579 years of CPU time had already been used on the project. For the first of these dates the breakdown of locations for the 10.9 million work units processed was:

- work 5 million work units;
- home 4.9 million work units;
- school 0.67 million work units;
- other 0.26 million work units.

It is not surprising that it is a major task to provide the data to keep all these clients supplied with work units.

Further information about the SETI@Home project can be found from its website – http://www.setiathome.ssl.berkeley.edu – and more general information can be obtained from the YAHOO Supercomputing and Parallel Computing

Table 14.1. Some SETI@Home statistics.

	30 July 1999	11 August 2000
Users	920,227	2,234,262
Results received	10,911,313	167,730,024
Floating-point operations	$2.182\,263 \times 10^{19}$	$3.354\,600 \times 10^{20}$
Average CPU time per work unit	30 hr 58 m 23 s	Not available

directory – http://dir.yahoo.com/Computers_and Internet/Supercomputing_and_Parallel_Computing/

14.5 POSSIBLE RELEVANCE TO REMOTE SENSING AND CLIMATE CHANGE

In contemplating the possible relevance of the general concept of the SETI@Home project to remote-sensing applications, our thoughts naturally turn to situations with intensive computing requirements. A possible example that we might think of is the processing of raw SAR (Synthetic Aperture Radar) data to generate an image. In the early days of spaceborne SAR, following the launch of *Seasat* in 1978, the lack of computing power was a problem. However, the speed and power of individual computers has now increased so much over the last 20 years that any attempt to set up a distributed supercomputing project along the lines of SETI@Home for SAR processing seems unlikely to be necessary or sensible. Another possibility is the question of the analysis of the archives of cloud images from meteorological satellites to attempt to determine whether there might be any change in the percentage of cloud cover or in the nature of clouds which might be indicative of climate change. We now have a few decades of cloud data, both from polar-orbiting spacecraft and from geostationary spacecraft. We also have had the ISCCP (International Satellite Cloud Climatology Project), but I do not know to what extent ISCCP has already reprocessed all the archived data to search for trends in the cloud patterns and, therefore, to what extent – if this has not been done – this might constitute an opportunity for a distributed supercomputing SETI@Home-type of approach or not. This should be investigated.

Another possibility is the question of the AVHRR (Advaced Very High Resolution Radiometer) archive of visible and near-infrared data (channels 1 and 2) and the vegetation indexes, particularly the Normalized Difference Vegetation Index (NDVI) which can be calculated from the calibrated channel-1 and channel-2 data. We are now in a position where this archive of data could be studied to look for evidence of local climate changes such as an increase in desertification.

It would seem that attempting to copy the SETI@Home project for a remote-sensing task to the extent of involving the general public is not likely to be necessary in terms of the amount of data needing to be processed or in terms of the organizational and publicity effort that would be involved. The other aspect is that the SETI@Home project is dependent on the goodwill of a large number of members of the public; this goodwill is unlikely to be forthcoming for many other projects. The search for extraterrestrial intelligence is something that captures the imagination of the public in a way that few other projects could.

However, although there is probably little case for attempting to involve the general public in a distributed supercomputing project along the lines of SETI@Home, there may still be scope for adopting the ideas of this project within an individual organization. Even within an organization (e.g. all the computers on a university campus), there is a very considerable useful amount of cheap processing power available. Moreover, within an organization, the whole task may be simpler for several reasons. There is likely to be a homogeneity of clients and therefore no need to support many different types of client. Within an organization there is likely to be more network bandwidth – enabling more data-intensive analysis – and better control of set-up on the users' computers.

15

Earth-observing systems and model-simulated datasets for climate study

V. Djepa, R. A. Vaughan and M. Menenti

15.1 INTRODUCTION

The objective of this chapter is to identify the main Earth-system factors and interactive processes affecting climate change and to point out the tools, datasets and models required to observe, predict and understand the changes of the climate on Earth. The definitions of the Earth system, feedback and interactive processes affecting climate change, and the basic principles for the retrieval of climate variables from remotely sensed or Earth-observation data are given. The temporal variability of the most important climate components and processes is discussed in order to evaluate the required timescale for observing and predicting the evolution of the Earth's ecosystems. The basic principles for the retrieval of climate components from Earth-observation data will be examined and the applications of some models for climate-variability prediction will be discussed.

International co-operation in Earth observation over the past few years has led to a series of space missions. *ERS-1*, *ERS-2* (European Space Agency [ESA]), *Meteosat* (EUMETSAT), NOAA, GOES, DMSP, *Topex-Poseidon*, CZCS on *Nimbus-7*, SeaWiFS, MODIS, SMMR, SSM/I, UARS, ERBE (NASA, etc) and *ADEOS* (NASDA) have all made significant contributions (see Chapter 3; for full names see Abbreviations and satellites, p. xv). Future important space missions are *Envisat* and *Earth Explorer* (ESA), MSG (EUMETSAT), *NOAA-K*, *-L*, *-M*, *EOS-COLOUR*, *EOS* (MOPIT, ASTER, EOSP, TES, AIRS, MIMR), *EOS-ALT* (*CHEM*, *SAGE III*, *ADEOS II*), and *Radarsat-2* (Canadian Space Agency [CSA]). Some of the international space programmes are co-ordinated by the United Nations and have high-governmental-level involvement; other programmes work through the International Council of Scientific Unions (ICSU) and the World Meteorological Organisation (WMO).

The main space centres and distributors of global datasets for climate study are listed in Table 15.1. EUMETSAT (European Organization for Meteorological

Table 15.1. Space centres and distributors of global datasets for climate study.

European Space Agency (ESA)
European Organization for Meteorological Satellites (EUMETSAT)
The EOS Data and Information System (EOSDIS)
Distributed Active Archive Centres (DAACs)
National Environmental Satellite Data and Information Service (NESDIS)
National Climatic Data Centre (NCDC)
National Geophysical Data Centre (NGDC)
National Oceanographic Data Centre (NODC)
NASA Langley Research Centre (NLRC)
Canadian Space Agency (CSA)
Japanese Space Agency (NASDA)

Satellites) is working closely with ESA in the development of new operational systems (*Meteosat* Second Generation [MSG], *METOP* and GCOS [Global Climate Observing System]). NOAA (National Oceanic and Atmospheric Administration) includes the National Weather Service, National Ocean Service, National Marine Fisheries Service, Office of Oceanic and Atmospheric Research and NESDIS (National Environmental Satellite Data and Information Service). The National Climate Data Center (NCDC) provides historic weather and climate datasets, including satellite imagery from the NOAA Satellite Active Archive (SAA). The NGDC (National Geophysical Date Center) maintains solar, marine and terrestrial geophysical information and the NODC (National Oceanographic Data Center) provides historic physical, chemical and biological data from the global oceans. The main objectives of Earth-observing systems and the space centres is to collect and process global datasets of the Earth and to assimilate the satellite products in Earth System Models (ESM) for climate change.

15.2 BASIC PRINCIPLES OF EARTH-OBSERVATION-RETRIEVED CLIMATE DATASETS

The retrieved climate components from Earth-observation datasets are available on a global scale from climate centres and distributors (see Table 15.1).

15.2.1 Climate components and feedback

Climate is usually defined as an average state of the weather of the Earth system over a finite time period of a number of years. Climate change is due to interactions of the Earth's ecosystems and climate forcing. The climate forcing includes external factors – such as solar activity – and internal factors – such as natural and human-induced changes (greenhouse gases, troposphere aerosols, land-surface changes etc; see Chapter 1). Climate change is also affected by negative or positive feedback into and between the Earth's ecosystems. The various components of the physical climate system have significant internal variability and fluctuations over a wide range of timescales. This involves a multidisciplinary study of the Earth as a system of interrelated processes between the atmosphere, oceans, land surface and polar ecosystems. The climate interactions are divided into fast components (atmospheric and weather related), medium-term components (atmospheric interactions with ocean, cryosphere and land biosphere) and long-term components (such as those from human activity). Three-dimensional distributions of the atmospheric constituents (clouds, aerosols, trace gases, wind field and water content) and their interaction with radiative and dynamic fields are needed to understand the internal variability of the short-term climate components. To evaluate the seasonal and inter-annual variability of the medium-term climate components and their impact on climate change, more accurate retrieval of fluxes between the ocean–atmosphere, atmosphere/land–biosphere and ocean–atmosphere–cryosphere interfaces are required. The main long-term components of the climate are the radiation budget of the Earth.

Systems and processes, including human activities and changes in CO_2 (carbon dioxide) level, contribute to climate variation over decades or longer timescales. Ocean mixing, deep- and bottom-water formation, the ocean carbon cycle and land–atmosphere interactive processes (heat and gas exchange) are also long-term factors. The biophysical and chemical processes of the land surface are determining factors for long-term climate change, but still are not adequately represented in climate models. The sea ice and snow cover have much shorter time scales than the ice sheets, but are important for climate conditions not only in the short term, but also over decades and longer time scales.

For long-term climate components, it is important to improve the datasets for temporal variability of ocean circulation, transport of energy and dissolved species, broadband radiation fluxes, sea-ice dynamics and deep-water formation.

Atmospheric datasets for climate study

The fast components of the climate are primarily atmospheric related and require an improved knowledge of the atmospheric constituents, processes and dynamic global energy, water and hydrological cycles. The main atmospheric constituents, controlling short-term climate changes are water vapour and trace gases (carbon dioxide, ozone [O_3] and nitrous oxide [N_2O]). The main atmospheric interactive processes affecting climate change are radiative, chemical and dynamical. The most important radiative processes are (i) atmospheric heating or cooling (ii) aerosol scattering and absorption (iii) cloud microphysics and radiative processes. The coupling between aerosols, clouds, precipitation and radiation is a powerful feedback mechanism in the climate system. The main chemical mechanism, controlling the interrelationship of the physical and chemical processes in the atmosphere, could be divided into five categories – photochemistry, heterochemistry, anthropochemistry, biochemistry and geochemistry.

Gas constituents

Cooling occurs by thermal emission to space primarily from water vapour (H_2O), carbon dioxide, (CO_2), ozone molecules (O_3), trace gases, clouds and the surface. The increase in concentration of infrared absorbers in the atmosphere would lead to a rise in the surface temperature and a reduction of the infrared flux to space, known as the enhanced greenhouse effect. Measurements in strong absorption lines are used to retrieve gas constituents. Water vapour plays a dominant role in the radiative balance and the hydrological cycle. It is a principal greenhouse gas and is involved in fundamental photochemical processes. Carbon dioxide, has strong absorption lines at 15 μm and its increasing concentration is a potential for greenhouse warming of the Earth. It affects the stratospheric ozone and is involved in geochemical and biochemical cycles of the land and marine biota. Ozone is the most photochemically active gas, important because of its strong absorption of ultraviolet and infrared radiation. Methane (CH_4) absorbs thermal infrared radiation at 2.22–2.41 μm, 3.1–3.7 μm, 3.85 μm and 7.3–8.3 μm, and modifies the energy balance. The first spectral range is free from water absorption and is used for satellite remote sensing of methane. Nitrous oxide absorbs at 8.6 μm and 7.8 μm and its concentration is increasing at about 0.2 per cent per year (Butcher 1992). It has a strong greenhouse effect through thermal infrared absorption in the troposphere. CFCs (chlorofluorocarbons) have a strong absorption band at 8–14 μm. Retrieval of stratospheric gases usually utilizes infrared limb sounders (LIMS), or solar-occultation measurements (SAMS on *Nimbus-7* for CH_4, NO [Nitric oxide], N_2O, CO [carbon monoxide], CO_2, or LIMS for HNO_3 [nitric acid]). Stratospheric and mesospheric gases, measured by the Upper Atmospheric Research Satellite (*UARS*) are given in Table 15.2. The impact of CFCs on the climate system involves direct radiative forcing in the long-wave spectrum and stratospheric-ozone depletion through the release of chlorine (Cl) at high altitudes.

It is expected that there will be increased destruction of stratospheric ozone in the next few years, as more chlorine is carried up into the stratosphere. Since 1978 daily global maps of total ozone have been obtained from *NOAA*'s High Resolution

Table 15.2. Stratospheric and mesospheric gases measured by *UARS*.

Instrument	Measured gases	Waveband
Cryogenic Limb Array	$O_3, NO, NO_2, N_2O, HNO_3$	IR, $\gtrsim$0.75 μm
Etalon Spectrometer	$H_2O, CH_4, CF_2, Cl_2, CF, Cl_3, HCl, ClONO_2, N_2O_5$	3.5–12.7 μm
Improved Stratospheric and Mesospheric Sounder (ISAMS)	$O_3, NO, NO_2, N_2O, HNO_3, N_2O_5, H_2O, CH_4, CO$	4.6–16.6 μm
Microwave Limb Sounder (MLS)	O_3, H_2O, ClO	63,183,205 mm
Halogen Occultation Experiment Solar occultation (HALOE)	O_3, NO, NO_2, H_2O	2.43–10.25 μm

Infrared Radiation Sounder (HIRS2) at 9.7 μm (Kidder and Van der Haar 1995). Ozone is also retrieved from LIMS in the 9.6-μm band. The best known method to retrieve ozone from satellite data makes use of backscattered ultraviolet radiation. Observations of the radiation reflected from the surface and backscattered by the atmosphere or clouds, at several wavelengths of the solar ultraviolet spectrum, are used for the retrieval of total ozone or the ozone profile. Total ozone is retrieved from the BUV (backscattered ultraviolet), which was flown on *Nimbus-4* from 1970 to 1977, and later from the Total Ozone Mapping Spectrometer (TOMS) on *Nimbus-7* and SBUV/2 on some of the *NOAA* satellites. Estimates of ozone profile use measurements at shorter, more opaque ultraviolet wavelengths, where the surface contribution is negligible. A third way to measure ozone takes advantage of the ozone absorption in the visible spectral range, using solar- or stellar-occultation experiments (Stratospheric Aerosol and Gas Experiment, SAGE/AEM2, SAGE II). The most accurate are the ultraviolet methods, but they cannot be applied in darkness – such as over the winter pole. Infrared methods, with either a downward-looking or a limb sounder, are perhaps a little less accurate than ultraviolet methods, but they do not depend on reflected sunlight.

The essential impact of greenhouse gases on climate state and ozone depletion requires further monitoring of ozone and physical and chemical processes that may affect the ozone layer. Chlorine, bromine (BrO) species, ozone and NO_2 (nitrogen dioxide) are retrieved from GOME (Global Ozone Monitoring Experiment), which allows investigation of the chemistry and dynamics of the troposphere and stratosphere, ozone and species trends and transport in the lower stratosphere. GOME on *ERS-2*, SCIAMACHY (Scanning Imaging Absorption Spectrometer for Atmospheric Cartography) on *Envisat*, TOMS and OMI on *METOP* and, after this, MLS, AIRS on *EOS-PM* and TES on *EOS-AM2* will ensure continuous monitoring of ozone for years to come.

Aerosols

A realistic representation of three-dimensional datasets of aerosol distribution is important for validation of climate models and to evaluate the impact of the increased concentration of aerosols on global warming and climate change. The aerosol measurements rely chiefly on the scattering of sunlight, using nadir-viewing techniques for tropospheric aerosols, or limb scanning for the stratosphere. *Landsat*, *GOES-1*, DMSP, CZCS on *Nimbus-7* and the *NOAA-5* scanning radiometers have been used for retrieving marine aerosols (Kidder and Van der Haar 1995). Sahara dust has been retrieved from AVHRR (Advanced Very High Resolution Radiometer) and GOES data. The aerosol optical thickness $\tau(0.55)$ is retrieved over water from AVHRR (Rao *et al.* 1989). In January 1989 NESDIS began to produce weekly maps of aerosol optical depth based on AVHRR channel-1 data. The maps have a 100-km grid over the world's ocean. Directional observations from ATSR-2 (Along-Track Scanning Radiometer) (Djepa and Vaughan 2000) and the future AATSR (Advanced Along-Track Scanning Radiometer) on *Envisat* allow accurate retrieval of aerosol optical thickness on a global scale with 1-km spatial resolution. MERIS (Medium Resolution Imaging Spectrometer) on *Envisat* will give aerosol optical thickness over water with higher spectral and spatial resolution. Solar-occultation experiments (SAM 1975, SAM II, 1978, SAGE I 1979 and SAGE II 1984) allow long-term stratospheric-aerosol-extinction profiles to be measured.

Temperature profile

For the initalization of global circulation models (GCMs) and numerical weather prediction (NWP) models, vertical profiles of temperature, humidity and wind speed are required on any grid point. Microwave Sounder Unit (MSU on *Nimbus-6*) channel-2 (52–55 GHz) brightness temperature is insensitive to the surface and has a weighting function that peaks in the middle troposphere. This is used to estimate the warming, or cooling, of the global atmosphere. TOVS (*TIROS* Operational Vertical Sounder) sounding (MSU, HIRS/2 and SSU) is used for temperature and water-vapour retrieval. The correlation of the 55.45-GHz brightness temperature anomaly with tropical cyclone central pressure is used for the retrieval of temperature profiles inside tropical cyclones by VAS and MSU (Kidder and Van der Haar 1995). The temperature and water vapour around 100 hPa are retrieved using statistical or regression analyses. TOVS soundings (troposphere temperature and water vapour, total ozone, cloud-top pressure and amount) are archived by NESDIS in the form of layer mean vertical temperatures for fifteen layers between the surface and 0.4 hPa. The precipitable water vapour is retrieved in three layers (surface–850, 850–700 and 700–500 hPa). VAS soundings have been used operationally at the NESDIS VAS Data Utilization Center since the summer of 1987. TOVS and GOES sounding have found great application for temperature in Numerical Weather Prediction Models (NWPMs) and GCMs.

Water vapour

Water vapour is a critical variable in atmospheric photochemistry, in the forecast of severe cyclones and weather and plays a key role in the Earth's climate, radiation budget and greenhouse effect. The atmospheric sensors for water vapour operate in strong absorption lines (6.3 μm, 23 GHz, 183 GHz, etc.). Total water vapour and liquid water content over the oceans are retrieved from Microwave Sounders (MWS/ATSRM) with resolution of 23 km. NASA's (National Aeronautics and Space Administration) Upper Atmosphere Research Satellite (*UARS*, launched 15 September 1991) is utilized for the retrieval of upper troposphere and stratosphere water vapour and ozone chemistry. Water-vapour distribution from 46 hPa to 0.2 hPa with an accuracy of 1.2–0.9 ppmv is retrieved from the MLS 183.3-GHz radiometer on board *UARS*. The temperature and pressure profiles, retrieved from the 63.3-GHz radiometer, are used as input variables for stratospheric water-vapour retrieval. The upper-troposphere water vapour is retrieved from the MLS 205-GHz radiometer when it is scanning down through the troposphere. TOVS, SMMR (Scanning Multi-channel Microwave Radiometer) soundings, NASA global distribution of precipitable water in three levels (surface–700 mb,

700–500 mb, 500–300 mb), total integrated precipitable water and liquid water path are available from DAACs (Distributed Active Archive Centres) and are widely used in global climatology.

Precipitation

Estimation of precipitation using remote-sensing techniques may be divided into three categories: those that use visible or infrared data, those that use passive microwave data and radar. The oldest and still most used precipitation-estimation technique is cloud indexing, assigning a rain rate to each cloud type. The precipitation over tropical oceans has been estimated by identifying High Reflective Clouds (HRCs). The Arkin technique (Kidder and Von der Haar 1995) for tropical precipitation uses GOES infrared data and the global-position index. The bispectral technique is applied for cloud precipitation, employing two-dimensional histograms (visible and infrared range), or clustering technique (Kidder and Von der Haar 1995). The advantage of microwave techniques is that microwave radiation penetrates clouds and the precipitation drops interact with the microwave radiation. The main disadvantage is the poor spatial and temporal resolution of passive microwave sounders. Below 22 GHz any ice above the rain is transparent and microwave radiometers respond directly to the rain layer. Above 60 GHz ice scattering is the dominant process and microwave radiometers sense only the ice and cannot see the rain below. The dual-polarization SMMR on board *Nimbus-7* is used for the retrieval of precipitation. The precipitation has been operationally retrieved also from SSM/I (Special Sensor Microwave/Imager) data. New possibilities for precipitation retrieval exist using active microwave satellite systems. A precipitation-sensing radar is planned to fly on the Tropical Rainfall Measuring Mission (TRMM). The main purpose of the Global Precipitation Climatology Project (GPCP), of the World Climate Research Programme (WCRP), is a comparison of different precipitation-estimation algorithms.

Winds

Three-dimensional wind profiles are important for improved weather and tropical cyclone forecast. In climate studies, wind profiles contribute to the world's global energy budget, atmospheric circulation, precipitation and distribution of atmospheric constituents. The advantage of geostationary satellites to take frequent images of the same area is used to estimate the horizontal wind at the level of the cloud, applying a cloud-tracking method. The vector difference of the location of a cloud in two successive images is divided by the time interval between images. A cross-correlation method is used for tracking clouds which are a few kilometres in diameter. Clouds with large vertical extent – such as cumulonimbus clouds, clouds associated with gravity waves, thunderstorm anvils – which are evaporating, or growing in a way not associated with the ambient wind, must be avoided. The advantages of the 6.7 μm water-vapour-tracked winds from *Meteosat* and 15-μm CO_2 VAS images are that it is possible to track winds in cloudless areas. As the emittance of cirrus clouds at 6.7-μm and 11 μm is about the same, the use of both water-vapour and window-infrared data to track high clouds may improve the accuracy of upper-level wind estimation. Cloud-drift winds from the two GOES satellites, *Meteosat* and *GMS* are used in short- and medium-range forecasting throughout the world. The High Resolution Doppler Imager (HRDI) and the Wind Imaging Interferometer (WINDII, on board *UARS*) view the limb of the atmosphere. The horizontal winds in the upper atmosphere are determined by measuring the Doppler shift in the emission and absorption lines in the visible and infrared parts of the spectrum (Kidder and Van der Haar 1995). The satellite infrared images are used also for estimating the location and wind speed of tropical cyclones. It is also possible to identify some mesoscale and sub-mesoscale atmospheric phenomena using active *EOS* SAR (Synthetic Aperture Radar) sensors. Winds from cloud motion are produced operationally by all the geostationary-satellite operating agencies (NESDIS produces winds from *GOES*, ESA produces winds from *Meteosat*, the Indian Space Research Organization [ISRO] produces winds from *Insat*, the Meteorological Satellite Centre [MSC] of Japan produces winds from *GMS*). Three-dimensional wind profiles will be received from the Doppler wind lidar, which is planned by ESA to fly on the *Earth Explorer* Atmospheric Dynamics Mission.

Clouds

Clouds affect climate by release and redistribution of sensible and latent heat, by reflection, absorption and emission of the radiation, by precipitation and modification of the radiation and turbulent transfer through the atmosphere. Physical modelling, histogram analysis, statistical, correlation or clustering methods are used for cloud classification. The first method uses the radiative properties (reflectance and emission) of different cloud types. For cloud detection, threshold methods in the visible or thermal spectral ranges are usually applied. The spatial coherence test is applied for cloud detection over oceans and the water-vapour absorption band at 6.3 μm is used for cirrus-cloud detection. The statistical clustering method is based on finding the concentrations of pixels (clusters) which are identified as cloud, land- or sea-surface types. The correlation technique also gives good results for cloud classification and is based on maximum correlation between the learning sets of particular cloud types and the studied area. The satellite-retrieved cloud variables are cloud-top temperature, cloud-top height (or pressure), reflectance, emittance, optical depth, liquid-water content, drop-size distribution and cloud amount (the fraction of the image which is covered by clouds).

Methods for the retrieval of cloud amount and cloud-top pressure from satellite sounders may be divided into two classes – radiance residual and radiance ratio methods – which are the basis of CO_2 slicing. Cloud-top temperature is determined from the thermal infrared-emitted radiance, measured usually in spectral ranges free from strong gas-absorption lines (AVHRR, ATSR, ATSR-2). Cloud albedo is calculated from optical satellite observations (AVHRR, ATSR-2) in the visible and near-infrared spectral range. Model simulations show that the cooling of the surface by low clouds exceeds that by higher clouds. It is possible to evaluate the impact of cloud type and top height on the cooling or heating rate of the surface from directional spectral observations of ATSR and ATSR-2 and after that from AATSR on *Envisat* (Djepa 1993). Thermal infrared satellite data from *GOES*, AVHRR and microwave observations at 19.35 GHz have found wide applications for weather forecasting and tornado and rainfall-intensity prediction. The main goal of the International Satellite Cloud Climatology Programme (ISCCP) is to collect global datasets from polar and geostationary satellites and to process them into cloud climatology. Data tapes of retrieved ISCCP parameters are available from the Satellite Services Division of NESDIS.

Accurate three-dimensional data fields of atmospheric constituents and proper representations of radiative and chemical processes are required to evaluate the impact of the atmospheric environment on climate change. For prediction of seasonal to inter-annual climate fluctuations it is necessary to evaluate the interaction of the atmosphere with the ocean, the cryosphere and the continental biosphere.

15.2.2 Ocean datasets for climate study

The most important Earth-observation ocean datasets for climate study are sea-surface temperature, wind speed, topography, salinity, phytoplankton concentration, sea ice and snow extent.

Sea ice and snow cover

Sea ice occupies vast areas of the world's oceans, about 10 per cent in the northern hemisphere and about 13 per cent in the southern hemisphere. Sea ice floats and is subject to large seasonal and annual variations. Sea ice and snow modify the surface albedo and energy balance at the surface. Prediction of seasonal to inter-annual climate fluctuations requires a proper description of ocean–atmosphere interactions and the impact of sea ice on turbulent heat exchange. The optical multispectral measurements from AVHRR, ATSR, ATSR-2 and the Thematic Mapper (TM) on *Landsat* have found wide application in sea-ice monitoring of polar regions. As the use of optical spectral data is limited by weather conditions and cloud-free days, and the emissivity from ice-covered oceans is higher than from ice-free water, microwave observations (SMMR, SSM/I) have also been used for ice monitoring and open-water detection. Brightness temperatures measured from AMSU (Advanced Microwave Sounding Unit) are used for the retrieval of snow cover, sea-ice concentration, cloud liquid water, rain rate and total precipitable water. Active-microwave remote sensing provides new possibilities for monitoring different features of sea ice. For example, K_u band (16.5 GHz) is used for the estimation of sea-ice concentration, X band (8.91 GHz) has the greatest potential for sea-ice age studies, L band (1.228 GHz) is best to map the extent of surface topography, while P band (0.428 GHz) is only able to detect large fractures. From sequential images, obtained from *Seasat* SAR, it is possible to measure ice motion with an accuracy of $\pm$150 m. SAR on *ERS* and *Radarsat* provides information about sea-ice drift with high spatial resolution. Monthly and seasonal anomalies of snow cover are available from CPC (Climate Prediction Center)/NOAA.

Sea-surface wind

Wind is the main driving factor of heat transport and ocean currents. Radar altimeters such as on *Topex-Poseidon*, passive-microwave radiometers (SMMR, SSM/I, AMSU) and microwave scatterometers on *ERS* and *Seasat* have been used for sea-surface wind retrieval. Wind data from the AMI (Active Microwave Instrument) on *ERS-2* are operationally assimilated in the NWPM. The advanced wind scatterometer is planned to fly on the *METOP* series of polar-orbiting meteorological satellites.

Sea state and topography

Accurate sea-surface heights are needed for ocean-dynamic topography, climate study, and Global Ocean Circulation Models (GOCMs). Since 1975 satellite altimeters have been used for the study of the geoid and gravity field of the Earth. Currently flying *Ceosat*, *ERS-2* and *Topex-Poseidon* altimeters are used for ocean topography. The NRA on *Topex* is the first spaceborne altimeter that uses two channels – 13.6 GHz (K_u band), 5.3 GHz (C band) – to compute the effect of the ionospheric free electrons in satellite range measurements. The NRA ground track has a spacing of 2.835° (314 km) at the equator and an along-track sampling rate of 7 km and sea-level measurements are obtained with a precision of ±2.4 cm. SAR is used to study the generation and propagation of long ocean waves. The distribution of short gravity waves, recorded by SAR on *Seasat*, shows correlation with a number of significant large-scale phenomena – such as local wind structure, long gravity waves, currents and local bathymetry. Internal and surface waves have been also studied using *Landsat* MSS band 7 (0.8–1.1 μm).

ERS-1, *ERS-2* and *Topex-Poseidon* altimeter missions will be continued with the *Jason-1* mission which is soon to be launched. An *Earth Explorer* gravity field and steady-state gravity circulation core mission (GOCE), planned for launch by ESA, will continue the studies of the gravity field of the Earth, the dynamics of the oceans and the interaction with the continents.

Salinity

Salinity and sea-surface temperature are the two major climatic factors controlling the circulation of the ocean. *Landsat* MSS data have been used for the retrieval of salinity of the sea water by applying regression models (Khorram 1982). Because of the essential impact of clouds on optical remote sensing, the microwave spectral range is used for temperature and salinity retrievals. Passive-microwave radiometers at L band (1.43 GHz) and dual-frequency microwave radiometers at L and S bands (2.65 GHz) have been used for remote sensing of water salinity and temperature with an accuracy of 0.91‰ and 0.56°C .

Ocean bio-optical climate datasets

Ocean colour is indirectly related to the chlorophyll concentration, affecting CO_2 exchange and global climate. Satellite colour images (CZCS, SeaWiFS [Sea-Viewing Wide-Field-of-view Sensor], AVHRR, TM) are used for the estimation of sea-surface phytoplankton, suspended particles and primary production. Data from CZCS (Level 1–3) are available from GSPC/DAAC or by FTP (File-Transfer Protocol) from http://daac.gsfc.nasa.gov/daac-bin/db_czcsbrs. Later, sea-surface phytoplankton and chlorophyll concentration will be retrieved with better spatial resolution from MODIS and MERIS satellite data.

Sea-surface temperature

Sea surface temperature (SST) controls the air temperature and moisture and depends on circulation patterns. Satellite observations in atmospheric windows of the microwave and infrared spectral ranges are used for SST retrieval. Infrared spectral measurements give better spatial resolution (<1 km), but are affected by atmospheric constituents and clouds. Microwave measurements in the spectral range >4 cm are transparent to clouds, but have very poor spatial resolution. SSTs have also been retrieved using HIRS/2, MSU and VAS data (Kidder and Van der Haar 1995) and, since 1991, have been retrieved from ATSR and ATSR-2 data. In cloudy conditions, SMMR on board the *Nimbus-7* satellite is used for SST retrieval. SST defines heat transport and evaporation, which affects precipitation and cloud formation. Coupling predicted SST anomalies with a GCM enables the impact of SST on atmosphere and ocean state to be evaluated. Monthly global temperature and precipitation anomalies are available from CPC/NOAA. Accurate global datasets of SST anomalies are needed in order to evaluate the impact of SST on climate change. ATSR, ATSR-2 and thereafter AATSR on *Envisat* will allow long-term monitoring of SST with an accuracy of better than 0.5 K.

15.2.3 Land–atmosphere interaction

The continental soil moisture, snow cover, temperature, roughness, Leaf Area Index (LAI) and vegetation cycles are critical variables impacting on climate change, land-surface evaporation and heat fluxes. The actual evaporation from the land determines the heat balance, the water balance of soil and air, the hydrological cycle and cloud formation and precipitation. For the retrieval of surface variables from satellite data, two main steps are employed – the atmospheric correction of measured spectral radiance and the estimation of surface variables.

Land-surface temperature

The surface temperature and emissivity components control the sensible and latent heat fluxes, weather and climate system and are input variables in NWP models and GCMs. The surface emissivity and temperature are retrieved from the atmospherically corrected surface radiances in the infrared, or microwave atmospheric windows. For an n-channel scanner in thermal infrared spectral range, there are $n + 1$ unknowns – the n values for emissivity and the surface temperature. Employing multi-viewing spectral measurements, we have $(n \times m)$ equations for n viewing angles and m spectral channels, allowing determination of spectral emissivities and aerodynamic temperatures of the components of the surface. As the surface within the whole pixel is generally heterogeneous and non-isothermal, the various parts of this surface may have different emissivities; the directional surface radiometric temperature and emissivity depend on the viewing angle, on the aerodynamic temperatures and emissivity of component surfaces, on the fractions occupied by different components and LAI (Plate 4(e,f)). The aerodynamic temperatures of vegetation and soil and their histograms, calculated from ATSR thermal data for a region of China, are presented in Plate 4(a–d)). It is shown that the aerodynamic temperature of vegetation is usually lower than that of the soil and the difference could be up to 20°C, which affects the heat fluxes of canopy and soil and, hence, weather and climate system. The emissivity of the heterogeneous surface, retrieved from ATSR-2 at 11-μm nadir is shown in Figure 15.1. The surface emissivity of the heterogeneous target depends on the emissivities of the component surfaces and viewing direction. ATSR, ATSR-2 and AATSR will allow retrievals with 1-km resolution of the aerodynamic and directional radiometric temperatures and emissivities over heterogeneous surfaces.

LAI and vegetation cover

Land-surface parametrizations in the NWP models and GCMs require specification of the vegetation type and amount along with surface temperatures and emissivities. The vegetation amount is parametrized through the fraction of the vegetation and LAI. Different methods (model simulations, measurements in the visible and thermal infrared spectral range) (Djepa and Vaughan 2000) have been applied for estimating LAI. Histograms of LAI, the fraction covered by vegetation (retrieved with a resolution of 1 km from thermal infrared data of ATSR-2) and the radiometric temperature during SGP '97 (Southern Great Plains hydrology experiment in 1997), are shown in Figures 15.2, 15.3 and 15.4 (this experiment originated from an interdisciplinary investigation titled 'Soil Moisture Mapping at Satellite Temporal and Spatial Scales', selected under NASA research announcement 95-MTPE-03). The continuing programme of ATSR, ATSR-2 and AATSR will permit the collection of long-term datasets of surface components, temperatures and emissivities, LAI and the fraction covered by vegetation, which are input variables for the initialization of NWPs and GCMs.

Albedo

The broadband hemispherical albedo is an important climate component, affecting turbulent heat exchange over the land surface. The two viewing angles of ATSR-2 and AATSR permit long-term monitoring of albedo of the component surfaces (vegetation, soil) and aerosol optical thickness (Djepa and Vaughan 2000) on a global scale with a resolution of 1 km.

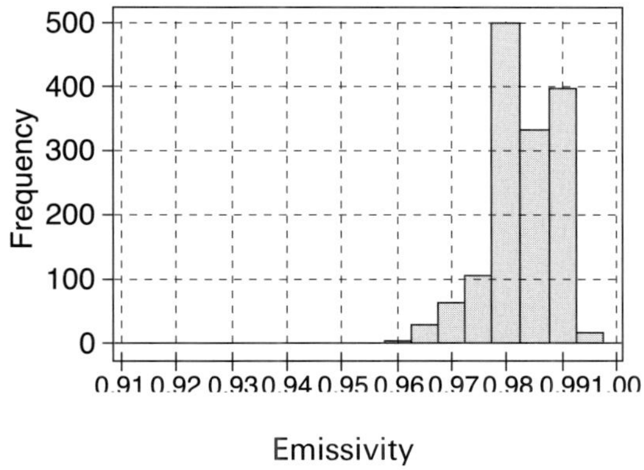

Figure 15.1. Surface emissivity ε_{n11}, at nadir, 11 μm, retrieved by ATSR-2, 1 July 1997.

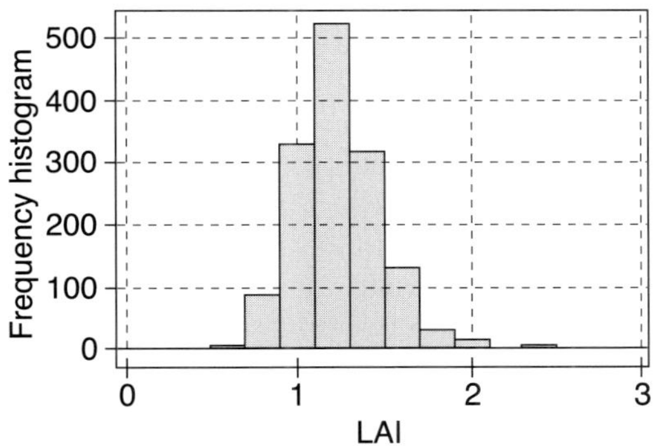

Figure 15.2. LAI, retrieved by ATSR-2, 1 July 1997, SGP '97.

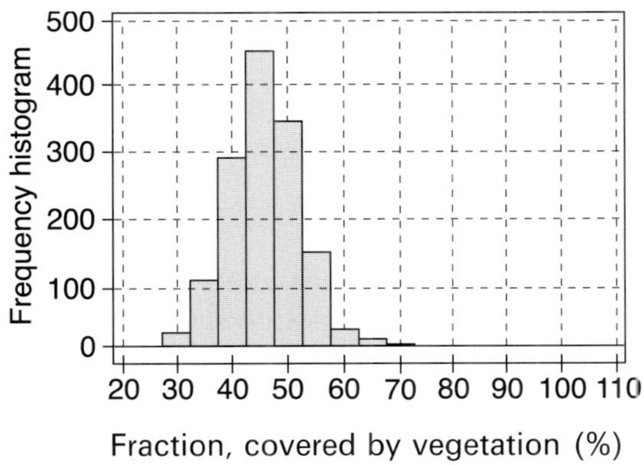

Figure 15.3. Fraction, covered by vegetation retrieved by ATSR-2, 1 July 1997, SGP '97.

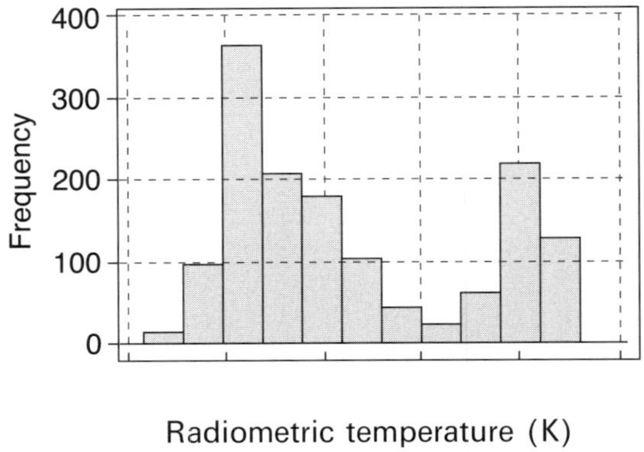

Figure 15.4. Surface radiometric temperature, at 11 μm, retrieved by ATSR-2, 1 July 1997.

Land-surface heat fluxes

The turbulent heat exchange on the surface depends on the wind speed, temperature, albedo, emissivity and soil moisture of the surface components. Changes in evaporation rate and sensible heat fluxes are closely related to carbon assimilation, atmospheric water balance, greenhouse effect, cloud formation, precipitation and the hydrological cycle. Quantitative estimation of turbulent heat exchange and evaporation on a global scale will help to evaluate the impact of evapotranspiration on cloud formation, weather and climate. It is possible to obtain datasets of evapotranspiration and the evapotranspiration fraction along the ATSR-2 track (Djepa and Vaughan 2000). Evaporation from the Earth's surface and evapotranspiration fraction are important variables for weather, climate study, irrigation, water resources and agriculture. The evapotranspiration fraction is a good indicator of the soil water supply and it is an important variable of the surface parametrizations of NWP models and GCMs. SAR on *ERS-2* and *Envisat* could also find application in soil-moisture and surface-roughness monitoring over vegetated regions. Further exploitation of AATSR, ASAR (Advanced Synthetic Aperture Radar), MSG (*Meteosat* Second Generation) and MODIS (Moderate Resolution Imaging Spectrometer) will help in the validation of land-surface retrieval algorithms and utilization of satellite datasets in climate study.

Net radiation of the Earth system

The measured heat fluxes at the top of the atmosphere are used to evaluate the impact of surface and atmospheric variables on the greenhouse effect (Djepa and Harwood 1998) and climate change, and for the validation of GCMs. The ERB (Earth Radiation Budget) experiment on *Nimbus-6* allows mapping of the seasonal and global distribution of the radiation budget of the Earth–atmosphere system. The Earth's net radiation is also measured from *NOAA*, *Meteosat* and *GOES*, and later by the GERBE/MSG mission.

15.3 CLIMATE MODELS AND DATASETS

Climate is a result of interactions between the component systems, where the land, ocean and atmosphere play significant roles. Climate models can be divided into one-dimensional, two-dimensional and general circulation models (GCMs). One-dimensional models are designed to determine average global temperatures. Two-dimensional climate models represent either two horizontal dimensions, or the vertical and one horizontal dimension. Calculation of the full three-dimensional character of the climate is the aim of GCMs. A great number of atmospheric (AGCM), ocean (OGCM), mesoscale models (MM4, MM5) and land biosphere/atmosphere models have recently been developed. Model simulations have shown not only climate sensitivity to cloud formation, but also greenhouse gases (CO_2 in particular), as well as land and ocean climate components. Results from climate models indicate that changes in surface temperature, vegetation cover and available moisture released to the atmosphere have significant feedback effects on cloud formation, which in turn impacts the radiation budget (evaporation) and precipitation fields on global and continental scales. Prognostic datasets, simulated by the UGAMP Global Circulation Model (UGCM), the Atmospheric General Circulation Model (AGCM) of the United Kingdom Meteorological Office (UKMO) and the ARPEGE Climate Model (ARPEGE-Climate, 1996), are used to evaluate the impact of the surface and atmospheric variables on climate change. The three-dimensional UGCM model is based on the forecast model of the European Centre for Medium-range Forecasts (ECMWF). The model consists of prognostic equations for temperature, specific humidity and surface pressure. It has nineteen levels in the vertical (from the surface to 10 mb) and the radiative scheme in the UGCM is the one developed by Morcrette (1989, 1991). The convection scheme used in the model is the Kuo convection scheme (Kidder and Von der Haar 1995) where the heating and moistening by cumulus clouds is determined by local moisture supply. UGCM datasets are used for chemistry and dynamic study of the upper troposphere and stratosphere and to evaluate their impact on long-term climate change. UGCM-simulated datasets allow the evaluation of ozone and species trends, cloud impact on the retrieved atmospheric constituents (Upper Troposphere Water Vapour [UTW], retrieved from MLS on board *UARS* in order to improve the retrieval algorithms for future sounding instruments.

ARPEGE is a climate version of the NWP model (Abel 1990). It includes improved surface parametrizations of snow, surface albedo, energy balance of the soil–canopy system and cloud radiative impact. The atmospheric model takes into account the dynamic relationship between large-scale variables – such as temperature and wind – and the effect of the subgrid-scale processes – including radiation, latent heat release and the exchanges of heat, moisture and momentum with the surface. The radiative scheme of the model is based on the ECMWF model. The radiative effects of trace gases, aerosols and clouds and the effect of the latent heat as a result of changes in water vapour are considered.

The impact of surface and atmospheric variables on the greenhouse effect has been examined by assimilating satellite retrieved datasets for water vapour into the

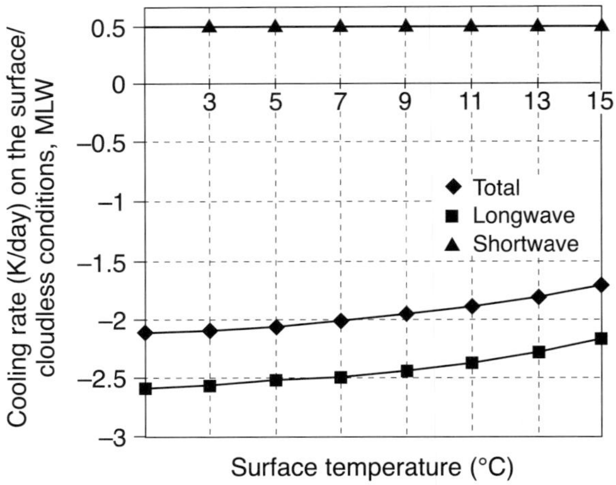

Figure 15.5. Impact of the surface temperature on the cooling rate of the surface, simulated by ECMWF radiative code, using UGCM profiles of temperature, water vapour and ozone mixing ratio.

ECMWF radiative code. The greenhouse effect is defined as $G = F^{u}_{ls} - F^{u}_{lt}$, where F^{u}_{ls} is the long-wave flux emitted from surface and F^{u}_{lt} is the long-wave flux on the top of the atmosphere, calculated by ECMWF. The net radiation F_s has been calculated using the radiative transfer code of Morcrette (1989) and is defined in terms of the net short-wave F_{ss} and long-wave F_{ls} radiation by $F_s = F_{ss} + F_{ls}$. The net short-wave and long-wave radiation are calculated from $F_{ls} = F^{d}_{ls} - F^{u}_{ls}$ and $F_{ss} = F^{d}_{ss} - F^{u}_{ss}$, where $F^{d}_{ls}, F^{u}_{ls}, F^{d}_{ss}, F^{u}_{ss}$ denote respectively downward and upward long-wave and short-wave fluxes. The short-wave fluxes are calculated solving the radiative transfer equation, assuming a plane-parallel atmosphere and considering the scattering by molecules, aerosols, cloud particles and molecular absorption in the 0.2–4-μm spectral range. The rate of atmospheric cooling by emission and absorption of long-wave radiation is calculated from $\mathrm{d}F_c/\mathrm{d}t = g\,\mathrm{d}F_{ls}/c_p\,\mathrm{d}p$, where c_p is the specific heat at constant pressure p and g is the acceleration due to gravity. The net atmospheric cooling or heating rate at the surface is estimated from $\mathrm{d}F/\mathrm{d}t = \mathrm{d}F_h/\mathrm{d}t - \mathrm{d}F_c/\mathrm{d}t$, where the rate of short-wave heating is defined as $\mathrm{d}F_h/\mathrm{d}t = g\,\mathrm{d}F_{ss}/c_p\,\mathrm{d}p$.The impact of SST on F_n, $\mathrm{d}F/\mathrm{d}t$ and G are examined for cloudy and cloudless conditions (Figure 15.5), using a mid-latitude winter (MLW) standard atmosphere profile.

Increasing the surface temperature by 12°C increases the greenhouse effect by $40\,\mathrm{W\,m^{-2}}$ and the surface cooling decreases by 0.7°C per day. This, and the recent results from climate models (Chahine 1992; Trenberth 1992), indicate the importance of having accurate datasets for surface variables (temperature, albedo, vegetation fraction, type and LAI) and atmospheric variables (wind speed, aerosol optical properties and gas concentration, cloud cover and type) for the initialization of climate and NWP models.

15.4 CONCLUSIONS

Current space missions provide global, quasi-synoptic and repetitive datasets of geophysical variables and many surface and atmospheric components of the Earth system on different space and timescales. To understand the internal variability in the components of the climate system, it is essential to advance our knowledge wth more accurate three dimentional global datasets of atmospheric climate components (wind field, clouds, aerosols) and their interactions with radiative and dynamic fields. More reliable estimates are needed of the impact of the greenhouse effect, land use and climate change on terrestrial ecosystems and the effect of biological processes and their components on renewable and non-renewable resources in the geosphere/biosphere–atmosphere interface.

It is also necessary to examine the seasonal and inter-annual processes and heat fluxes at the ocean–atmosphere, atmosphere–land–biosphere and ocean–atmosphere–cryosphere interfaces and their energy impacts. It is important to evaluate the variations in ocean–atmosphere exchange processes, the watercycle of the ecosystem and their response to climate change, sea-ice dynamics and deep-water formation. Of geat importance is to quantify the current mass balance of the polar ice sheets and their response to climate change. This requires more accurate and new Earth-observation datasets and space technology (laser, multi-angle systems [AATSR], high spectral and spatial resolution passive and active systems). The currently flying ATSR and ATSR-2 confirmed the advantages of multi-viewing systems for atmospheric three-dimensional data fields and land-surface climate components. Using advanced Earth-observation datasets in GCMs will help to quantify the global budgets, detect evolution of Earth's ecosystems and validate integrated high-resolution models.

References

Abel, P. (1990) *Report of the Workshop on Radiometric Calibration of Satellite Sensors of ARPEGE-Climate*, 1996.

Abel, P., Guenther, B., Galimore, R. N. and Cooper, J. W. (1993) Calibration results for NOAA-11 AVHRR channels 1 and 2 by congruent path measurements. *Journal of Atmospheric and Oceanic Technology* **10**, 493–508.

Agrawala, S. (1999) Early science-policy interactions in climate change: lessons from the Advisory Group on Greenhouse Gases. *Global Environmental Change* **9**, 157–169.

Allen, M. R., Mutlow, C. T., Blumberg, G. M. C., Christy, J. R., McNider, R. T. and Llewellyn-Jones, D. T. (1994) Global change detection. *Nature* **370**, 24–25.

Arrhenius, S. (1896) On the influence of carbonic acid in the air upon the temperature of the ground. *Philosophical Magazine* **41**, 237–276.

Avissar, R. and Verstraete, M. M. (1990) The representation of continental surface processes in atmospheric models. *Reviews of Geophysics* **28**, 35–52.

Baer, F. (1977) Adjustments of initial conditions required to suppress gravity oscillations in non-linear flows. *Beiträge zur Physik der Atmosphäre* **50**, 350–366.

Baily, R. G. and Hogg, H. C. (1986) A world ecoregions map for resource reporting. *Environmental Conservation* **13**, 195–202.

Barbosa, P. M., Pereira, J. M. C. and Grégoire, J.-M. (1998) Compositing criteria for burned area assessment using multitemporal low resolution satellite data. *Remote Sensing of Environment* **65**, 38–49.

Barkstrom, B. (1973) A comparison of Minnaert reflectance law and the reflectance from a non-conservative isotropic scattering atmosphere. *Journal of Geophysical Research* **78**, 6370–6372.

Barrett, J. (1996) Do CO_2 emissions pose a global threat? In: J. Emsley (ed.), *The Global Warming Debate*. The European Science and Environment Forum, London, pp. 60–70.

Barry, R. G. and Chorley, R. J. (1998) *Atmosphere, Weather and Climate* (7th edn). Routledge, London.

Bate, R. (1998) *Global Warming – the Continuing Debate*. The European Science and Environment Forum, London.

Bate, R. and Morris, J. (1994) *Global Warming – Apocalypse or Hot Air?* IEA Studies on the Environment No. 1. Institute of Economic Affairs, London.

Belward, A. S. (ed.) (1996a) *The IGBP-DIS Global 1 km Land Cover Data Set 'DISCover' Proposal and Implementation Plans*, Report of the Land Cover Working Group of IGBP-DIS, IGBP-DIS Working Paper No. 13. IGBP DIS Office, Toulouse, France.

Belward, A. S. (1996b) AVHRR data sets for global terrestrial ecosystem monitoring. In: G. D'Souza, A. S. Belward and J-P. Malingreau (eds), *Advances in the use of AVHRR data for Land Applications*. Kluwer Academic, Dordrecht, pp. 455–470.

Belward, A. S. (1999) International co-operation in satellite sensor calibration; the role of the CEOS working group on calibration and validation. *Advances in Space Research* **23**, 1443–1448.

Berk, A., Bernstein, L. S. and Robertson, D. C. (1989) *MODTRAN: A moderate resolution model for LOWTRAN 7*, Report GL-TR-89-0122. Hanscom Air Force Base, MA, Geophysics Laboratory.

Berreen, J. and Meyer, A. (1992) A package marked 'Return to sender'. Some problems with the Climate Convention. *Network '92* **18**, 6–7.

Boehmer-Christiansen, S. (1996) Political pressure in the formation of scientific concensus. In: J. Emily (ed.) *The Global Warming Debate*. The European Science and Environment Forum, London, pp. 234–248.

Bowyer, S., Zeitland, G., Tarter, J., Lampton M. and Welch, J. (1983) The Berkeley Parasitic SETI program. *Icarus* **53**, 147–155.

Brest, C. L. and Rossow, W. B. (1992) Radiometric calibration and monitoring of NOAA AVHRR data for ISSCP. *International Journal of Remote Sensing* **13**, 235–273.

Brest, C. L., Rossow, W. B. and Roiter, M. (1997) Update of radiance calibrations for ISCCP. *Journal of Atmospheric and Oceanic Technology* **14**, 1091–1109.

Brown, J. F., Loveland, T. R., Merchant, J. W., Reed, B. C., and Ohlen, D. O. (1993) Using multisource data in global land characterization: concepts, requirements, and methods. photogrammetric engineering and remote sensing **59**, 977–987.

Brown, J. W., Brown, O.B. and Evans, R.H. (1993) Calibration of advanced very high resolution radiometer: a new approach. *Journal of Geophysical Research* **98**, 18257–18268.

Brown, L. R., Renner, M. and Halweil, B. (1999) *Vital Signs 1999*. W.W. Norton, New York.

Brown, O. B., Brown, J. W. and Evans, R. H. (1985) Calibration of advanced very high resolution radiometer infrared observations. *Journal of Geophysical Research* **90**, 11667–11677.

Buckley, R. (1994) *World Population: The Biggest Problem of All?* European Schoolbooks Publishing Limited, Cheltenham.

Butcher, S. S., Charlson, R. J. Orians G. H. and Wolf, G.V. (1992) *Global Biogeochemical Cycles*. Academic Press, London, p. 380.

CEOS (1995) *Leadership to Ensure High-quality Earth Observation Data: A Strategic Vision*. Ispra, Varese, Italy, The CEOS Working Group on Calibration and Validation, Secretariat Space Applications Institute, DG-JRC, 32 pp.

Chahine, M. T. (1992) The hydrological cycle and its influence on climate. *Nature* **359**, 373–380.

Charney, J. G. (1975) Dynamics of desert and drought in the Sahel. *Quarterly Journal of the Royal Meteorological Society* **101**, 193–202.

Che, N. and Price, J. C. (1992) Survey of radiometric calibration results and methods for visible aand near-infrared channels of NOAA-7, -9, and -11 AVHRRs. *Remote Sensing of Environment* **41**, 19–27.

CIESIN (Center for International Earth Science Information Network) (1998) Environmental Treaties and Resource Indicators (ENTRI) [online]. Palisades, NY, CIESIN. URL: http://www.ciesin.org

Cihlar, J., Belward, A. S., and Govaerts, Y. (eds) (1999) *METEOSAT Second Generation Opportunities for Land Surface Applications and Research.* EUMETSAT Scientific Publications, Darmstadt, 63 pp.

Cobb, J., Donnelly, C., Bowyer, S., Werthimer, D. and Lampton, M.(1997) The SERENDIP IV interference rejection and signal detection system. In: C. B. Cosmovici, S. Bowyer and D. Werthimer (eds), *Astronomical and Biochemical Origins and the Search for Life in the Universe* (*Proceedings of the Fifth International Conference on Bioastronomy, Capri, 1–5 July, 1996*), IAU Colloquium No. 161. Editrice Compositori, Bologna, pp. 677–682.

Congalton, R. G. (1991) A review of assessing the accuracy of classifications of remotely sensed data. *Remote Sensing of Environment* **37**, 35–46.

Corbyn, P. (1996) Carbon dioxide fluctuations resulting from climate change. In: J. Emsley (ed.), *The Global Warming Debate.* The European Science and Environment Forum, London, pp. 71–78.

Cosnefroy, H., Briottet, X. and Leroy, M. (1993) Characterization of desert areas with METEOSAT 4 data for the calibration of optical satellite sensors. In: P. S. Chavez and R. A. Schowengerdt (eds), *Proceedings of SPIE Conference on Recent Advances in Sensors, Radiometric Calibration, and Processing of Remotely Sensed Data.* SPIE, Bellingham, Washington, DC, pp. 203–211.

Courtier, P., Thépaut, J.-N. and Hollingsworth, A. (1994) A strategy for operational implementation of 4D-Var, using an incremental approach. *Quarterly Journal of the Royal Meteorological Society* **120**, 1367–89.

Cracknell, A. P. (1994a) Climate change – the background. In: R. A. Vaughan and A. P. Cracknell (eds), *Remote Sensing and Global Climate Change.* Springer, Berlin, pp. 1–33.

Cracknell, A. P. (1994b) Basis and structure of climate models. In: R. A. Vaughan and A. P. Cracknell (eds), *Remote Sensing and Global Climate Change*, Springer, Berlin, pp. 135–176.

Cracknell, A. P. (1997) *The Advanced Very High Resolution Radiometer.* London, Taylor and Francis, 534 pp.

Crutzen, P. J. and Andreae, M. O. (1990) Biomass burning in the tropics: impact on atmospheric chemistry and biogeochemical cycles. *Science* **250**, 1669–1678.

Daley, R. (1991) *Atmospheric Data Analysis.* Cambridge University Press, Cambridge.

Danko, D. M. (1992) The digital chart of the world. *Geoinfosystems* **2**, 29–36.

De Boissezon, H., Gonzales, G., Pous, B. and Sharman, M. (1993) Rapid estimates of crop acreage and production at a European scale using high resolution imagery – operational review. *Proceedings of the International Symposium on Operationalization of Remote Sensing, held at Enschede, The Netherlands, 19–23 April 1993.* International Institute for Aerospace Survey and Earth Sciences, Enschede, vol. 2, pp. 94–105.

De Grandi, G. F., Mayaux, P., Malingreau, J-P., Rosenqvist, Å., Saatchi, S. and Simard, M. (2000) New perspectives on global ecosystems from wide-area radar mosaics: flooded forest mapping in the Tropics. *International Journal of Remote Sensing* **21**, 1235–1249.

Dee, D., Cohn, S., Dalcher, A. and Ghil, M. (1985) An efficient algorithm for estimating noise covariances in distributed systems. *IEEE Transactions on Automatic Control* **30**, 1057–1065.

DeFries, R. S. and Belward, A. S. (2000) Global and regional land cover characterization from satellite data: an introduction to the Special Issue. *International Journal of Remote Sensing* **21**, 1083–1092.

Dignon, J. and Penner, J. E. (1991) Biomass burning: a source of nitrogen oxides in the atmosphere. In: J. S. Levine (ed.), *Global Biomass Burning, Atmospheric Climatic and Biospheric Implications.* The MIT Press, Cambridge, MA, pp. 370–375.

Diner, D. J., Gregory, P., Asner, P., Davies, R., Knyazikhin, Y., Muller, J-P., Nolin, A.W., Pinty, B., Schaaf, C.B. and Stroeve, J. (1999) New directions in Earth observing: scientific applications of multi-angle remote sensing. *Bulletin of the American Meteorological Society* **80**, 2209–2228.

Djepa, V. (1993) Determination of clouds parameters from ATSR/ERS1 data. *Proceedings of the 13th Annual EARSeL Symposium, Dundee, UK, 29 June–1 July 1993*. Springer, Budapest, pp. 393–400.

Djepa V. and Harwood, R. (1998) Impact of atmospheric and surface variables on sea surface heat fluxes, based on modelling and satellite information, COSPAR, Birmingham. *Advances in Space Res*earch **22**(5), 697–700.

Djepa V. and Vaughan, R. A. (2000) Retrieval of land-atmosphere interactive processes and variables from the Along Track Scanning Radiometer (ATSR-2) on board ERS-2. *International Journal of Remote Sensing*, in press.

Dwyer, E., Pinnock, S., Grégoire, J-M. and Pereira, J. M. C. (2000) Global spatial and temporal distribution of vegetation fire as determined from satellite observations. *International Journal of Remote Sensing* **21**, 1289–1302.

Easterling, D.R., Evans, J.L,, Groisman, P.Ya., Karl, T.R., Kunkel, K.E. and Ambenje, P., 2000: Observed variability and trends in extreme climate events, a brief review. Bulletin of the American Meteorological Society, 81, 417-425.

EC (1999) *Official Journal of the European Communities*, **128**, 92–96.

Eidenshink, J. E., and Faundeen, J. L. (1994) The 1 km AVHRR global land data set: first stages in implementation. *International Journal of Remote Sensing* **15**, 3443–3462.

Emsley, J. (1996) *The Global Warming Debate*. The European Science and Environment Forum, London.

Emsley, J. (1998) Introduction. In: R. Bate (ed.), *Global Warming – the Continuing Debate*. The European Science and Environment Forum, London, pp. 22–26.

EOS (1999) Science plan. In: R. Greenstone and M. D. King (eds), *The State of Science in the EOS Program*, NASA Goddard Space Flight Center, Greenbelt, MD, 397 pp.

ESA (1998a) *The Living Planet Programme – Call for Core Missions*, SP-1196. European Space Agency, Noordwijk.

ESA (1998b) *The Living Planet Programme – Call for Earth Explorer Opportunity Missions*, SP-1226. European Space Agency, Noordwijk.

ESA (1998c) *The Earth Explorers: The Science and Research Elements of ESA's Living Planet Programme*, SP-1227. Noordwijk, European Space Agency.

ESA (1998d) *Further Achievements of the ERS Missions*, SP-1228. European Space Agency, Noordwijk.

ESA (1998e) *Envisat Mission – Opportunities and Applications*, SP-1218. European Space Agency, Noordwijk.

ESA (1998f) *Taking Nature's Pulse – Europe's ERS Satellites*, BR-134. European Space Agency, Noordwijk.

ESA (1999) *Introducing the ESA 'Living Planet Programme': The ESA Strategy for Earth Observation*, SP-1234. Noordwijk, European Space Agency.

Estes, J. E., and Mooneyhan, W. (1994) Of maps and myths. *Photogrammetric Engineering and Remote Sensing* **60**, 517–524.

Fisher, M. and Lary, D. J. (1995) Lagrangian four-dimensional variational data assimilation of chemical species. *Quarterly Journal of the Royal Meteorological Society* **121**, 1681–1704.

Fourier, J. B. J. (1824) Remarques générales sur les températures du globe terrestre et des espaces planétaires. *Annales de Chimie et de Physique* **27**, 136–167.

Geitelson, A., and M.N. Merzlyak, 1994: Quantitative estimation of chlorophyll-a using reflectance spectra. *Journal of Photochemistry and Photobiology: B (Biology)* **22**, 247–252.

Ghil, M., Cohn, S., Tavantzis, J., Bube, K. and Isaacson, E. (1981) Applications of estimation theory to numerical weather prediction. In: L. Bengtsson, M. Ghil and E. Kèallen (eds.), *Dynamic Meteorology: Data Assimilation Methods*, Springer Verlag, New York, pp. 139–224.

Gibson, J. K., Kallberg, P., Uppala, S., Hernandez, A., Nomura, A. and Serrano, E. (1997) *ERA Description*, ECMWF re-analysis project report series, 1. European Centre for Medium Range Weather Forecasts, Reading.

Gille, J. C. (1984) *Middle Atmosphere Science*, NASA Technical Memorandum 86129, vol. 1 (part. 2). Goddard Space Flight Center, Greenbelt, MD, pp. 49–52.

Gobron, N., Pinty, B., Verstraete, M. and Govaerts, Y. (1999) The MERIS Global Vegetation Index (MGVI): description and preliminary application. *International Journal of Remote Sensing* **20**, 1917–1927.

Goodess, C.M., Palutikof, J.P. and Davies, T.D., 1992: The Nature and Causes of Climatic Change: Assessing the Long-term Future. Belhaven Press, London, 248pp.

Gordon, C., Cooper, C., Senior, C.A., Banks, H., Gregory, J.M., Johns, T.C., Mitchell, J.F.B. and Wood, R., 2000: The simulation of SST, sea ice extents and ocean heat transports in a version of the Hadley Centre coupled model without flux adjustments. Climate Dynamics, 16, 147-168.

Gore, A. (1993) *Earth in the Balance. Ecology and the Human Spirit*. Dutton Plume, New York.

Gribbin, J. (1978) *The Climate Threat*. Fontana/Collins, Glasgow.

Gribbin, J. (1988) *The Hole in the Sky – Man's Threat to the Ozone Layer*. Corgi, London.

Gribbin, J. (1990) An assault on the climate consensus. *New Scientist*, 15 December 1990, 26–31.

Grubb, M., Koch, M., Thomson, K., Munsen, A. and Sullivan F. (1993) *The Earth Summit Agreements: A Guide and Assessment*. Royal Institute of International Affairs/Earthscan, London.

Grubb, M., Vrolijk, C. and Brack, D. (1999) *The Kyoto Protocol – A Guide and Assessment*. Royal Institute of International Affairs/Earthscan, London.

Gruber, A. and Winston, J. S. (1978) Earth-atmosphere radiative heating based on NOAA scanning radiometer measurements. *Bulletin of the American Meteorological Society* **59**, 1570–1573.

Gutman, G. (1999a) On the use of long-term global data of land reflectances and vegetation indices derived from the Advanced Very High Resolution Radiometer. *Journal of Geophysical Research* **104**, 6241–6255.

Gutman, G. (1999b) On the monitoring of land surface temperatures with the NOAA/AVHRR: Removing the effect of satellite orbit drift. *International Journal of Remote Sensing* **20**, 3407–3413.

Hanson, M. C., Defries, R. S., Townshend, J. R. G. and Sohlberg, R. (2000) Global land cover classification at 1 km spatial resolution using a classification tree approach. *International Journal of Remote Sensing* **21**, 1331–1364.

Hay, A. M. (1988) The derivation of global estimates from a confusion matrix. *International Journal of Remote Sensing* **9**, 1395–1398.

Henry, P., Dinguirard, M. and Bodilis, M. (1993) Multispectral calibration over desert areas. *Bulletin of the American Meteorological Society* 67–76.

Holdgate, M. (1995) How can development be sustainable? *RSA Journal* **143**(5464), November, 15–29.

Houghton, J. T. (ed.) (1984) *The Global Climate*. Cambridge University Press, Cambridge.

Houghton, J. T. (1991) The Bakerian Lecture 1991, the predictability of weather and climate. *Philosophical Transactions of the Royal Society of London* **A337**, 521–572.

Houghton, J. T. (1997) *Global Warming – the Complete Briefing*. Cambridge University Press, Cambridge.

Houghton, J. T., Callander, B. A. and Varney, S. K. (1992) *Climate Change – the Supplementary Report to the IPCC Scientific Assessment*. Cambridge, Cambridge University Press.

Houghton, J. T., Jenkins, G. J. and Ephraums, J. J. (1990) *Climate Change – the IPCC Scientific Assessment*. Cambridge University Press, Cambridge.

Houghton, J. T., Meira Filho, L.G., Bruce, J., Lee Hoesung, Callander, B. A., Haites, E., Harris, N. and Maskell, K. (1994) *Climate Change 1994 – Radiative Forcing of Climate Change and an Evaluation of the IPCCIS92 Emission Scenarios*. Cambridge University Press, Cambridge.

Houghton, J. T., Meira Filho, L. G., Callander, B. A., Harris, N., Kattenberg, A and Maskell, K. (1996) *Climate Change 1995 – the Science of Climate Change*. Cambridge University Press, Cambridge.

Houghton, R. A., Skole, D. L. and Lefkowitz, D. S. (1991) Changes in the landscape of Latin America between 1850 and 1985. II. A net release of CO_2 to the atmosphere. *Journal of Forest Ecology and Management* **38**, 173–199.

Hulme, M., Mitchell, J., Ingram, W., Lowe, J., Johns, T., New, M. and Viner, D., 1999: Climate change scenarios for global impacts studies. Global Environmental Change, 9, S3-S19.

Ignatov, A and Gutman, G. (1999) Monthly mean diurnal cycles in surface temperatures over land for global climate studies. *J. Climate* **12**, 1900–1910.

IGOS (1999) *An Integrated Global Observing Strategy*. Online at http://www.igospartners.org

IPCC (1992) Climate Change 1992: the Supplementary Report to the IPCC Scientific Assessment. J.T. Houghton, B.A. Callandar and S.K. Varney (eds.), Cambridge University Press, Cambridge.

IPCC (1996) *Greenhouse Gas Inventory Reference Manual, Revised 1996 IPCC Guidelines for National Greenhouse Gas Inventories*, vol. 3, J. T. Houghton, L. G. Meira Filho, B. Lim, K. Tréanton, I. Mamaty, Y. Bonduki, D. J. Griggs and B. A. Callander (eds). Intergovernmental Panel on Climate Change, United Nations Environment Programme, Organisation for Economic Cooperation and Development, 425 pp.

Jeanjean, H. and Achard, F. (1997) A new approach for tropical forest area monitoring using multiple resolution data, *International Journal of Remote Sensing* **18**, 2455–2461.

Johns, T. C., Carnell, R. E., Crosley, J. F., Gregory, J. M., Mitchell, J. F. B., Senior, V. A., Tett, S. F. B. and Wood, R. A. (1997) The second Hadley Centre coupled ocean-atmosphere GCM: spinup and validation. *Climate Dynamics* **13**, 103–134.

Jones, P. D., New, M., Parker, D. E., Martin, S. and Rigor, I. G. (1999) Surface air temperature and its changes over the past 150 years. *Reviews of Geophysics and Space Physics* **37**, 173–199.

Kaufman, Y. and Holben, B. N. (1993) Calibration of the AVHRR visible and near-IR bands by atmospheric scattering, ocean glint, and desert reflection. *International Journal of Remote Sensing* **14**, 21–52.

Keeling, C. D. and Whorf, T. P. (1998) *Atmospheric Carbon Dioxide Record from Mauna Loa*. Online at http://www.cdiac.esd.ornl.gov/trends/co2/sio-mlo.htm

Keeling, C. D., Backastow, R. B. and Whorf, T. P. (1982) Measurements of concentration of carbon dioxide at Mauna Loa Observatory, Hawaii. In: W. C. Clarke (ed.), *Carbon Dioxide Review*. Oxford, Oxford University Press, pp. 377–385.

Khorram, S. (1982) Remote sensing of salinity in the San Francisco Bay delta. *Remote Sensing of Environment* **12**, 15–22.

Kidder S. Q. and Von der Haar, T. H. (1995) *Satellite Meteorology*. Academic Press, San Diego, 466 pp.

Kidwell, K, (1995) *NOAA Polar Orbiter Data Users Guide*. NOAA/NESDIS Climatic Data Center, Satellite Data Services Division, Washington, DC.

Kondratyev, K. Ya. (1998) *Multidimensional Global Change*. Wiley, Chichester.

Kondratyev, K. Ya. (1999) *Climatic Impact of Aerosols and Clouds*. Springer-Praxis, Chichester.

Kondratyev, K. Ya. and Cracknell A.P. (1998) *Observing Global Climate Change*. Taylor and Francis, London.

Kondratyev, K. Ya. and Donshenko, V. K. (1999) *Ecodynamics and Geopolicy. Volume 1. Global Problems*. SPB Science Centre Publications, St Petersburg [in Russian].

Kondratyev, K. Ya. and Pokrovsky, O. M. (1977) Planning multipurpose experiments in remote sensing of the parameters of the environment and natural resources. *Izvestiya of USSR Academy of Sciences, Series Geography* **3**, 83–89.

Kondratyev, K. Ya. and Pokrovsky, O. M. (1989) The International Geosphere-Biosphere Programme. Key aspects of requirements to Earth satellite observation data. *Izvestiya of USSR Academy of Sciences, Series Geography* **1**, 20–28.

Kramer, H. J. (1996) *Observation of the Earth and its Environment – Survey of Missions and Sensors*. Springer, Heidelberg,

Lawford, R. G. (1999) A midterm report on the GEWEX continental-scale international project (GCIP). *Journal of Geophysical Research-Atmospheres* **104**(D16), 19279–19292.

Leakey, R. and Lewin, R. (1995) *The Sixth Extinction: Biodiversity and its survival*. Doubleday, New York.

Lillesand, T. M. and Kiefer, R.W. (2000) *Remote Sensing and Image Interpretation*, 4th edn. Wiley, Chichester.

Loeb, N. J. (1997) In-flight calibration of NOAA AVHRR visible and near-infrared bands over Greenland and Antarctica. *International Journal of Remote Sensing* **18**, 477–490.

Lorenc, A. C., Bell, R. S. and Macpherson, B. (1991) The Meteorological Office analysis convection data assimilation scheme. *Quarterly Journal of the Royal Meteorological Society* **117**, 59–89.

Loveland, T. R., Merchant, J. W., Ohlen, D. O. and Brown, J. F. (1991) Development of a land cover characteristics data base for the conterminous US. *Photogrammetric Engineering and Remote Sensing* **57**, 1453–1463.

Loveland, T. R., Zhu, Z., Ohlen, D. O., Brown, J. F., Reed, B. C. and Yang, L. (1999) An analysis of the IGBP global land cover characterization process. *Photogrammetric Engineering and Remote Sensing* **65**, 1021–1032.

Machenauer, B. (1977) On the dynamics of gravity oscillations in a shallow water model with application to normal mode initialization, *Contributions to Atmospheric Physics* **50**, 253–271.

Mantripp, D. (1994) *Radar Altimetry*. In: *The Determination of Geophysical Parameters from Space*. Institute of Physics, Bristol.

Martonchik, J. V., Diner, D. J., Pinty, B., Verstraaete, M. M., Nyneni, R. B., Knyazikhin, Y. and Gordon, H. R. (1998) Determination of land and ocean reflective, radiative and biophysical properties using multiangle imaging. *IEEE Transactions on Geoscience and Remote Sensing* **36**, 1266–1281.

Mathews, E. (1983) Global vegetation and land use: new high resolution data bases for climate studies, *Journal of Climate and Applied Meteorology* **22**, 474–487.

Matson, M. and Dozier, J. (1981) Identification of subresolution high temperature sources using a thermal IR sensor. *Photogrammetric Engineering and Remote Sensing* **47**, 1311–1318.

Menaut, J. C. (1983) The vegetation of African Savannas. In: F. Bourlière (ed.), *Ecosystems of the World. 13. Tropical Savannas*. Elsevier Scientific, Amsterdam, pp. 109–149.

Miller, K. and Barber, C. (1992) Biodiversity after Earth Summit: prospects for the Convention on Biodiversity. *Network '92*, No. 18, p. 5.

Mitchell, R. M., O'Brien, D. M. and Forgan, B. W. (1992) Calibration of the NOAA AVHRR shortwave channels using split-pass imagery: 1. Pilot study. *Remote Sensing of Environment* **40**, 57–65.

Morcrette, J. J. (1989) Description of the radiation scheme in the ECMWF model. *Technical Memorandum*, 165. ECMWF, Reading.

Morcrette, J. J. (1991) Radiation and cloud radiative properties in the European Centre for Medium Range Weather Forecasting System. *Journal of Geophysical Research* **96**, 9121–9132.

Muirhead, K. and Cracknell, A. P. (1984) Identification of gas flares in the North Sea using satellite data. *International Journal of Remote Sensing*, **6**, 827–833.

Myneni, R. B., Keling, C. D., Tucker, C. J., Asrar, G. and Nemani, R. (1997) Increased plant growth in the northern high latitudes from 1981–1991. *Nature* **386**, 698–701.

NASA/NOAA (1989) Memorandum of Understanding between the National Aeronautics and Space Administration and the National Oceanic and Atmospheric Administration for Earth Observation Remotely Sensed Data Processing, Archiving, and Related Science Support (also Addendum dated 1990). NASA, Washington, DC.

National Research Council (2000) *Reconciling Observations of Global Temperature Change*. National Academy Press, Washington, DC.

Neckel, H. and Labs, D. (1984) The solar radiation between 3300 and 12500 A. *Solar Physics* **90**, 205–208 (addenda to this work, and contributions of various groups to the extension to longer wavelengths have also been used).

Nguyen, L., Minnis, P., Kirk Ayers, J., Smith, Jr W. L. and Shu Peng Ho (1999) Intercalibration of geostationary and polar imager data using AVHRR, VIRS, and ATSR-2 data. Preprint volume, *10th Conference on Atmospheric Radiation, Madison, Wisconsin*. American Meteorological Society, pp. 405–408.

Nobre, C. A., Sellers, P. J. and Snukla, J. (1991) Amazonian deforestation and regional climate change. *Journal of Climate, American Meteorological Society* **4**, 957–988.

Noilham, J. and Planton, S. (1989) A simple parametrization of land surface processes for meteorological models. *Monthly Weather Reviews* **117**, 536–549.

Nuttall, N. (1999) Oceans to rise 20ft if ice sheet melts. *The Times (London)*, 18 October 1999.

Olson, J. S., Watts, J. A. and Allison, L. J. (1985) *Major World Ecosystems Ranked by Carbon in Live Vegetation: A Database*, NDP-017. Carbon Dioxide Information Center, Oak Ridge National Laboratory, Oak Ridge, TN.

Parrish, D. and Cohn, S. (1985) *A Kalman Filter for a Two-dimensional Shallow-water Model: Formulation and Preliminary Experiments*, National Meteorological Center Office Note 34. US Department of Commerce, NOAA, National Weather Service, Washington, DC.

Pearson, E. S. and Hartley, H. O. (eds) (1966) *Biometrika Tables for Statisticians*, vol. 1, 3rd edn. Cambridge University Press, Cambridge (see table 41 and p. 84).

Pinnock, S. and Grégoire, J-M. (1999) *The World Fire Web, Global Vegetation Monitoring Unit*. Online at http://www.gvm.sai.jrc.it

Pokrovsky, O. M. (1981) *Problems of the Optimisation of Remote Sensing Systems for Determining the Parameters of the Atmosphere, Ocean and Natural Environment.* VINITI, Russian Institute of Scientific and Technical Information, Moscow.

Posmentier, E. S, Soon, W. H. and Baliunas, S. L. (1998) Relative impacts of solar irradiance variations and greenhouse changes on climate, 1880–1993. In: R. Bate (ed.), *Global Warming – the Continuing Debate.* The European Science and Environment Forum, London, pp. 159–172.

Rabier, F., Courtier, P., Pailleux, J. Talagrand, O. and Vasiljevic, D. (1993) A comparison between four-dimensional variational assimilation relying on three-dimensional variational analyses. *Quarterly Journal of the Royal Meteorological Society* **119**, 845–880.

Rao, C. R. N. (1987) *Pre-launch Calibration of Channels 1 and 2 of the Advanced Very High Resolution Radiometer*, NOAA Technical Report NESDIS 36. Department of Commerce, National Oceanic and Atmospheric Administration, National Environmental Satellite, Data, and Information Service.

Rao, C. R. N. and Chen, J. (1993) Calibration of the visible and near-infrared channels of the Advanced Very High Resolution Radiometer (AVHRR) after launch. In: P. S. Chavez and R. A. Schowengerdt (eds), *Proceedings of SPIE Conference on Recent Advances in Sensors, Radiometric Calibration, and Processing of Remotely Sensed Data.* SPIE, Bellingham, Washington, DC, pp. 56–66.

Rao, C. R. N. and Chen, J. (1995) Inter-satellite calibration for the visible and near-infrared channels of the Advanced Very High Resolution Radiometer on the NOAA-7, -9, and -11 spacecraft. *International Journal of Remote Sensing* **16**, 1931–1942.

Rao, C. R. N., and Chen, J. (1999) Revised post-launch calibration of the visible and near-infrared channels of the Advanced Very High Resolution Radiometer (AVHRR) on the NOAA-14 spacecraft. *International Journal of Remote Sensing* **20**, 3485–3491.

Rao, C. R. N., Chen, J. Staylor, F.W. Abel, P. Kaufman, Y.J. Vermote, E. Rossow, W.R. and C. Brest (1993a) *Non-linearity Corrections for the Thermal Infrared Channels of the Advanced Very High Resolution Radiometer: Assessment and Recommendations*, NOAA Technical Report NESDIS 69. Department of Commerce, National Oceanic and Atmospheric Administration, National Environmental Satellite, Data, and Information Service.

Rao, C. R. N., Chen, J. Staylor, F.W. Abel, P. Kaufman, Y.J. Vermote, E. Rossow, W.R. and C. Brest (1993b) *Degradation of the Visible and Near-Infrared Channels of the Advanced Very High Resolution Radiometer on the NOAA-9 Spacecraft: Assessment and Recommendations for Corrections*, NOAA Technical Report NESDIS 70. Department of Commerce, National Oceanic and Atmospheric Administration, National Environmental Satellite, Data, and Information Service, Washington, DC.

Rao, C. R. N., Stowe, L. L. and McClain, E. P. (1989) Remote sensing of aerosols over the oceans using AVHRR data: theory, practice and applications. *International Journal of Remote Sensing* **10**, 743–749.

Rees, W. G. (1999) *The Remote Sensing Data Book*. The University Press, Cambridge.

Reynolds, R. W. and Smith, T. M. (1994) Improved global sea surface temperature analyses. *J. Climate* **7**, 929–948.

Richards, T. S., Gallego, J. and Achard, F. (2000) Sampling for forest cover change assessment at the Pan-Tropical Scale. *International Journal of Remote Sensing* **21**, 1473–1490

Robinson, J. W., Gunther, F. J. and Campbell. W. J. (1983) *Ground Truth Sampling and LANDSAT Accuracy Assessment*, document N83-26161. National Technical Information Services.

Rosenfeld, G. H. (1986) Analysis of thematic map classification error matrices. *Photogrammetric Engineering and Remote Sensing* **52**, 681–686.

Rosenfield, G. H. and Fitzpatrick-Lins, K. (1986) A coefficient of agreement as a measure of thematic classification accuracy. *Photogrammetric Engineering and Remote Sensing* **52**, 223–227.

Rosenfield, G. H., Fitzpatrick-Lins, K. and Ling, H.S. (1982) Sampling for thematic map accuracy testing. *Photogrammetric Engineering and Remote Sensing* **48**, 131–137.

Rosenqvist, A., Taylor, V., Chapman, B., Shimada, M., Freeman, A., De Grandi, G., Saatchi, S. and Rauste, Y. (2000) The Global Rain Forest Mapping Project – A review. *International Journal of Remote Sensing* **21**, 1375–1387.

Rouse, J. W. Jr, Haas, R. H., Schell, J. A., Deering, D. W. and Harlan, J.C. (1974) *Monitoring the Vernal Advancement and Retrogradation (Green Wave Effect) of Natural Vegetation*, NASA type III, Final report, RSC 1978–4. Greenbelt, MD, Goddard Space Flight Center.

Running, S. W., Justice, C. O., Salomonson, V., Hall, D., Barker, J., Kaufmann, Y. J., Strahler, A. H., Huete, A. R., Muller, J.-P., Vanderbilt, V., Wan, Z. M., Teillet, P. and Carneggie, D. (1994a) Terrestrial remote sensing science and algorithms planned for EOS/MODIS. *International Journal of Remote Sensing* **15**, 3587–3602.

Running, S. W., Loveland, T. R. and Pierce, L. L. (1994b) A vegetation classification logic based on remote sensing for use in global biogeochemical models. *Ambio* **23**, 77–81.

Salati, E. (1987) The forest and the hydrological cycle. In: R. E. Dickinson (ed.), *The Geophysiology of Amazonia: Vegetation and climate interactions*. J. Wiley & Sons, New York, pp. 273–296.

Scepan, J. (1999) Thematic validation of high resolution global land cover data sets. *Photogrammetric Engineering and Remote Sensing* **65**, 1051–1060.

Schiffer, R. A. and Rossow, W. B. (1983) The International Satellite Cloud Climatology Project: the first project of the World Climate Research Programme. *Bulletin of the American Meteorological Society* **64**, 779–784.

Segalstad, T. V. (1996) In: J. Emsley (ed.), *The Global Warming Debate*. The European Science and Environment Forum, London, pp. 41–50.

Singer, S. F. (1996) A preliminary critique of IPCC's Second Assessment of Climate Change. In: J. Emsley (ed.), *The Global Warming Debate*. The European Science and Environment Forum, London, pp. 146–157.

Slater, P. (1997) Sensor requirements for the radiometric consistency of long-term, global, land data sets. In: B. Guerther, J. Butler and P. Ardanuy (eds), *Workshop on Strategies for Calibration and Validation of Global Change Measurements, Arlington, VA, 10–12 May 1995*, NASA Reference Publication 1397. NASA, Washington, DC, pp. 69–72.

Slater, P. N., Biggar, S.F. Thome, K.J. Gilman, D.L. and Spyak, P.R. (1996) Vicarious calibration of EOS sensors. *Journal of Atmospheric and Oceanic Technology*. **13**, 343–359.

Smith, G. R., Levin, R.H. Abel, P. and Jacobowitz, H. (1988) Calibration of the solar channels of NOAA-9 AVHRR using high altitude aircraft measurements. *Journal of Atmospheric and Oceanic Technology* **5**, 631–639.

Soon, W., Baliunas, S. L., Robinson, A. B., Robinson Z. W. and Idso, S. B. (1999) *Environmental Effects of Increased Atmospheric Carbon Dioxide*, publication number 4809. Smithsonian Centre for Astrophysics, Harvard.

Staylor, W. F. (1990) Degradation rates of the AVHRR visible channel for the NOAA-6, -7, and -9 spacecraft. *Journal of Atmospheric and Oceanic Technology* **7**, 411–423.

Stehman, S. V. (1996) Estimating the kappa coefficient and its variance under stratified random sampling. *Photogrammetric Engineering and Remote Sensing* **62**, 401–407.

Steyn-Ross, D.A. and Steyn-Ross, M. (1992) Radiance calibrations for advanced very high resolution radiometer infrared channels. *Journal of Geophysical Research* **97**, 5551–5568.

Strong, A., Barrrientos, C., Duda, C. and Sapper, J. (1997) Improved satellite techniques for monitoring coral bleaching. In: H. A. Lessios and I. G. Macintyre (eds), *Proceedings. of 8th International Coral Reef Symposium, Panamá, 24–26 June 1996*. Smithsonian Tropical Research Institute, Balboa, Panamá, pp. 1495–1498.

Stroppiana, D., Pinnock, S. and Grégoire, J. M. (2000) The Global Fire Product: daily fire occurrence from April 1992 to December 1993, derived from NOAA-AVHRR data. *International Journal of Remote Sensing* **21**, 1279–1288.

Sullivan, W. T., Werthimer, D., Bowyer, S., Cobb, J., Gedye, D. and Anderson, D. (1997) A new Major SETI project based on Project SERENDIP data and 100,000 personal computers, In: C. B. Cosmorovici, S. Bowyer and D. Werthimer (eds), *Astronomical and Biochemical Origins and the Search for Life in the Universe (Proceedings of the Fifth International Conference on Bioastronomy, Capri, 1–5 July 1996)*, IAU Colloquium No. 161. Editrice Compositori, Bologna, pp. 729–734.

Tahnk, W. R. and Coakley, J. A. (2001) Improved calibration coefficients for NOAA-14 AVHRR visible and near-IR channels. *International Journal of Remote Sensing* **22**, 1269–1283.

Talagrand, O. and Courtier, P. (1987) Variational assimilation of meteorological observations with the adjoint vorticity equations. Part 1. Theory. *Quarterly Journal of the Royal Meteorological Society* **113**, 1311–1328.

Tarantola, A. (1987) *Inverse Problem Theory*. Elsevier, Amsterdam.

Teillet, P. M., Slater, P. N., Ding, Y., Santer, R. P., Jackson, R. D. and Moran, M. S. (1990) Three methods for the absolute calibration of the NOAA sensors in-flight. *Remote Sensing of Environment* **31**, 105–120.

Tickell, C. (1977) *Climatic Change and World Affairs*. Center for International Affairs, Harvard.

Tolba, M. K. and El-Kholy, O. A. (1992) *The World Environment 1972–1992*. Chapman and Hall, London.

Townshend, J. R. G. (ed.) (1992) *Improved Global Data for Land Applications*, IGBP report number 20. IGBP Secretariat, Royal Swedish Academy of Sciences, Stockholm, Sweden, 87 pp.

Townshend, J. R. G. and Williams, D. (1998) *The Concept of an Integrated Global Observing Strategy*. Online at: http://www.igospartners.org

Townshend, J. R. G., Justice, C. O., Skole, D., Malingreau, J.-P., Cihlar, J., Teillet, P. M., Sadowski, F. and Ruttenberg, S. (1994) The 1-km AVHRR global data set: needs of the International Geosphere Biosphere Program. *International Journal of Remote Sensing* **15**, 3319–3332.

Trenberth, K. E. (1992) *Climate System Modeling*. Cambridge University Press, Cambridge.

Tucker, C. J. and Sellers, P. J. (1986) Satellite remote sensing of primary production. *International Journal of Remote Sensing* **7**, 1395–1416.

Tyndall, J. (1863) On radiation through the Earth's atmosphere. *Philosophical Magazine and Journal of Science*, Series 4, **25**, 200–206.

Ulbrich, U. and Christophe, M. (199) A shift of the NAO and increasing storm track activity over Europe due to anthropogenic greenhouse gas forcing. *Climate Dynamics* **15**, 551–559.

UNEP (1999) *Earthwatch Strategic Framework for Environmental Observing, Assessment and Reporting*, UN System-Wide Earthwatch Coordination document. UN System-wide

Earthwatch United Nations Environment Programme, Geneva, Switzerland. Online at: http://www.unep.ch/earthw.html

UNFCCC (1992) *United Nations Framework Convention on Climate Change*, adopted in New York on 9th May 1992. United Nations Climate Change Secretariat, Bonn. Online at: http://www.unfccc.de/

UNFCCC, (1997) *The Kyoto Protocol to the United Nations Framework Convention on Climate Change (Addendum)*, FCCC/CP/1997/L7/Add. 1, December 10. United Nations, New York.

Unninayer, S. and Schiffer, R. A. (1997) *In-situ Observations for the global Observing Systems: A Compendium of Requirements and Systems*, NP-1997(01)-002-GSFC. NASA Office of Mission to Planet Earth, Greenbelt, MD, 286 pp.

Vaughan, R. A. (1994) Remote sensing, systems and data. In: R. A. Vaughan and A. P. Cracknell (eds), *Remote Sensing and Global Climate Change*. Springer, Heidelberg.

Vaughan, R. A. (1998) The role of remote sensing in climate monitoring. In: R. Bate (ed.), *Global Warming – the Continuing Debate*. The European Science and Environment Forum, Cambridge.

Vaughan, R. A. and Cracknell, A. P. (eds) (1994) *Remote Sensing and Global Climate Change*. Springer, Berlin.

Vaughan, R. A. and Karamjavan, Z. (2000) Topographic mapping using Topex satellite imagery. *Journal of Surveying Engineering* **126**(3), 106–122.

Vermote, E. and Kaufman, Y. J. (1995) Absolute calibration of AVHRR visible and near-infrared channels using ocean and cloud views. *International Journal of Remote Sensing* **16**, 2317–2340.

Verstraete, M. M. and Dickinson, R. E. (1986) Modeling surface processes in atmospheric general circulation models. *Annales Geophysicae* **4**, 357–364.

Verstraete, M. M. and Pinty, B. (1996) Designing optimal spectral indices for remote sensing applications. *IEEE Transactions on Geoscience and Remote Sensing* **34**, 1254–1265.

Walton, C. C., Sullivan, J. T., Rao, C. R. N. and Weinreb, M. P. (1998) Corrections for detector nonlinearities and calibration inconsistencies of the infrared channels of the advanced very high resolution radiometer. *Journal of Geophysical Research* **103**, 3323–3337.

Warrick, R. A., Le Provost, C., Meier, M. F., Oerlemans, J. and Woodworth, P. L. (1996) Changes in sea level. In: J. T. Houghton, L. G. Meira Filho, B. A. Callandar, N. Harris, A. Kattenberg and K. Maskell (eds), *Climate Change 1995: The Science of Climate Change*. Cambridge University Press, Cambridge, pp. 364–405.

Washington, W. M. and Parkinson, C. L. (1986) *An Introduction to Three Dimensional Climate Modelling*. University Science Books, Mill Valley.

Weber, G. (1996) European temperature variations since 1525. In: J. Emsley (ed.), *The Global Warming Debate*. The European Science and Environment Forum, London, pp. 113–138.

Weinreb, M. P., Hamilton, G. Brown, S. and Koczor. R. J. (1990) Nonlinearity corrections in the calibration of the Advanced Very High Resolution Radiometer infrared channels, *Journal of Geophysical Research* **95**, 7381–7388.

Werthimer, D., Bowyer, S., Ng, D., Donnelly, C., Cobb, J., Lampton, M. and Airieau, S. (1997) The Berkeley SETI Program: SERENDIP IV Instrumentation. In: C. B. Cosmovici, S. Bowyer and D. Werthimer (eds), *Astronomical and Biochemical Origins and the search for Life in the Universe (Proceedings of the Fifth International Conference on Bioastronomy, Capri, 1–5 July 1996)*, IAU Colloquium No. 161. Editrice Compositori, Bologna, pp. 683–688.

WGBU (1998) *The Accounting of Biological Sinks and Sources under the Kyoto Protocol: A Step Forwards or Backwards for Global Environmental Protection?* German Advisory Council on Global Change special report. Wissenschaftlicher Beirät der Bundesregierung Globale Umweltveränderungen, Bremerhaven, 75 pp. Online at: http://www.wbgu.de/wbgu_sn1998_engl.html

Whitlock, C. H., Staylor, W. F. Suttles, J. T. Smith, G. Levin, R. Frouin, R., Gautier, C. Teillet, P. M. Slater, P. N. Kaufman, Y. J. Holben, B. N , Rossow, W. B., Brest., C. and Lecroy, S. R. (1990) AVHRR and VISSR satellite instrument calibration results both for cirrus and marine stratocumulus IFO periods, FIRE Science Results 1988, NASA Conference Proceedings CP 3083.

Wigley, T. M. L. and Raper, S. C. B. (1992) Implications for climate and sea level of revised IPCC emissions scenarios. *Nature* **357**, 293–300.

World Commission on Environment and Development (WCED) (1987) *Our Common Future*. Oxford University Press, Oxford.

Wu, A. and Zhang, Q. (1994) A method for determining the sensor degradation rates of the NOAA AVHRR channels 1 and 2. *Journal of Applied Meteorology* **33**, 118–122.

Zupanski, D. (1997) A general weak constraint applicable to operational 4DVAR data assimilation systems. *Monthly Weather Review* **125**, 2274–2292.

Index